ORGANIC CHEMISTRY

ORGANIC CHEMISTRY

AN INTRODUCTION

M. F. GRUNDON

Professor of Chemistry
The New University of Ulster

and

H. B. HENBEST

Professor of Organic Chemistry
The Queen's University of Belfast

Second Edition

OLDBOURNE
LONDON

OLDBOURNE BOOK CO. LTD.
2 Portman Street, London, W.1

First published 1962
Second edition 1968

SBN 356 02231 5

Printed in Great Britain by
Spottiswoode, Ballantyne & Co. Ltd.
London and Colchester

CONTENTS

Chapter		Page
	Preface	vii
1	The Ground State	1
2	Alkanes	12
3	Alkyl Halides	31
4	Alcohols and Related Compounds	43
5	Alkenes and Alkynes	55
6	Aldehydes and Ketones	73
7	Carboxylic Acids and their Derivatives	86
8	Lipids	106
9	Amines	114
10	Organometallic Compounds	126
11	Polyhalides, Polyalcohols and Related Compounds	132
12	Substituted Alkenes. Polymers	138
13	Substituted Carboxylic Acids. Enantiomers. Proteins	149
14	Carbohydrates	172
15	Benzene and Related Hydrocarbons	182
16	Aryl Halides, Phenols and Related Compounds	202
17	Aryl Aldehydes, Ketones, Carboxylic Acids and Sulphonic Acids	214
18	Aryl Nitro-compounds, Amines and Related Compounds	237
	Answers to Questions	251
	Index	255

PREFACE TO SECOND EDITION

This book provides an introductory course in organic chemistry suitable for students in their last years at school or in their first year at university.

Chapters 1 and 2, which form the basis for the remainder of the book, are concerned with the ground state properties of molecules, including structure and stereochemistry, and with the nature of chemical reactions and the associated changes in energy. The chemistry of monofunctional aliphatic compounds is discussed in Chapters 3–10. This is followed by an account of bifunctional and more complex compounds, including proteins, and carbohydrates (Chapters 11–14), and of benzene and its derivatives (Chapters 15–18). A course based on this book can terminate at Chapters 10, 14 or 18, depending on the standard required.

Within each chapter, an account of nomenclature and ground state properties of one class of compound is followed by a discussion of certain reactions. The reaction products generally provide links with succeeding chapters. This arrangement minimises the mention of compounds and reactions not already discussed. General methods of preparation of each class of compound are summarised near the beginning of a chapter together with cross-references to more detailed accounts elsewhere in the book.

The text has been arranged so that new concepts and methods are introduced gradually. Resonance stabilisation, for instance, is not considered until Chapter 7, but receives further discussion in Chapters 15–18. The application of modern physical techniques to the determination of molecular structure is described briefly and exemplified often; a detailed account of these methods does not seem appropriate until the reader has access to the necessary instruments.

As it is difficult to express natural phenomena in terms of symbols, a few practical procedures are described in the book and some of the questions are concerned with experimental methods. This also reminds the reader that theory is based on experiment.

The reactions that have been chosen for discussion are either generally important in synthesis or are of particular value in an analytical, technological or theoretical context. Those included usually proceed in yields of 50–100%. Situations where the reaction pathways are more complicated, due, for example, to the intervention of rearrangements, are better dealt with in a more advanced book.

The authors have tried to provide only the basic ingredients of a modern course in organic chemistry, allowing the teacher to add material of his own choosing, and leaving the student to expand his knowledge of the subject and relate it to other disciplines by further reading. Articles or books on the history of science are included in this counsel.

January 1968

M. F. G.
H. B. H.

The honour of kings is to search out a matter.

PROVERBS 25:2

CHAPTER 1

THE GROUND STATE

The names given to the early ages of man (stone, bronze, iron) indicate the importance of these materials for the development of human civilisation. Apart from food, and later, clothing (linen was used in Egypt in 1500 B.C.), the utilisation of organic materials and compounds made relatively slow progress until comparatively recently. The emergence of scientific inquiry in the Renaissance period led to work that may be regarded as the forerunner of modern chemistry. During 1772–1777, Lavoisier in France developed improved methods for the determination of the proportions of elements in substances. Although a few organic compounds (sucrose, benzoic acid, and methyl alcohol) had been isolated previously from biological sources, the work of Scheele in Sweden (*ca.* 1770–1786) showed that many more compounds (e.g. malic acid, tartaric acid, acetaldehyde and glycerol) could be obtained when appropriate experimental procedures were used.

Let us consider one organic substance in more detail. Methionine, a colourless solid, is one of the products from the hydrolysis of many proteins (p. 160). We now know that a sample of this substance consists of a very large number of distinct, yet identical, molecules. A *molecule* is a stable assemblage of atoms. Each molecule of methionine contains five carbons, eleven hydrogens, one nitrogen, two oxygens and one sulphur, a composition which is more briefly expressed as $C_5H_{11}NO_2S$, the *molecular formula* for methionine. Methionine is classified as an organic compound because it contains the element carbon.

In the period around 1800 the belief was held that organic compounds could be formed only by living organisms with the aid of a 'vital force'. Berzelius suggested (1807), therefore, that substances isolated from biological sources should be called organic compounds in contrast to inorganic substances obtained from mineral sources. The definition of organic compounds had soon to be changed, since in 1828 Wöhler found that evaporation of an aqueous solution of ammonium cyanate gave urea, a compound previously obtained only from urine:

$$NH_4^{\oplus}\ CNO^{\ominus} \longrightarrow \begin{matrix} H_2N \diagdown \\ \quad\quad C{=}O \\ H_2N \diagup \end{matrix} \quad \text{urea}$$

As most of the biological products were being found to contain carbon, organic chemistry was redefined as the *chemistry of carbon compounds*.

Occurrence and Isolation. Many new organic compounds have been obtained from chemical and from biological sources in the last 130 years. A biological cell contains water and a wide range of organic substances, a few of which are referred to in this book. The elements nitrogen, sulphur and phosphorus as well as hydrogen and oxygen are present in many biological compounds. Cells also contain inorganic ions which are essential for normal health (Na, Mg, Cl, K, Ca, Mn, Fe, Co, Cu, Zn). The function of a living cell depends on a delicately poised complex of chemical reactions (pp. 111, 168) the detailed arrangement of which has evolved over a period of approximately four thousand million years.

The ease of isolation of a single organic compound from a chemically or biologically derived mixture can depend on (*a*) the relative abundance of the compound, (*b*) its chemical stability, (*c*) its physical properties relative to those of other substances present, and (*d*) the separation techniques that are available and suitable for use in the problem. The antibiotic, penicillin, is produced by micro-organisms of the *Penicillium* species when they are grown in water containing nutrient materials. Penicillin was difficult to isolate (*a*) because of its chemical reactivity and (*b*) because of its low concentration in the aqueous phase containing many other substances. Use can, however, be made of the fact that organic compounds distribute themselves in different proportions between two immiscible solvents. The following figures show that the ester, isopropyl acetate (cf. p. 93) is more effective than the other solvents listed for extracting the antibiotic from an aqueous solution. Separation and evaporation of the ester layer gives penicillin in a more concentrated form, for most of the other substances in the aqueous layer are not extracted by the organic solvent.

Distribution of penicillin G (p. 234) between organic and aqueous phases at 0°

Solvent	*Concentration in solvent* / *Concentration in water at pH 2·5*
isopropyl acetate	67
chloroform	39
ether	14
toluene	0·059
carbon tetrachloride	0·015

Chromatography, another valuable technique for the separation of substances, was first developed for studying the composition of mixtures of pigments present in plants. Various modifications of the technique are now available, for example, thin layer chromatography for the separation of non-volatile compounds, and gas (or vapour phase) chromatography for the analysis of mixtures of more volatile substances (pp. 196, 221). A solid compound can often be obtained in a purer state by crystallisation, in which a hot saturated solution of the substance in an appropriate solvent or solvent mixture is cooled. The method has the elegance of simplicity but the disadvantage that some of the compound is retained in solution. This occurs even with a pure compound: when a solution of methionine (0·30 g.) in water (3 ml.) at 95° is cooled to 5°, 0·24 g. of the compound is recovered as colourless crystals.

Composition and Structure. When a new compound is obtained it is first necessary to find out what elements are present in it. The origin of the substance may provide some evidence. Thus a product is likely to contain carbon if it is obtained from the reaction of a known organic compound and if it is also soluble in organic solvents. The presence of nitrogen, sulphur or the halogens can be detected by strongly heating the substance with sodium when the organic molecules are disrupted and identifiable salts (sodium cyanide, sodium sulphide, sodium halides) are obtained.

The proportion of each element in the compound is then determined quantitatively by analysis. Values for carbon and hydrogen are obtained by combustion analysis: an accurately weighed sample (*ca.* 3 mg.) of the compound is heated (700°) in a stream of oxygen, when carbon dioxide and water are produced. The amounts of these products are determined by passing the effluent gas in succession through weighed tubes containing magnesium perchlorate (to absorb water) and soda lime (to absorb carbon dioxide).

$$C_xH_y(E)_z + O_2(\text{excess}) \longrightarrow xCO_2 + \frac{y}{2}H_2O$$

In this general equation $(E)_z$ represents elements other than carbon and hydrogen.

In the early procedure for combustion analysis, a 0·5 g. (500 mg.) sample of compound and a working time of up to one month was required. Following pioneering work by Andrews (Belfast) and Ernich (Graz) on microanalytical procedures, these figures were reduced to 3–30 mg. and one hour by Pregl (Vienna, 1911). More recent developments include improvements in balances for weighing samples (less than 0·1 mg. can be used), and automated methods for combustion of the sample and for the quantitative estimation of products by gas chromatography.

If nitrogen is present in the original compound, a separate analysis is carried out. A sample is heated with copper oxide, and the volume of nitrogen gas is measured. Further analyses may be performed for sulphur and the halogens if they are present.

The use of these analytical methods may be considered by reference to a compound containing carbon, chlorine, hydrogen (probably) and oxygen (possibly). Combustion analysis of a sample (3·124 mg.) of the compound gave 5·383 mg. of carbon dioxide and 1·842 mg. of water. The *weights* of carbon and hydrogen in the original sample are therefore 1·468 mg. (5·383 × 12·01/44·01) and 0·206 mg. (1·842 × 2·016/18·016) respectively (atomic and molecular weights: carbon, 12·01; carbon dioxide, 44·01; hydrogen, 1·008; water, 18·016); the corresponding percentage weights are 47·0 (1·468 × 100/3·124) for carbon and 6·6 (0·206 × 100/3·124) for hydrogen.

Analysis of another sample (3·665 mg.) of the substance gave 6·850 mg. of silver chloride, equivalent to 1·695 mg. of chlorine in the sample. The compound therefore contains 46·3% chlorine. Addition of the percentage weights for carbon, hydrogen and chlorine gives a figure (99·9) that shows that only these elements are present in the compound. The relative atomic proportions of the three elements are then obtained by dividing their percentage weights by their atomic weights:

carbon = 3·92 (47·0/12·01); hydrogen = 6·55 (6·6/1·008)
chlorine = 1·306 (46·3/35·457)

As molecules contain integral numbers of atoms, the atomic proportions have to be adjusted arithmetically to give the simplest integral ratio. In the present case the ratio 3:5:1 for carbon:hydrogen:chlorine is obtained, and the compound has the *empirical formula*, $(C_3H_5Cl)_n$ where n is an unknown integer. [The following example may be worked through in order to give practice in the procedure.

Analysis gave the following percentage weights: C, 40·3; H, 7·4; N, 9·4; S, 21·45; O, 21·45% (by difference from 100%). Empirical formula, $(C_5H_{11}NO_2S)_n$.]

The empirical formula only expresses the simplest ratio of atoms present in the molecule, but when n is known the actual molecular formula is obtained. One method of finding the value of n is to determine the molecular weight of the substance, e.g. by the cryoscopic (freezing point depression) method. In the present example, the experimental molecular weight was 151. Since this value is close to that (153·05) calculated for $n=2$, the *molecular formula* of the original compound must be $C_6H_{10}Cl_2$. The calculation simply involves adding together the weights of the atoms present: $153{\cdot}05 = 6 \times 12{\cdot}01 + 10 \times 1{\cdot}008 + 2 \times 35{\cdot}456$.

The molecular formula expresses the numbers of atoms present in a molecule, but it does not give any information about the specific way in which the atoms are linked together; the *structure* of the compound. Numerous arrangements that satisfy the valence rules can be drawn for a molecule of formula, $C_5H_{11}NO_2S$, for example, (A) and (B), but only one represents the structure of a molecule of methioninc.

$CH_3CONHCH_2CH(OH)CH_2SH$ (A)

$CH_3SCH_2CH_2CH(NH_2)COOH$ (B)

Chemical tests often supply useful information. Methionine, for example, is more soluble in aqueous acid than in water and is therefore more likely to contain an amino-group (p. 114) as in structure (B) than a non-basic amide group (p. 94) present in structure (A).

Molecular structure is now studied by one or more of the following physical methods. The figures against each item refer to the cost of an instrument and the approximate date of the introduction of the technique for general use.

(*a*) Ultraviolet and visible absorption spectroscopy (p. 139); £700–3000; 1930;
(*b*) infrared absorption spectroscopy (see below); £1500–3000; 1945;
(*c*) dipole moment (p. 32); £550; 1960;
(*d*) nuclear magnetic resonance spectroscopy (p. 186); £10,000–20,000; 1960;
(*e*) mass spectroscopy (see below); £15,000–30,000; 1965;
(*f*) X-ray and electron diffraction (see below); £2000–20,000; 1965;
(*g*) atomic absorption spectroscopy (for detection and estimation of metals); £1200; 1965.

Methods (*a*), (*b*), (*d*) and (*e*) give information fairly quickly (10 min. to a few hours) and are used, sometimes in conjunction with chemical and other evidence, for elucidating the structures of molecules of simple to medium complexity. X-ray procedures are more time-consuming, but are particularly useful for investigating the structures of very complex molecules provided that the substance can be obtained in a suitably crystalline form. The structure of vitamin B_{12} (cyanocobalamin), $C_{63}H_{88}O_{14}N_{14}PCo$, (insufficient absorption of which from food gives rise to pernicious anaemia) was established by X-ray crystallography, and much progress is currently being made towards solving the structures of crystalline proteins (p. 163), especially those that contain a dense atom such as iron (in haemoglobin), which can serve as a reference point in the analysis. Computers are used to perform the complicated calculations required in this work.

Diffraction methods are also invaluable for determining more detailed features of molecular structure, for example, bond lengths and angles. X-ray analysis of one of the two crystalline forms of methionine shows that each molecule has the bond lengths indicated in (A) and the bond angles shown in (B):

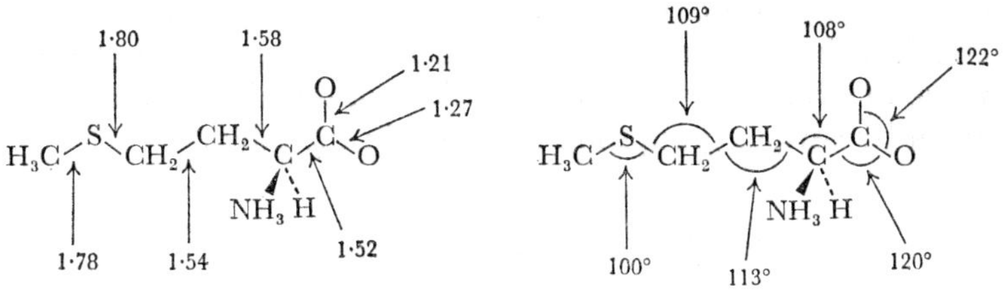

(A) (B)

distances in Ångstrom units (1Å = 10^{-8} cm.)

When an organic compound is bombarded by a stream of electrons under high vacuum in a mass spectrometer, an electron is lost from some of the molecules and a positive radical ion (the *molecular ion*) results:

$$\text{M:} \longrightarrow \text{M}\cdot^{\oplus} + \text{e}$$

The molecular ion may then dissociate, for example into positive ions and radicals (p. 19):

$$\text{d}\cdot + \text{c}^{\oplus} \longleftarrow \text{M}\cdot^{\oplus} \longrightarrow \text{a}^{\oplus} + \text{b}\cdot$$

An analysing magnet deflects the positive ions to different extents and the resultant spectrum reveals their mass and relative abundance. The peak of highest unit mass is usually attributable to the molecular ion, and mass spectrometry therefore is a precise method for establishing the molecular weight of an organic compound provided that it is sufficiently volatile. The mode of cleavage of the molecular ion is dependent on the functional groups present, and the fragmentation patterns (mass and relative abundance of ions $a^{\oplus}$, $c^{\oplus}$, etc.) may provide valuable structural information (pp. 84, 95).

The infrared region of the electromagnetic spectrum lies between the visible and the microwave regions (12,000–300 cm.$^{-1}$). When an organic molecule absorbs infrared radiation, the energy temporarily stretches or bends one or more bonds. The absorption of the radiation is selective, so that each absorption band covers only a narrow region of the spectrum. The position of these absorption bands are often characteristic of particular groups or structural features. Thus a carbon–oxygen single bond normally gives a band near 1100 cm.$^{-1}$ whereas a carbon–oxygen double bond absorbs selectively near 1700 cm.$^{-1}$. Infrared spectroscopy is therefore of great diagnostic value in structural work; the positions of the absorption bands given by various groups are recorded in the table:

Absorption Bands in the Infrared Region

Positions of absorption bands (in cm.$^{-1}$)*	*Structural group responsible for the absorption band or bands*
3600–3100	O—H in alcohols and phenols N—H in primary and secondary amines and amides
3250	≡C—H in alkynes
3000	=C—H in alkenes and benzene compounds
2900	—C—H in alkanes
2800	O=C—H in aldehydes
2600	S—H in thiols
2300	C≡N in nitriles
2100	C≡C in alkynes
1750–1700	C=O in aldehydes, ketones and esters
1650	C=O in amides
1550, 1350	NO_2 in nitro-compounds
1300, 1150	S=O in sulphones
1100	C—O in alcohols and ethers
950–700	C—H in alkenes and benzene compounds

* These are given to the nearest 50 or 100 cm.$^{-1}$ The precise location of an absorption band depends on the detailed structure of the molecule.

Several other physical properties are very relevant to the practice of organic chemistry, for example, melting and boiling points, densities, and solubilities. Two useful generalisations concerning solubility and miscibility are (*a*) that similar substances are usually mutually soluble, and dissimilar substances are often but slightly soluble in each other, and (*b*) that a solid having a higher m.p. is usually less soluble in an organic solvent than one having a lower m.p. Dissolution in a solvent and melting are similar in that both involve breaking down of the crystal lattice of the solid.

Covalent Bonds. The structure of an atom is normally such that the number of negative electrons surrounding a positively charged nucleus is equal to the value of the nuclear charge (the *atomic number* of the element). The electrons are grouped in concentric principal shells and sub-shells; the maximum capacity of the first two principal shells are two and eight respectively. The electronic configurations of sodium and chlorine are shown in the diagram below, in which the usual notation is employed. A particularly stable electronic arrangement arises when the principal shells are filled, as in the noble gases; for example, in the reaction between sodium and chlorine to give the ionic compound $Na^{\oplus}Cl^{\ominus}$, both ions attain such electronic configurations:

sodium $1s^2/2s^2, 2p^6/3s^1$ ⟶ sodium$^{\oplus}$ $1s^2/2s^2, 2p^6/$ (neon configuration)

chlorine $1s^2/2s^2, 2p^6/3s^2, 3p^5$ ⟶ chloride$^{\ominus}$ $1s^2/2s^2, 2p^6/3s^2, 3p^6/$ (argon configuration)

The energy needed to remove the $3s$ electron from sodium (118 kcal./mole) is more than compensated by the combined liberation of energy resulting from the acquisition of an electron by chlorine (84·9 kcal./mole) and from the electrostatic force between the positive and negative ions (180·4 kcal./mole*).

An alternative attractive force is the *covalent* bond (term suggested by Lewis in 1916), which results from the *sharing* of electrons between two atoms.

The hydrogen molecule provides the simplest example:

2H· two hydrogen atoms ⟶ 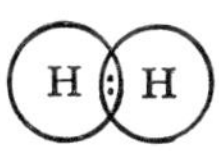 or H:H or H—H or H_2 one hydrogen molecule

A single covalent bond, indicated by a line, is formed by the sharing of two electrons. Energy is evolved when an electron is attracted by two positive nuclei instead of one; the particular internuclear distance at which this gain in energy is maximal corresponds to the bond length (p. 7).

Carbon has the electronic configuration $1s^2/2s^2, 2p^2$ or (He)$2s^2, 2p^2$ [(He) represents helium configuration]. Thus, there

* This value corresponds to the lattice energy of sodium chloride derived from the reaction: $Na^{\oplus}$(gas) + $Cl^{\ominus}$(gas) ⟶ NaCl(solid).

are four electrons available for covalent bond formation in the outer or valence shell; at the same time four electrons are required to complete the shell (to give a neon-like configuration). A carbon atom gains these electrons when it combines with four atoms of hydrogen to form a molecule of methane, CH_4. The carbon now has eight electrons in its valence shell (cf. neon), each of the four hydrogens having contributed an electron. At the same time, due to electron sharing, each hydrogen is associated with two electrons (stable helium configuration):

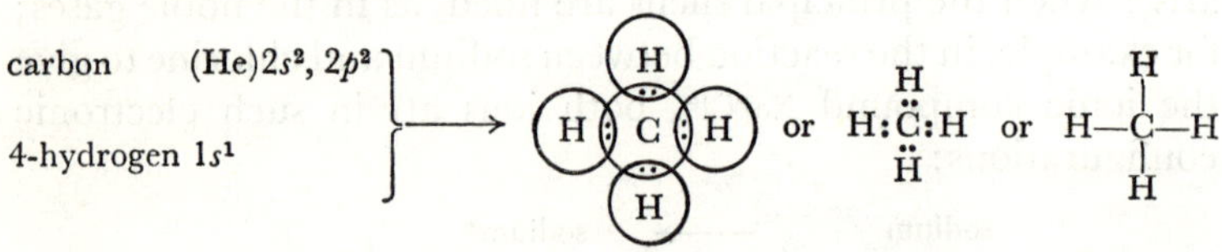

A neon-like configuration with eight electrons in the outer shell is also adopted when possible by other elements (boron, nitrogen, oxygen, fluorine) in the first row of the periodic classification. The special stability of this arrangement is the basis of the 'octet rule' for these elements.

Dimethyl ether is an example of a compound that contains carbon–oxygen single covalent bonds:

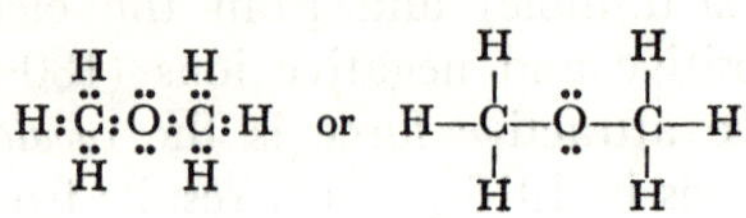

In accord with the octet rule the oxygen and each of the carbons has an outer shell of eight electrons. Oxygen has the configuration $1s^2$, $2s^2$, $2p^4$ and gains a share in two extra electrons in forming two covalent bonds. Four electrons around oxygen remain unbonded; these are arranged in two pairs and are usually referred to as 'lone' (or unbonded) pairs of electrons. Lone pairs of electrons often play important rôles in the chemistry of the elements that carry them.

Covalent bonds can be alternatively described in terms of molecular orbitals (p. 18).

This chapter has been concerned with the structure of molecules in their most stable or *ground states*. A molecule can be transformed temporarily into an *excited state* (p. 139) when it acquires energy insufficient to break one or more bonds, but for a structural change to occur in a chemical reaction the reactants must move through a state of maximum energy (the *transition state*, p. 24).

QUESTIONS

1. (*a*) What is the colourless natural product 3 mg. of which when burnt in oxygen yields 11 mg. of carbon dioxide?

(*b*) Comment on the fact that when the colourless substance is heated to 3600° (under nitrogen) it changes into a black material which contains the same proportion of carbon.

2. (*a*) A dyestuff gave an analysis of 73·3% carbon, 3·8% hydrogen, 10·7% nitrogen, the remainder being oxygen. Molecular weight determinations gave values in the range 255–268. Calculate the molecular formula of the compound.

(*b*) Work out the empirical formula of a compound that was isolated from the treatment of sour milk with a silver salt. Analytical results for the compound: C, 18·3; H, 2·5; O, 24·4; Ag, 54·8%.

(*c*) A selective herbicide (MCPA) contains 53·9% carbon, 4·5% hydrogen, 17·7% chlorine, remainder oxygen. A molecular ion peak of approximate mass 200·5 was produced in a mass spectrometer. What is the molecular formula?

3. When insulin (p. 160) was first isolated in a crystalline state, it was found to contain a small amount of zinc (in a combined form). How could the presence of this metal be established in a sample of insulin?

4. Write a balanced equation for the thermal decomposition of glyceryl trinitrate (p. 136) assuming that nitrogen, water, oxygen and carbon dioxide are the only products. Calculate the volume of gas produced (after cooling to N.T.P.) by exploding 1 g. of the compound.

5. (*a*) How may the fatty substances be removed from a sample of biological material?

(*b*) Equal weights of a vegetable oil, potassium hydroxide, and water were heated at 100° for 30 min., a clear solution being formed. Addition of hydrochloric acid gave an upper oily layer that solidified on cooling. Crystallisation of the solid from ethanol gave crystals, m.p. 62° which, by analysis, contained 74·95% carbon, 12·6% hydrogen, remainder oxygen. The molecular weight was *ca.* 255. Calculate the molecular formula of the solid.

(*c*) How might you find out if a sample of a meteorite contains (i) carbon and (ii) material of biological origin?

6. What evidence is there that (*a*) atoms and (*b*) molecules exist?

Whatever is now establishment was once innovation.

SYDNEY SMITH (1771–1845)

CHAPTER 2

ALKANES

A molecule composed only of carbon and hydrogen atoms is called a hydrocarbon. An *alkane* is a hydrocarbon that contains only single covalent bonds.

The simplest alkane, methane, has been known for a long time in the form of 'marsh gas', which is produced by the bacterial decomposition of organic matter under anaerobic conditions. Methane is also present in coal seams; when released during mining operations, it can accumulate in sufficient amount to form an explosive mixture with air ('fire damp'). Natural gas consists of methane with decreasing amounts of ethane, propane and the two butanes. Petroleum (rock oil), associated with natural gas and asphalt (or bitumen), seeps through the earth's crust at many places. Asphalt was used in the near East for the construction of roads and buildings from about 3000 B.C. onwards. It was also used for preserving the dead, the word 'mummy' coming from the Coptic 'mum' meaning bitumen. 'Greek fire', believed to have been invented by the Byzantines, consisted of petroleum thickened by addition of resin or sulphur. In naval warfare in the period A.D. 650–1200, this material was ejected as a flaming liquid on the enemy's ships.

More recent aspects of petroleum technology are discussed at the end of this Chapter.

Structure and Nomenclature. Alkanes have the general formula, C_nH_{2n+2}, each member of the series of compounds differing from the next by a CH_2 unit. The three simplest alkanes are:

methane

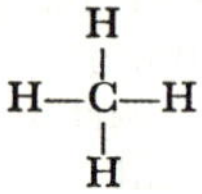

or CH_4

ethane

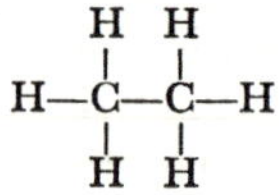

or CH_3CH_3 or C_2H_6

propane

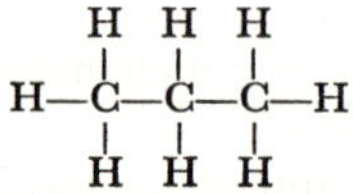

or $CH_3CH_2CH_3$ or C_3H_8

For each of these compounds the fully condensed formula is equivalent to the structural formula. Thus starting from the C_3H_8 formula of propane, there is only one way of joining the carbon and hydrogen atoms together, and at the same time keeping to the valence rules.

Lengthening the chain of carbon atoms gives the following compounds (up to C_{10}).

		b.p.
$CH_3CH_2CH_2CH_3$	butane	−0·55°
$CH_3CH_2CH_2CH_2CH_3$	pentane	36
$CH_3CH_2CH_2CH_2CH_2CH_3$	hexane	68·3
$CH_3CH_2CH_2CH_2CH_2CH_2CH_3$	heptane	98·4
$CH_3(CH_2)_6CH_3$	octane	125·7
$CH_3(CH_2)_7CH_3$	nonane	150·8
$CH_3(CH_2)_8CH_3$	decane	174

The names, methane to decane, are the basis for naming other compounds which are derived from the alkanes by replacement of one or more hydrogens by other atoms or groups of atoms.

Structural Isomerism. The above alkanes, from butane to decane, are the first members of the series of compounds known as the normal (or n-) alkanes. The term 'normal' signifies that the compound contains an unbranched chain of carbon atoms. The distinction is made because there are alternative 'branched chain' structures possible for butane and the higher alkanes. In the case of butane, two structures can be written both of which are derived from the molecular formula, C_4H_{10}.

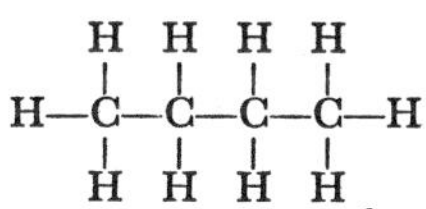

or $CH_3CH_2CH_2CH_3$

butane

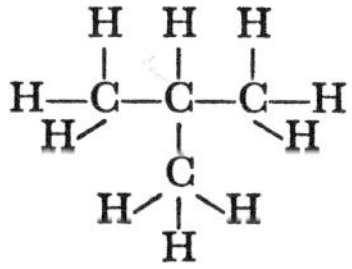

or $CH_3CH(CH_3)_2$

isobutane or
2-methylpropane (systematic)

Butane and isobutane are said to be *structural isomers*, the term referring generally to compounds that have the same molecular formula but different structural formulae. Isobutane has a 'branched chain' structure and butane (or n-butane) has a 'straight chain' structure. The carbon atoms of butane have been drawn in a straight line in the above diagram, but alternative diagrams can be drawn provided that the four carbons still remain joined in a sequence:

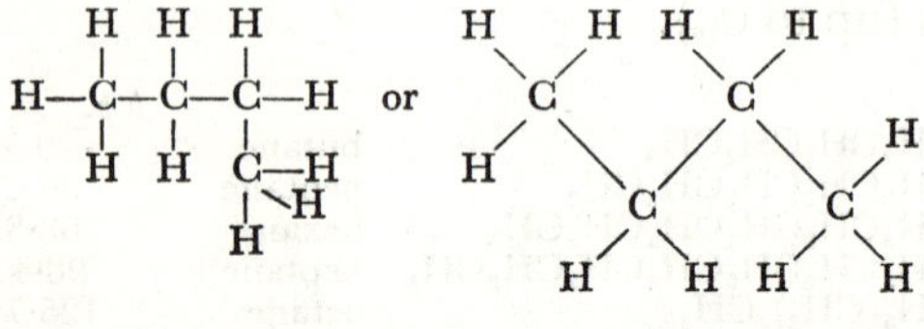

More structural isomers become possible when additional carbon and hydrogen atoms are present in a molecule. Thus there are three isomeric pentanes, those additional to n-pentane being:

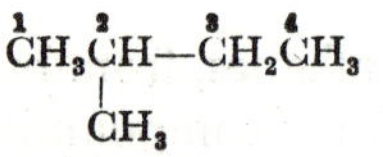

isopentane or
2-methylbutane (systematic)

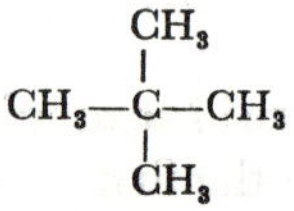

neopentane or
2,2-dimethylpropane (systematic)

As long as the valence rules are obeyed, an exact forecast of the number of possible isomers of each alkane can be made. Thus there are 18 octanes, all of which are known, and 75 isomeric decanes, more than half of which have been characterised. If the higher, multibranched alkanes are required, their syntheses by rational methods could pose some formidable problems.

Systematic names for the isomers containing branched chain structures are obtained by regarding the longest sequence (chain) of carbons as the parent structure, and additional carbons as substituents; thus, in the above compounds, one or two additional carbons (as CH_3 groups) replace one or two hydrogens in the parent structures. For example, isopentane contains a sequence of four carbon atoms and is therefore a derivative of butane. If these four carbons are numbered in sequence, the additional CH_3 (methyl group; see below) in isopentane is seen to be attached to the carbon labelled 2. The resultant systematic name is 2-methylbutane. The systematic name, 2,2-dimethylpropane, for neopentane signifies that the longest chain of carbon atoms is three (propane); the term *di*methyl means that there are two extra CH_3 groups and the prefix 2,2- signifies that each is joined to carbon atom 2 of the propane chain.

Substituents derived from alkanes by removal of a hydrogen atom are called *alkyl* groups. Individual alkyl groups are named by replacing -ane of the corresponding alkane by -yl, e.g. methyl (CH_3) from methane and ethyl (CH_3CH_2 or C_2H_5) from ethane. Alkanes are given the general formula, RH, indicating that they are composed of an alkyl group, R, and a hydrogen atom. *Methylene* is a useful term for denoting the divalent CH_2 group. Propane, $CH_3CH_2CH_3$, contains one methylene and two methyl groups.

The number of alkyl groups joined *directly* to a particular carbon (its degree of substitution) is indicated by one of the words, primary, secondary, tertiary or quaternary. 2,2,4-Trimethylpentane is a compound that contains carbon atoms with different degrees of substitution:

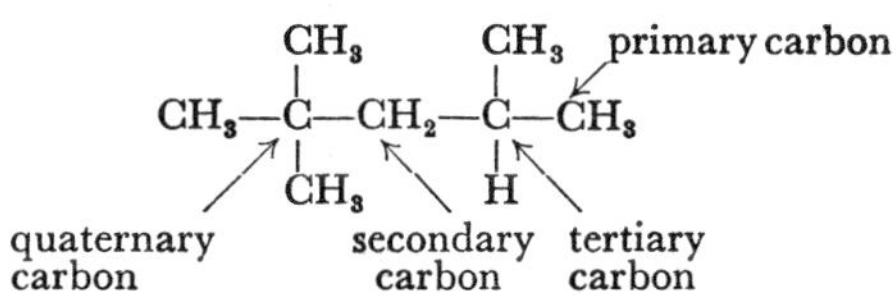

Tetrahedral Carbon. The structures of hydrocarbons and other organic molecules are usually represented for convenience in two dimensions, but this does not mean that the bonds involved lie in one plane. In fact, when a carbon atom is singly bonded to four atoms or univalent groups, experimental evidence (p. 157; X-ray studies of crystalline organic compounds) has established that the four bonds from carbon are directed in space towards the corners of a tetrahedron, so that the angle between two bonds is approximately 109°. If the four substituents are the same, the bond angles are equal (109° 28′) as are the bond lengths. A molecule of methane is represented in diagram (1); the part in lines of normal thickness (the carbon atom and two hydrogen atoms) is in the plane of the paper, while the black and dotted spheres (representing the remaining hydrogen atoms) are in front and behind the plane, respectively. Diagrams (2) and (3)

are alternative representations of a methane molecule. In (2), the four hydrogen atoms are placed at the corners of a regular tetrahedron with the carbon atom (not shown in the diagram) at the centre of the tetrahedron. In (3), the wedge represents a bond projecting in front of the plane of the paper, and the dotted line a bond behind the plane; the other two bonds lie in the plane of the paper.

The tetrahedral arrangement of bonds around carbon is

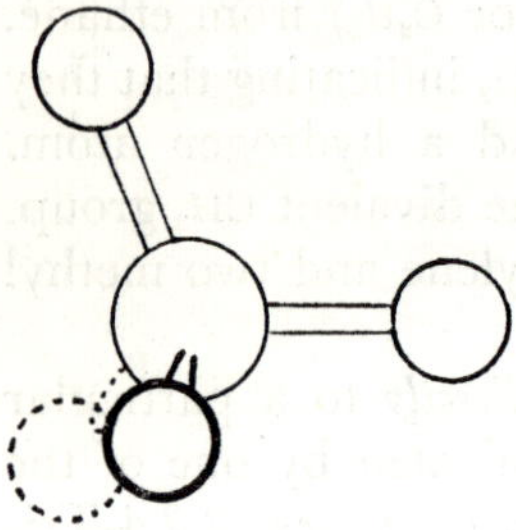

(1)

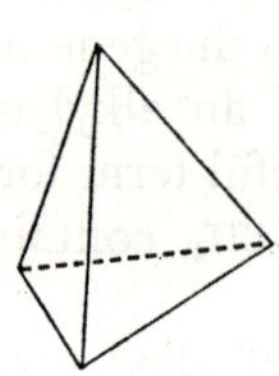

(2)

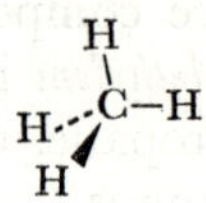

(3)

maintained in hydrocarbons higher in the series than methane, and two representations of an ethane molecule are given in diagrams (4) and (5).

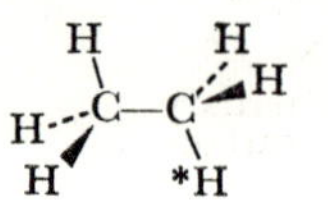

(4)

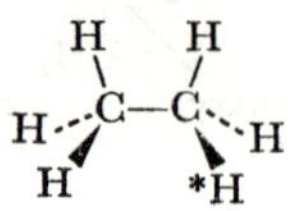

(5)

In diagram (4) the hydrogens are as far apart as possible, whereas in (5) the hydrogens are in the closest possible positions. Arrangements (4) and (5) contain the same bond angles and lengths; the change from one to the other results from a 60° rotation of one of the methyl groups about the C—C bond (see starred hydrogen). These two forms must be readily interconvertible because ethane exists as a single homogeneous substance. The ease of rotation about the C—C bond gives rise to an infinite number of arrangements (*conformations*) which differ only in the spatial relationship of the two methyl groups.

Actually rotation about the C—C bond in ethane is not entirely unrestricted, because the energy of each molecule is not the same in all conformations. The situation may be depicted by an *energy profile*, a graph in which the energy of a molecule is plotted against the angle through which one methyl group has rotated.

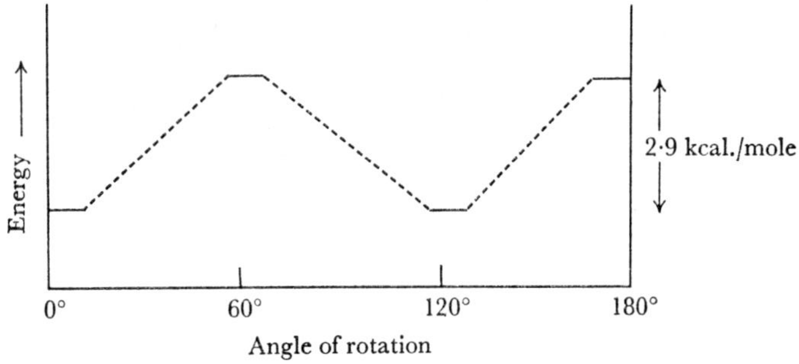

Each minimum in the graph corresponds to a conformation of type (4), and each maximum to an arrangement (5). The height of the barrier to rotation is 2·9 kcal./mole, which is very much less than the value required (*ca.* 17 kcal./mole) to give isomers capable of separate existence at room temperature. A molecule of ethane therefore adopts all possible conformations, but in the substance in bulk there will be a predominance of those arrangements (4) in which the hydrogens are farthest apart. The energy barriers to rotation in other σ bonds (see below) are usually small, for example 1·95 in CH_3—NH_2, 1·15 in CH_3—CHO, and 0·006 kcal./mole in CH_3—NO_2, and again rotational isomers cannot be isolated.

The above discussion is one aspect of *stereochemistry*, a subject concerned with specific ways in which atoms and groups are arranged in space within molecules, including those undergoing reaction.

The use of molecular models can be a valuable aid for the appreciation of stereochemical problems.*

The Electronic Structure of Covalent Bonds. The simple view of the formation of covalent bonds by the sharing of electrons between atoms (p. 9) does not predict that there will be a

* Various types of models are available and include (*a*) Atomic Models (ball and spring type) obtainable from A. Gallenkamp and Co., Technico House, London, W.C.2. (*b*) Courtauld Atomic Models (space filling) from Griffin and George Ltd., Alperton, Wembley, Middlesex. (*c*) Dreiding Stereomodels (metal links) from Koch-Light Laboratories Ltd., Colnbrook, Bucks. (*d*) Framework Molecular Models (representation of molecular orbitals) from Prentice-Hall Inc., 20 Wellbeck St., London, W.1.

tetrahedral arrangement of bonds around a carbon atom. For a fuller understanding of stereochemistry we need to consider the electronic structure of bonds in more detail.

There is a high probability that the single ($1s^1$) electron of the hydrogen atom will be found within a spherical area surrounding the nucleus. This area is described as an *s* atomic *orbital*, and such orbitals can contain a maximum of two electrons of opposing spin. A hydrogen molecule is formed by the overlap of the atomic orbitals of two hydrogen atoms:

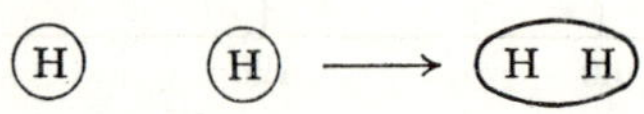

$$H\cdot + \cdot H \longrightarrow H : H$$

This amalgamation leads to the formation of a molecular orbital, which like an atomic orbital, contains a maximum of two electrons. When formed from two *s* electrons in this way the filled molecular orbital (constituting a bond) is elliptically symmetrical about the line joining the nuclei, with the highest electron density between the nuclei.

It will be recalled that, of the eight electrons that form the four bonds in methane, four ($2s^2$, $2p^2$) are derived from carbon and four ($1s^1$ each) from hydrogen. In methane, one of the $2s$ electrons of carbon is promoted to the $2p$ level giving an 'sp^3' configuration; these four atomic orbitals simultaneously become equivalent (by 'hybridisation'), and four covalent bonds are formed by overlap of these atomic orbitals with the *s* orbitals of the four hydrogen atoms, as indicated in diagram (6); diagram (7) gives a more accurate picture of the molecular orbitals.

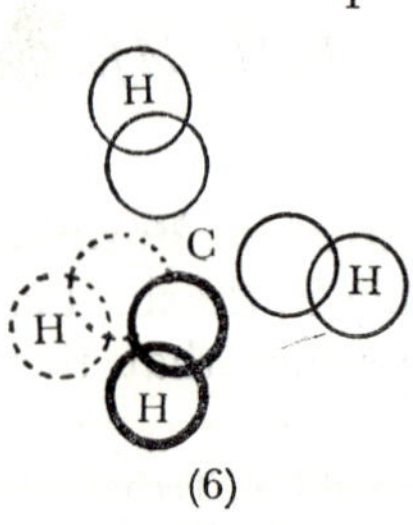

(6)

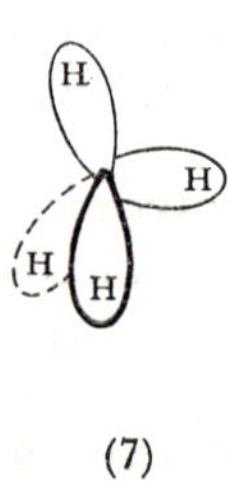

(7)

Repulsion between the filled molecular orbitals favours the tetrahedral arrangement in which the 'electron clouds' are as far apart as possible. The preference for the staggered conformation (4) in a molecule of ethane (see above) may originate from repulsion between the two groups of three electron pair C—H bonds.

Physical Properties of Alkanes. At room temperature the n-alkanes range from gases (CH_4 to C_4H_{10}), to liquids (C_5H_{12} to $C_{16}H_{34}$), and then to solids ($C_{17}H_{36}$ upwards). Physical properties of compounds depend on molecular weight (p. 5) and on molecular shape. Structural isomers have different molecular shapes and therefore different melting and boiling points.

	m.p.	*b.p.*
pentane	−130°	36°
2-methylbutane	−160°	28°
2,2-dimethylpropane	−17°	9·5°

The three isomeric pentanes can be separated therefore by fractional distillation.

The cross-section of an n-alkane is smaller than that of a branched chain hydrocarbon, and this difference is the basis of their separation using 'molecular sieves'. These materials are aluminosilicates, of natural or synthetic origin, which contain pores or channels of regular size and shape. When a mixture of n-butane (cross-section *ca.* 4·9 Å) and isobutane (*ca.* 5·6 Å) is passed over a sieve of intermediate porosity, the former hydrocarbon is adsorbed preferentially. It can be subsequently recovered by warming the sieve under reduced pressure.

Halogenation of Alkanes. Alkanes are usually unaffected by acidic and alkaline reagents and by reducing agents. Their most important reactions are those with suitably reactive radicals,* which lead to the introduction of new groups by replacement of one or more hydrogens. The halogenation of methane provides the simplest illustration of this type of reaction.

An equimolar mixture of methane and chlorine remains unchanged when kept at 20° in the dark. When the mixture is heated (above 200°) or irradiated (ultraviolet or strong visible light)

* The following definition will be used: a *radical* is an atom or molecule that contains an unpaired electron. Originally, the word 'radical' was used to denote a group of atoms within a normally stable molecule; 'group' is now generally preferred for this purpose.

rapid, sometimes explosive, reaction occurs and chloromethane and hydrogen chloride are formed:

$$CH_4 + Cl_2 \longrightarrow \underset{\substack{\text{chloromethane} \\ \text{(methyl chloride)}}}{CH_3Cl} + HCl$$

The action of the light or the heat imparts enough energy to a small fraction of the chlorine molecules, so that fission into pairs of chlorine radicals (chlorine atoms) occurs.

(*a*)

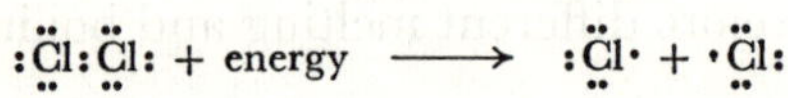

These chlorine atoms, each of which contains seven electrons in the outer shell, are more reactive than chlorine molecules. Reaction between a chlorine atom and a molecule of methane gives a stable molecule of hydrogen chloride together with a methyl radical, which is unstable (reactive) because the carbon only has seven electrons in its outer shell.

(*b*)

(methyl radical)

Each methyl radical then reacts with a *molecule* of chlorine to give the stable methyl chloride and a chlorine atom.

(*c*)

Thus, for every chlorine atom consumed in reaction (*b*), another is generated in reaction (*c*). Therefore reactions (*b*) and (*c*) proceed recurrently until all of the original methane and chlorine molecules are used up (an example of a *chain* reaction).

Process (*a*) is termed an *initiation* step as it provides the few chlorine atoms that are required to start the main reaction. Stages (*b*) and (*c*) are *propagation* steps in which the reaction of the two substances proceeds to completion. Each of the steps (*a*), (*b*) and (*c*) is a chemically significant change of a simple type: in (*a*) one bond has been broken, and in (*b*) and (*c*) an atom has been transferred from one species to another. Such steps are called *elementary processes*. Some organic reactions involve only one elementary step (cf. Chapter 3), whereas others are more complicated than the formation of methyl chloride from methane.

The description of the pathway by which a chemical change proceeds is referred to as the *mechanism*; it is concerned with the question 'how does the reaction occur'? The description is never complete, because it is always possible to study the situation further.* In the discussion of the chlorination of methane, answers to the questions 'what is the role of the light', and 'why does the reaction proceed explosively' were considered. We might now ask 'would light of a particular wavelength be especially effective for the initiation step (*a*)', and then 'how could this question be answered in practice'?

Chlorination proceeds further if a mixture of methyl chloride and chlorine is heated or irradiated. Chlorine atoms and organic radicals ($\cdot CH_2Cl$) are then intermediates in the reaction mechanism, the overall equation being:

$$CH_3Cl + Cl_2 \longrightarrow \underset{\text{(methylene dichloride)}}{\underset{\text{dichloromethane}}{CH_2Cl_2}} + HCl$$

Further hydrogens can be replaced by chlorine to give, successively, $CHCl_3$ (trichloromethane or chloroform) and CCl_4 (tetrachloromethane or carbon tetrachloride).

In practice, the chlorination of methane gives a mixture of products, the composition of which depends on the initial methane:chlorine ratio. The mixture can be separated by fractional distillation:

CH_3Cl	b.p. −24°	$CHCl_3$	b.p. 61·3°
CH_2Cl_2	b.p. 40·7°	CCl_4	b.p. 78·8°

The first product from the chlorination of ethane (at 200° or in the presence of light) is chloroethane or ethyl chloride, CH_3CH_2Cl. Reactions of higher alkanes with chlorine (one molecular equivalent) give mixtures of isomeric mono-chloro-compounds, arising from attack at the various hydrogens in each starting

* Finding out how a chemical reaction takes place is a 'black box' type of problem. Reactants and products are usually stable enough to make their identification possible, but what happens in between, in the black box of the reaction, needs to be found out by experiments that lead to reasonable inferences.

material; for example, the following compounds are formed from 2-methylbutane (isopentane):

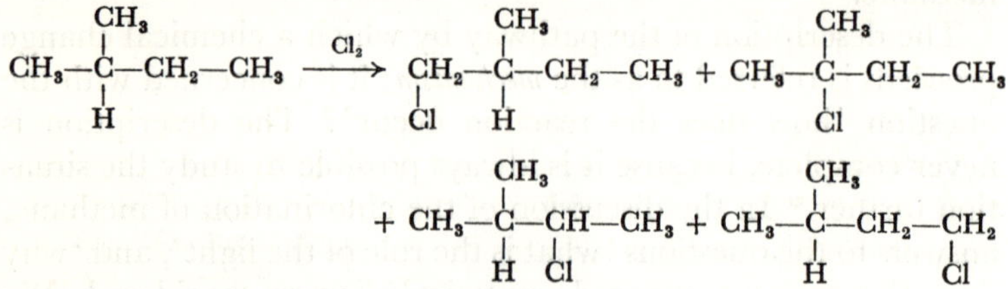

Alkyl bromides can be prepared by heating or irradiating a mixture of an alkane and bromine, but the reaction does not proceed as readily as chlorination. Fluorine reacts vigorously with alkanes by a radical mechanism even at 20° in the absence of light, but iodine does not normally react with alkanes. The order of halogen reactivity is therefore F > Cl > Br > I.

Energy Changes in the Halogenation of Alkanes. Some features of the mechanism of the chlorination of methane have been discussed above (p. 19), but no reference was made to the energy factors involved. This problem may, in general, be dissected into two parts: (i) an equilibrium aspect concerned with the direction and the extent of reaction, and (ii) a kinetic aspect concerned with the rate (velocity) of the reaction or, in more detail, with the rate of each of the three elementary processes.

Let us first consider the possibility of bringing the reaction to *equilibrium*:

$$\text{methane} + \text{chlorine} \rightleftharpoons \text{chloromethane} + \text{hydrogen chloride}$$

In this situation, the speed of the forward reaction is equal to that of the reverse reaction, and the *equilibrium constant*, K, is given by the following expression, where the terms in square brackets represent the concentrations of the components:

$$K = \frac{[\text{chloromethane}][\text{hydrogen chloride}]}{[\text{methane}][\text{chlorine}]}$$

If $K > 1$, there is a greater concentration of chloromethane and hydrogen chloride present at equilibrium and, *vice versa*, if $K < 1$. The difference in *free* (or chemically available) *energy*, ΔG, between the two pairs of components of the system is related to the value of K by the following equation, where R is the gas constant and T the absolute temperature:

$$\Delta G = -2{\cdot}303\, RT \log K$$

When $K=1$, $\Delta G=0$, and there is no difference in free energy between the groups of components on either side of the equation. A negative ΔG value ($K > 1$) signifies that the components on the right-hand side predominate at equilibrium, whereas the converse applies when ΔG is positive. ΔG is composed of two energy terms: the *heat of reaction*, ΔH, and a temperature-entropy contribution, $T\Delta S$ (see p. 37). The latter term will be ignored for the further discussion of the halogenation of alkanes, because the heat change (ΔH) is the main type of energy factor involved in each of the recurrent elementary processes (*b*) and (*c*). Heats of reaction can be measured experimentally by calorimetric methods; for the chlorination of methane the value is -27 kcal./mole. (The corresponding value for the combustion of methane, $CH_4 + 2O_2 \rightarrow CO_2 + 2H_2O$ is -192 kcal./mole). This means that the reaction is exothermic or, in other words, the equilibrium given above will lie strongly to the right-hand side.

From an analysis of thermochemical evidence, values of *bond dissociation energies* can be assigned to covalent bonds, each link giving a characteristic value. Some relevant bond dissociation energies (in kcal./mole) are

CH_3—H	101	Cl—Cl	57
CH_3—Cl	80	H—Cl	103

Such values are especially useful for discussing reactions that are not easy to study directly: for example, ΔH values for the suggested stages (*b*) and (*c*) are estimated to be -2 (101–103) and 23 kcal./mole (57–80) respectively. A possible alternative mechanism for the chlorination of methane could involve the recurrent steps (*d*) and (*e*):

$$(d)\quad Cl\cdot + CH_4 \longrightarrow Cl{-}CH_3 + H\cdot$$
$$(e)\quad H\cdot + Cl_2 \longrightarrow H{-}Cl + Cl\cdot$$

Although step (*e*) is energetically more favourable ($\Delta H = -46$ kcal./mole) than either (*b*) or (*c*), reaction (*d*) gives a positive

ΔH value (+21 kcal./mole), and effective propagation *via* this step would therefore be impossible.

Activation, and the Transition State. A favourable ΔG value for a chemical equation signifies that the reaction *can* take place, but it does not reveal if it *will* occur under a particular set of conditions: the fact that methane and chlorine (or hydrogen and oxygen) are stable at 25° in the absence of light shows that

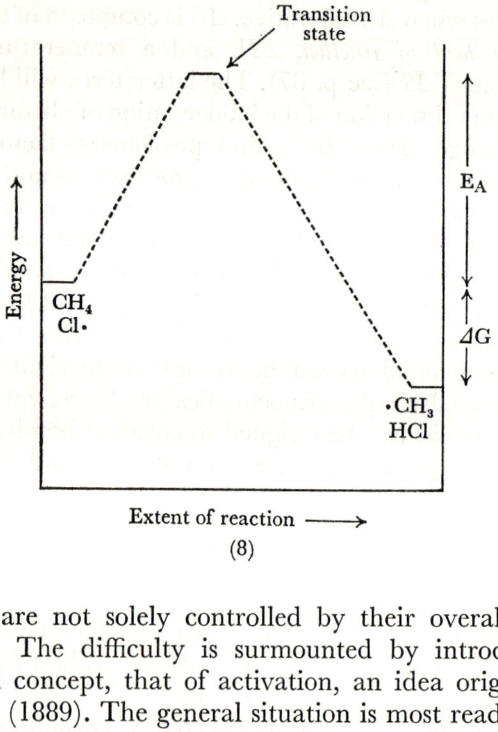

(8)

reactions are not solely controlled by their overall thermodynamics. The difficulty is surmounted by introducing an additional concept, that of activation, an idea originated by Arrhenius (1889). The general situation is most readily visualised in terms of an energy profile,* exemplified in diagram (8) by step (*b*) of the methane-chlorine reaction.

ΔG, the fall in free energy during the complete reaction, is represented by the difference in the levels of the initial and final states. If ΔG were the only factor involved, the reaction would

* Dotted lines instead of continuous lines or curves are used in energy profiles of reactions in this book because, in general, very little information is available concerning the regions in between minima and maxima.

simply reduce to the equilibrium, $CH_4 + Cl\cdot \rightleftharpoons \cdot CH_3 + HCl$; in other words it would become *equilibrium controlled.*

However chemical changes are frequently not 'spontaneous' as implied by the term, equilibrium. They occur at finite rates and thus are *kinetically controlled.* This is because the required bond alterations will not take place until the reactants acquire a certain minimum amount of energy (become sufficiently activated) before they collide. This energy increment is called the *energy of activation*, symbolized as E_A. When these conditions are fulfilled a maximum in the energy profile is reached. This particular energy level is called the *transition state* or the activated complex. At this point the reaction products can be formed by completion of the bond changes that have already partly occurred at the transition state. This is a downhill pathway, the surplus energy being dissipated as kinetic energy. The partial formation and breaking of covalent bonds at the transition state are conveniently represented by dotted lines:

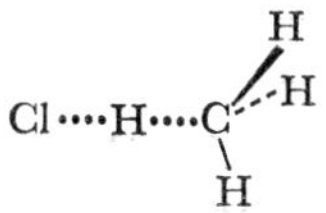

The energy increment, E_A, required by the reactants in order to reach the transition state is usually derived from the kinetic energies of the species involved. At any one moment the particles in the system are moving at various speeds; the fraction of them containing enough energy to be able to react will be proportional to $e^{-\frac{1}{2}Mv^2/RT}$ (Maxwell-Boltzmann distribution), when e is the base of natural logarithms, M is the molecular weight, v the average velocity, R the gas constant, and T the absolute temperature. If the rate depends on the proportion of such energy-rich particles, and if the energy of activation is simply related to the kinetic energy term $\frac{1}{2}Mv^2$, the equation can be rewritten as

$$\text{rate} = ke^{E_A/RT} \quad \text{(Arrhenius equation)}$$

The expression is temperature dependent, and the observed increase of reaction rates with temperature is important evidence for the activation theory.

Many important concepts have been introduced in the above sections. A deeper understanding of these ideas will develop from consideration of further examples in subsequent Chapters.

Petroleum. Methane the major component of natural gas is available for chemical syntheses, because it can be separated readily from ethane and other alkanes, but alkanes in bulk are obtained from petroleum. Crude petroleum contains straight chain and branched chain hydrocarbons ranging from methane up to about $C_{70}H_{142}$. Cyclic hydrocarbons, especially methyl and ethyl derivatives of cyclopentane and cyclohexane, are often present, and Borneo petroleum contains 30–40% of aromatic hydrocarbons, e.g. toluene (p. 187). Organic acids ('naphthenic acids') and organosulphur compounds are present in most petroleums, but organonitrogen compounds are found only in trace amounts.

Petroleum occurs underground in the interstices of porous rocks such as sandstone. Wells to reach such underground reservoirs of oil were first drilled in Pennsylvania, U.S.A., in 1859, and in the Baku district of Russia in 1873. Large reservoirs of oil have been since discovered in Venezuela, in Texas, in the Sahara Desert and around the Persian Gulf. Natural gas has recently been found in Holland and under the North Sea.

The origin of petroleum has been a subject of some controversy, but it is now apparent that hydrocarbons are present in most, perhaps all, forms of plant and animal life in small amounts (e.g. 5 parts in 100,000 in fish), and that most petroleum is derived from biological material, although some may also be formed from inorganic substances such as metal carbides. Marine deposits of moderate age (10,000–15,000 years) contain hydrocarbon and asphaltic material (3 parts per million), and it appears that over longer periods of time droplets of chemically stable hydrocarbons have accumulated to give underground pools of oil. The saturated hydrocarbons, pristane, $C_{19}H_{40}$, and phytane, $C_{20}H_{42}$, occur in petroleum; the fact that the structures of these compounds are based on isoprene units (p. 145) strongly suggests a biological origin. Evidence for the existence of life on our planet more than 10^9 years ago comes from the detection of pristane and phytane in rocks of the Precambrian period. These investigations were assisted by the introduction of techniques for the separation and identification of small amounts of material; gas chromatography and mass spectrometry (p. 7) are particularly effective.

Crude petroleum is a very dark liquid, often admixed with water and sand. Distillation is used to separate the crude oil into fractions of different boiling point ranges:

Boiling range	*Name*		*Uses (apart from cracking)*
20–90°	light petroleum		solvent
90–120°	ligroin		solvent
100–200°	petrol, gasoline	naphtha	fuel
200–300°	paraffin, kerosene	naphtha	fuel
above 300°	gas oil		fuel
above 300°	lubricating oil		lubricant
residue	asphalt, bitumen		road construction

Further stages in refining depend on the chemical composition of the distillation fractions and their ultimate uses. The petrol fraction obtained directly from distillation of natural petroleum often contains a large proportion of straight chain hydrocarbons; it will then have a low 'octane number', and not be suitable as a fuel in modern high compression motor engines. The octane number is a measure of the performance of the fuel. 2,2,4-Trimethylpentane (octane number = 100) is a good fuel because it burns smoothly and does not cause the engine to 'knock'. At the other end of this arbitrary scale, the straight-chain n-heptane is given the octane number of 0. The octane number of a petrol fraction can be improved by the addition of either branched chain hydrocarbons (synthesised separately from lower alkanes and alkenes) or other suitable compounds, e.g. benzene and ethyl alcohol (octane numbers of 106 and 95 respectively). The performance of fuels in internal combustion engines is also improved by the addition of tetraethyl-lead (p. 130).

Some of the lower alkanes are separated and purified from natural gas and petroleum and then used in chemical syntheses, but the complexity of the natural material makes the isolation of the higher alkanes unattractive. Higher boiling petroleum fractions are now converted into a mixture of simpler molecules by a process known as *cracking*. The C—C and C—H bond dissociation energies in alkanes are high (*ca.* 83 and 103 kcal./mole respectively), and temperatures above 400° are necessary to promote the endothermic reactions involved in cracking. At such

SOME PRODUCTS FROM PETROLEUM

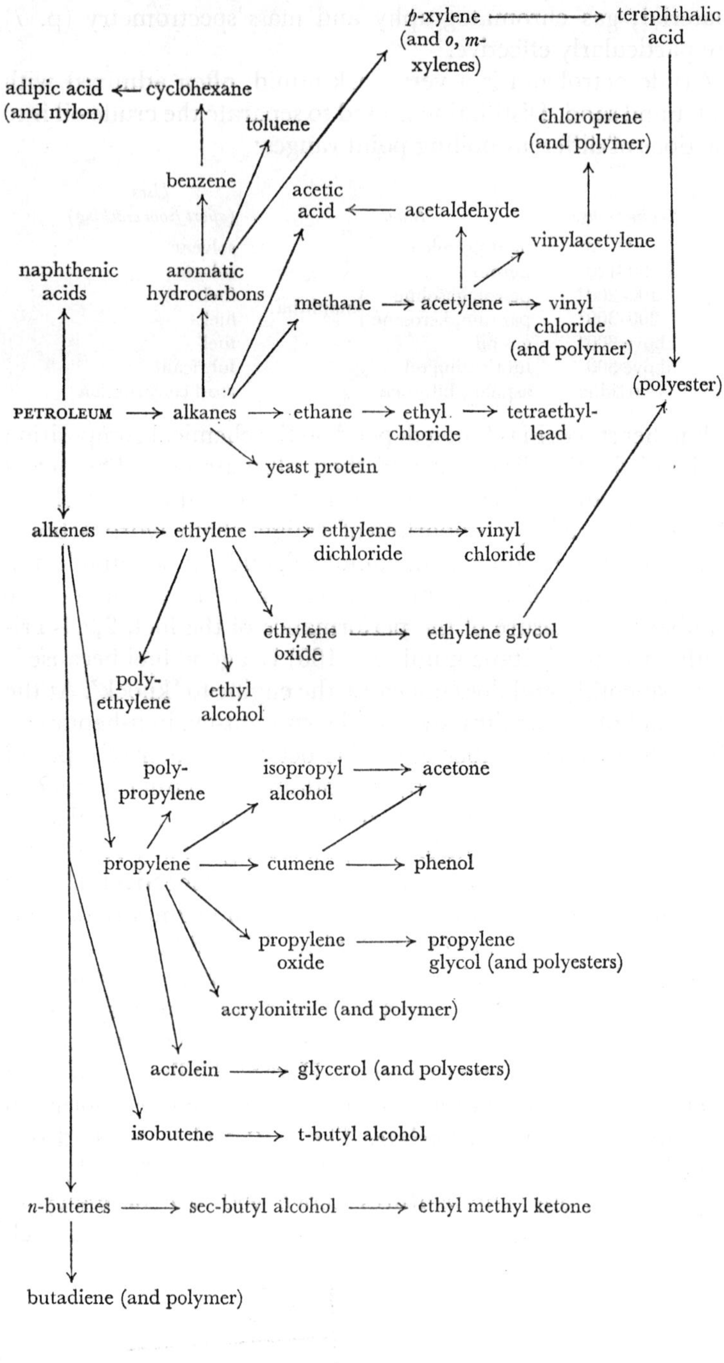

temperatures, the products most required, ethylene and propylene (Chapter 5), can undergo further changes to give coke and hydrogen. Optimum amounts of the two compounds are obtained by using a short reaction time (2 sec. at 825°), and by having steam present as a diluent; typical yields are ethylene (30 %) and propylene (15 %). Adsorption procedures are used to separate the two compounds from alkanes and hydrogen, but fractional distillation at low temperatures (−110° to −50°) is proving to be more economical as the scale of cracking increases.

Complex chemical changes are involved in the microbiological, enzymic conversion of alkanes into protein (p. 160). A yeast is grown in a dispersal of the hydrocarbon in water containing ammonium salts and other nutrients, the rate of protein formation being about a thousand times that in synthesis in animals. The proportion of lysine (p. 158) in yeast protein is more than adequate for human diet, whereas cereal grains are often deficient in this essential amino-acid.

The flow chart on the opposite page outlines routes from petroleum to various products.

U.K. chemical industry; some production figures, 1965

	Value–£ millions	*Organic chemicals–thousand tons*	
dyestuffs	73	acetone	118
fertilisers	135	butadiene	109
inorganic chemicals	267	ethylene	529
oils and fats	159	ethylene oxide	97
organic chemicals	366	formaldehyde	88
paints and varnishes	179	phenol	48
pharmaceuticals	247	phthalates	139
plastic materials	281	propanols	144
soap and detergents	139	propylene	260

QUESTIONS

1. Write structural formulae for hydrocarbons of molecular formula C_6H_{14}, and give a systematic name to each isomer.

2. Write structural formulae for the five *mono*-chloro-compounds obtained by chlorination of 4-methylheptane. Suggest procedures for finding out (*a*) if hydrogen chloride is also formed, and (*b*) when all the chlorine has reacted.

3. Comment on the following observations.

(*a*) A gaseous substance (analysis: C, 80%; H, 20%) was mixed with nitric acid vapour and passed through a tube at 400°. The product contained a compound (b.p. 114°) that gave the analytical results: C, 32%; H, 6·7%; N, 18·7%.

(*b*) A mixture of n-butane and deuterium was heated at 66° for 70 min. in the presence of platinum. Examination of the gaseous product by mass spectrometry showed that compounds with molecular weights from 58 to 68 were present.

4. Point out errors in the following:

(*a*) Bromination of methane proceeds according to the equation:

$$CH_4 + Br_2 \longrightarrow CH_2Br_2 + 2HBr$$

(*b*) By analysis the compound had the molecular formula $C_5H_{10}Cl$.

5. Suggest a mechanism for the formation of hydrogen chloride from hydrogen and chlorine. The reaction proceeds extremely rapidly at 20° when light is admitted to a mixture of the two pure starting materials.

6. (*a*) How many primary, secondary, tertiary and quaternary carbons are there respectively in 2-methylbutane?

(*b*) In the photo-initiated chlorination of 2-methylbutane the following yields of monochloro-compounds are obtained:

$(CH_3)_2C(Cl)CH_2CH_3$	22%	$ClCH_2CH(CH_3)CH_2CH_3$	30%
$(CH_3)_2CHCH(Cl)CH_3$	33%	$(CH_3)_2CHCH_2CH_2Cl$	15%

Calculate the percentage yield of each of the four products on the assumption that the rate of chlorination of each C—H group in 2-methylbutane is the same. Remember to take account of the fact that some carbons are bound to more than one hydrogen.

7. From the following bond dissociation energies (in kcal./mole) calculate the heat of reaction (ΔH) for each of the three stages in the fluorination and bromination of methane, state whether each is exothermic or endothermic, and compare with chlorination (p. 19).

C—H	101	H—Br	87	Br—Br	46	C—Br	67
		H—F	134	F—F	36	C—F	107

8. (*a*) Which of the butanes has a zero dipole moment, and which shows a moment of 0·13 D? (see pp. 32, 58)

(*b*) How could the alkane be recovered from a gaseous mixture of butane and air?

(*c*) What minimum volume of air (at N.T.P.) is theoretically required to enable the combustion of 1 g. of octane to proceed to completion.

There is nothing so practical as a good theory.

KURT LEWIN (1890–1947)

CHAPTER 3

ALKYL HALIDES

Alkyl halides, compounds derived from alkanes by replacement of one or more hydrogens, are useful for the synthesis of organic compounds because the halogen can be replaced by other groups. Ethyl chloride, CH_3CH_2Cl, was first obtained from the action of sulphur chloride or metal chlorides ($FeCl_3$, $AlCl_3$) on ethyl alcohol.

Structure and Nomenclature. Alkyl halides are named as derivatives of alkanes by indicating the presence and position of the halogen by prefixes. Many of the simpler compounds also have semisystematic names in which the alkyl and halide groups are written as pairs. Examples of both methods are:

	Systematic name	*Semisystematic name*
CH_3CH_2I	iodoethane	ethyl iodide
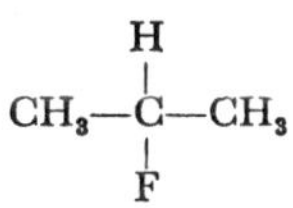 or CH_3CHFCH_3	2-fluoropropane	isopropyl fluoride
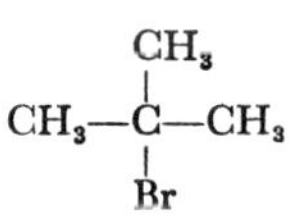 or $(CH_3)_3CBr$	2-bromo-2-methylpropane	t-butyl bromide (t- short for tertiary)
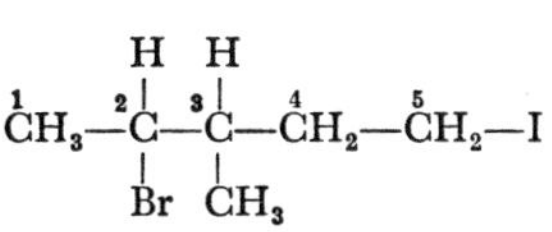 or $CH_3CHBrCH(CH_3)CH_2CH_2I$	2-bromo-5-iodo-3-methylpentane (4-bromo-1-iodo-3-methylpentane if the pentane chain is numbered in the reverse direction)	

In the last two compounds, each set of prefixes is arranged in alphabetical order. Alkyl halides (general symbol, RX, where R is alkyl and X is halogen) can be divided into three groups depending on the number of carbons attached to the C-halogen group. Compounds containing the group CH_2-halogen (including methyl halides) are called *primary* alkyl halides, abbreviated to RCH_2X. Compounds of general formulae, R_2CHX, and R_3CX, are referred to as *secondary* and *tertiary* alkyl halides respectively.

Physical Properties. The boiling points of alkyl halides depend on the complexity of the alkyl groups, and increase as the atomic weight of the halogen constituent is increased:

	b.p.	*dipole moment*, D
fluoroethane (ethyl fluoride)	−32°	1·95
chloroethane (ethyl chloride)	13°	2·0
bromoethane (ethyl bromide)	38°	2·0
iodoethane (ethyl iodide)	72°	1·8

Ethyl chloride is sometimes employed for minor operations, local anaesthesia being produced by rapid cooling of the tissues when the low boiling halide is applied. Methyl iodide is a liquid, b.p. 42°, but the other methyl halides are gases at room temperature.

The electrons of a carbon–carbon bond are distributed symmetrically, and the bond is 'non-polar'. Halogens, however, have a greater power of attracting electrons than carbon (are more *electronegative*), and in a carbon-halogen bond there is displacement of the bonding electrons towards halogen giving a $\delta^{\ominus}$ charge at halogen and leaving a $\delta^{\oplus}$ charge at carbon. The polar character of a carbon–halogen bond can therefore be expressed:

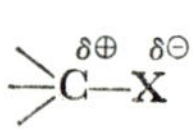

The degree of charge separation in a molecule can be measured, and this *dipole moment* is usually expressed in Debye units (D). The dipole moments of the ethyl halides given in the table above largely result from the polar nature of the carbon–halogen bonds, since the carbon–carbon and carbon–hydrogen bonds in the ethyl groups are essentially non-polar.

Alkyl halides are only slightly soluble in water in which they form upper or lower layers depending on their specific gravities. The following figures for the 1-butyl halides show that specific gravities and refractive indices, both of which are useful in the identification of alkyl halides, increase in the order fluoride, chloride, bromide and iodide.

	b.p.	*Density**	*Refractive index, n_D*
1-fluorobutane	32°	0·775	1·342
1-chlorobutane	77·8°	0·885	1·402
1-bromobutane	101·6°	1·275	1·440
1-iodobutane	130°	1·616	1·500

* Alkanes have densities in the 0·6–0·75 region.

Primary Alkyl Halides and the S_N2 Reaction. 1-Bromopropane, a liquid, is almost insoluble in an aqueous solution of sodium hydroxide. When the mixture of the two phases is stirred at 20°, the layer of organic halide *slowly* diminishes in volume due to its conversion into the water-soluble propan-1-ol.

$$\underset{\text{1-bromopropane}}{CH_3CH_2CH_2—Br} + \underset{\text{(solvated ions)}}{{}^{\oplus}Na^{\ominus}OH} \xrightarrow[\text{water}]{\text{in}} \underset{\text{propan-1-ol}}{CH_3CH_2CH_2—OH} + \underset{\text{(solvated ions)}}{{}^{\oplus}Na^{\ominus}Br}$$

In the reaction the halogen in the alkyl halide has been replaced by a hydroxyl (OH) group; the process is termed *hydrolysis*. Study of the kinetics of this type of reaction reveals that the rate is second order and proportional to the concentrations of both reactants:

$$\text{rate} = k[CH_3CH_2CH_2Br]\,[HO^{\ominus}\ \text{solvated}]$$

k is the *specific rate constant* usually expressed in moles per litre per second at the temperature of the reaction. A picture of the reaction which is consistent with these results is that hydrolysis of a molecule of alkyl bromide only occurs when it suffers a sufficiently energetic collision with a hydroxide ion. However the latter needs to collide with the carbon carrying the bromine substituent; if it collides with other parts of the molecule hydrolysis will not occur. When the bromine is at another position in the molecule a different product is formed:

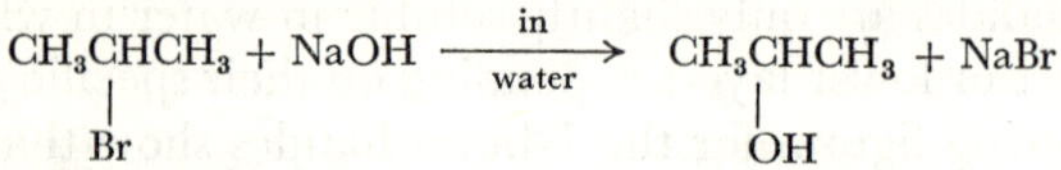

2-bromopropane propan-2-ol

Energy profiles can be drawn similar to that for the reaction of methane with a chlorine atom (p. 24) with the central formula below representing the structure of the transition state. Despite this analogy there is a fundamental difference between the two

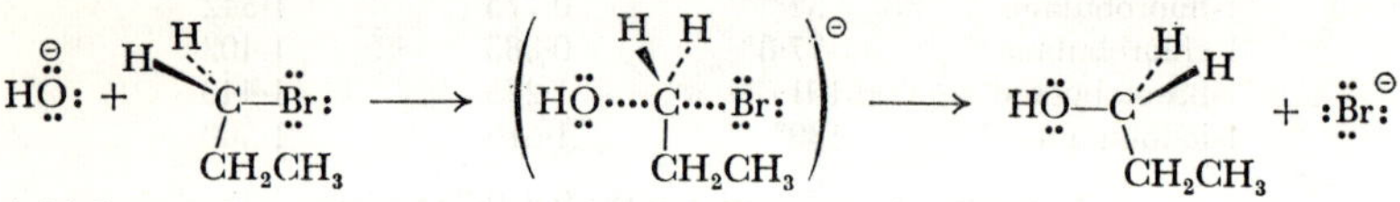

initial ground states transition state final ground states

reactions. In the halogenation of alkanes cleavage of a bond yields neutral radicals (a *homolytic* process), whereas ionic species are involved in the hydrolysis of an alkyl bromide. The latter is an example of a *heterolytic* reaction in which both electrons of the new covalent bond are supplied by one species ($HO^{\ominus}$), and both electrons of the bond that breaks are acquired by bromine (to give $Br^{\ominus}$). Heterolytic reactions are much more likely to take place in polar solvents (e.g. water, alcohols, acetone, dimethylformamide) than in the gas phase, because ions associate with such solvents (by *solvation*) (p. 88) and are thus brought into solution and stabilised.

In the hydrolysis of an alkyl bromide the hydroxide ion is behaving as a *nucleophile* because, according to Ingold's definition of this term, it acts by donating electrons to or sharing them with another atomic nucleus (carbon in the present case). In a complementary manner the alkyl bromide is functioning as an *electrophile* in the activation step, because it acquires electrons from another species. Nucleophilic reagents must contain at least one pair of unbonded electrons so that a new bond can be formed. A second order substitution reaction effected by an incoming nucleophile is often referred to as an S_N2 reaction. One elementary process only is involved in an S_N2 reaction, as indicated by the simple contour of the energy profile, because bond making and breaking occur simultaneously.

Other, more condensed ways of representing the alkaline hydrolysis of 1-bromopropane are:

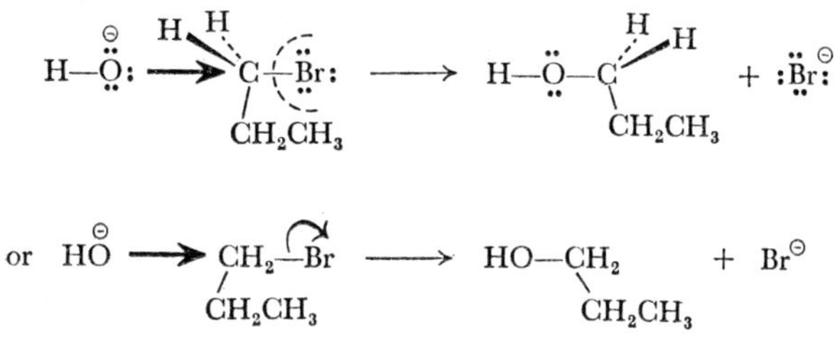

The bold arrows denote movement of electron pairs, and this procedure is generally useful for representing the mechanisms of heterolytic reactions (cf. pp. 97, 134). The diagrams also indicate that when reaction occurs the nucleophile approaches carbon from the direction remote from the bromine atom, thus causing an 'umbrella-like' inversion of the molecule (cf. p. 157).

The vast difference in reactivity between organic and inorganic (ionic) bromides when treated with sodium hydroxide should be noted. Whereas the alkaline hydrolysis of 1-bromopropane proceeds at a slow and measurable rate, reaction between magnesium bromide and sodium hydroxide in water proceeds at an exceedingly rapid rate under normal conditions and magnesium hydroxide is precipitated:

$$2NaOH + MgBr_2 \xrightarrow{\text{in } H_2O} 2NaBr + Mg(OH)_2 \downarrow$$

An aqueous solution of magnesium bromide contains hydrated metal cations and halide anions which are distributed randomly throughout the solution. When a solution containing hydrated sodium and hydroxide ions is added, magnesium and hydroxide ions very rapidly meet by diffusion, and precipitation occurs almost immediately. By contrast a molecule of an alkyl bromide retains its covalent structure in the presence of water until a sufficiently activated collision leads to reaction.

Many anions and neutral molecules contain an unbonded electron pair and fulfil the structural requirements of a nucleophile. We may expect them to react with alkyl halides in an analogous way to hydroxide ions. Such reactions proceed in the best yields (above 75%) using *primary* alkyl halides, and the following chart illustrates how 1-bromobutane, for example, can be transformed into several different types of compound.

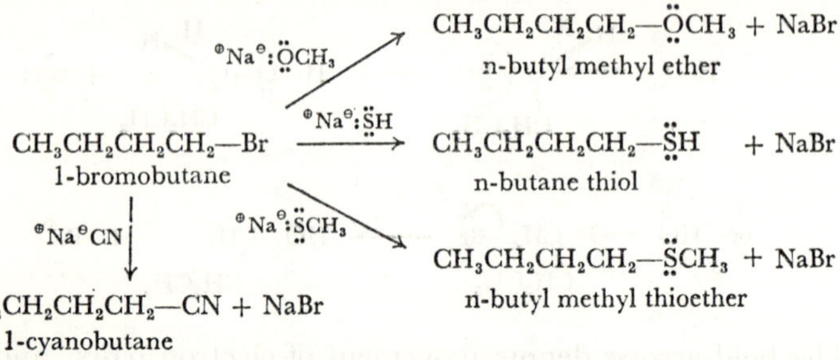

For these reactions to proceed to completion at convenient rates it is necessary to find a liquid or liquid mixture that will act both as a diluent and as a solvent for the covalent organic halide and the ionic sodium salt. For one of the reactions the following procedure was developed.

A mixture of 1-bromobutane (68·5 g.; 0·5 mole*), sodium cyanide (27 g.; 0·55 mole) and ethylene glycol (p. 133) (150 ml.) was stirred and heated in a 500 ml. two-necked flask (condenser and thermometer). The b.p. was close to that of the halide initially (102°), but rose rapidly when almost all of it had reacted (after about 1 hour). The product, 1-cyanobutane, was isolated by distillation, its b.p. being much lower (138–142°) than that of ethylene glycol (197°). The yield was 37·4 g. (90% of that possible).

An earlier method, in which aqueous ethanol was used as solvent, gave a lower yield (80%), and was less convenient (reflux time of 25 hours, and a more difficult fractional distillation).

Although ammonia, $:NH_3$, is a neutral molecule it contains an unbonded pair of electrons and therefore can react as a nucleophile with an alkyl halide:

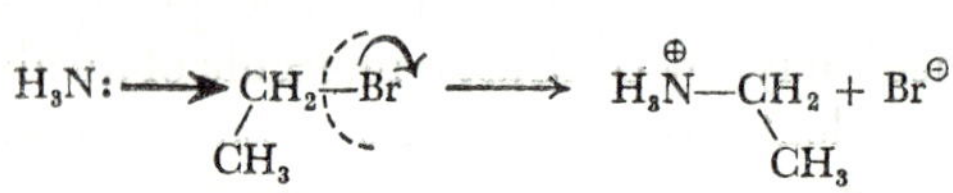

In this process the two unbonded electrons of ammonia are used to form a nitrogen–carbon covalent bond, in which the two

* 1 mole is defined as the formula weight in grams, while the term 'mol.' indicates relative molar quantities. An alternative expression for this experiment is '... 1-bromobutane (68·5 g.; 1 mol.), sodium cyanide (27 g., 1·1 mol.)...

electrons are shared equally. The nitrogen now only has a 'half-share' of its two original electrons. As a consequence, the charge on nitrogen changes from neutral to a unit positive value as the bond from nitrogen to carbon is formed, Other nucleophilic reagents undergo the same alteration of charge (loss of one negative unit) when they react with alkyl halides. Thus, in the hydrolysis of alkyl halides, the charge on oxygen changes from negative ($H\ddot{\underset{..}{O}}:^{\ominus}$) to neutral ($H{-}\ddot{\underset{..}{O}}{-}R$) as the covalent bond to carbon develops.

In practice, products other than alkylammonium halides are also formed when alkyl halides are treated with ammonia. This feature of the reaction is discussed later (p. 116).

Entropy. We should now consider the concept of *entropy* (S), which refers to the extent of disorder in a substance. Earlier the relation between the equilibrium constant and the free energy difference in an equilibrium was presented, and mention was made of the idea that free energy is composed of two terms: the relevant equation is $\Delta G = \Delta H - T\Delta S$, where T is the absolute temperature and ΔS is the change in entropy. A definition of an entropy change is given by $\Delta S = q/T$, where q is the amount of heat supplied to the substance at the absolute temperature T. A crystalline solid consists of an ordered array of atoms or molecules, but when it melts (at T) a quantity of heat (q) is required in order to convert it all into the liquid form. In the liquid state the atoms or molecules are arranged more randomly than in the original solid, and the transformation therefore results in an increase in entropy. Positive entropy changes also occur in reactions of the type $A \rightarrow B + C$, where there is an increase in the number of molecules. A consequence of a positive ΔS value is that an increase in temperature of the system will make the $T\Delta S$ energy term more negative in value; this can become sufficiently large to override a positive (unfavourable) ΔH value for the change. Molecules represented by A can be more stable at lower temperatures, while those represented by B and/or C become more stable than A at higher temperatures. Examples of this effect are given on pp. 29, 70, 122, 145.

From the point of view of thermodynamics, a system of molecules is always striving to reach the position of equilibrium under the prevailing conditions. Equilibrium represents the position at which the two opposing tendencies of lowest energy and highest entropy are balanced.

By studying the kinetics of a reaction at different temperatures, the energy of activation (E_A) can be separated into terms that represent the heat ($\Delta H^{\ddagger}$) and the temperature-entropy ($T\Delta S^{\ddagger}$) contributions. For the following reaction (see below), the values of 19 kcal./mole for the heat (enthalpy) of activation, and -3 kcal./mole for the entropy of activation were obtained:

$$CH_3CH_2Br + I^{\ominus} \xrightarrow[\text{at } 20\text{–}50^{\circ}]{\text{in acetone}} CH_3CH_2I + Br^{\ominus}$$

The values, which are typical for an S_N2 displacement, show that unfavourable heat and entropy changes are both involved in the activation step (the minus $T\Delta S^{\ddagger}$ value becoming positive from the $-T\Delta S^{\ddagger}$ term in the expression for E_A).

Preparation of Alkyl Halides. The various methods for preparing alkyl chlorides and bromides include the halogenation of alkanes, the replacement of the hydroxyl group of alcohols (p. 45), and the addition of hydrogen chloride and hydrogen bromide to alkenes (p. 61). The less generally available alkyl fluorides and iodides can be made from their chloro- or bromo-analogues by halogen exchange reactions. For fluorides, the alkyl bromide is heated with mercuric fluoride:

$$\underset{\text{1-bromopropane}}{2CH_3CH_2CH_2Br} + HgF_2 \longrightarrow \underset{\text{1-fluoropropane}}{2CH_3CH_2CH_2F} + HgBr_2$$

An adaptation of this method is used for the preparation of dichlorodifluoromethane from carbon tetrachloride:

$$CCl_4 + 2HF \xrightarrow[SbF_5]{SbF_3} Cl_2CF_2 + 2HCl$$

Dichlorodifluoromethane, b.p. $-30°$, is a relatively non-toxic compound used as a propellent for aerosol sprays and as a fluid in refrigerators.

Alkyl iodides are formed when alkyl chlorides or bromides are treated with sodium iodide, e.g.

$$\underset{\text{2-chloropentane}}{CH_3CH_2CH_2\underset{\substack{|\\ Cl}}{C}HCH_3} + NaI \xrightarrow[\text{acetone}]{\text{in}} \underset{\text{2-iodopentane}}{CH_3CH_2CH_2\underset{\substack{|\\ I}}{C}HCH_3} + NaCl$$

Sodium iodide is soluble in acetone but, as the substitution reaction proceeds, the less soluble sodium chloride is precipitated.

Nucleophilic exchange of halogen is used for the introduction of another isotope of the halogen. For example, a mixture of ordinary ethyl bromide and lithium bromide in which the bromine is radioactive (^{82}Br; the natural stable isotopes of bromine are ^{79}Br and ^{81}Br) can be brought to equilibrium. Recovery of the ethyl bromide gives material containing 'labelled' as well as 'unlabelled' halide.

$$CH_3CH_2Br + Li^{82}Br \underset{\text{acetone}}{\overset{\text{in}}{\rightleftharpoons}} CH_3CH_2{}^{82}Br + LiBr$$

Atomic Charge and Structure. Many heterolytic reactions involve ions or molecules with positively or negatively charged groups. The general structures of cations, neutral molecules and anions derived from carbon, nitrogen, oxygen and fluorine are given in the following table, where R represents an alkyl group or hydrogen.

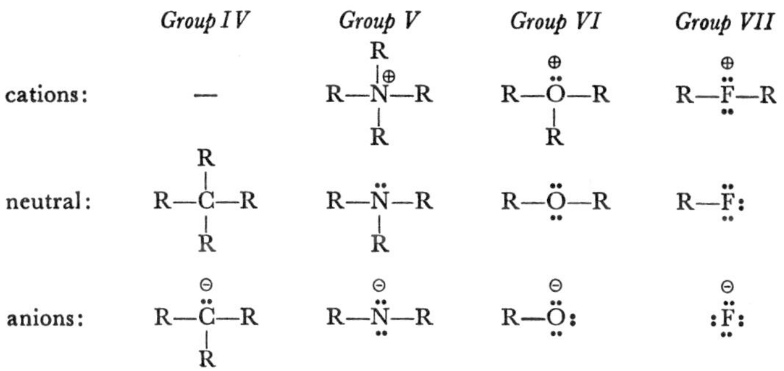

It should be noted that for the species in the table the valence shell of each central atom has eight electrons, but the number of covalent bonds to that atom depends on its charge in addition to its position in the periodic classification. These relationships can be illustrated by considering the three oxygen-containing species. In a neutral molecule, R—Ö—R, the six valence electrons of oxygen (four in two unbonded pairs, and two from two shared pairs) effectively compensate for the nuclear charge. The species, $R_3O:^{\oplus}$, bears a unit positive charge since there are now only five electrons (two in an unbonded pair and three in shared pairs) available to neutralise the nuclear charge on oxygen. In the case of the anion, $R\ddot{\underset{..}{O}}:^{\ominus}$, oxygen possesses six electrons in three unbonded pairs and a 'half-share' in the bonded pair; the total of seven electrons indicates that the atomic charge has a unit negative value. The relation between ammonia and ammonium ion has been discussed already (p. 36). The species in the above table all have the same orbital configuration at the central atom and in this respect are *isoelectronic.*

Elimination of Hydrogen Halide. *Formation of alkenes.* Although propan-1-ol is the main product when 1-bromopropane is treated with aqueous sodium hydroxide, a competing (or 'side') reaction gives propylene, formed by *elimination* of hydrogen bromide from the alkyl bromide.

main reaction:

$$CH_3CH_2CH_2Br + NaOH \longrightarrow CH_3CH_2CH_2OH + NaBr$$

side reaction:

$$CH_3CH_2CH_2Br + NaOH \longrightarrow CH_3CH{=}CH_2 + NaBr + H_2O$$

Elimination also competes with substitution in the other reactions of alkyl halides discussed above. The ratio of elimination to substitution depends on the number of alkyl groups attached to the carbon of the carbon–halogen bond; elimination occurs to the greatest extent with tertiary halides, and to the least extent with primary halides. For example, in the reactions of alkyl halides with metal alkoxides, ethers are only formed in good yields starting from primary halides. Elimination becomes the predominant reaction when a *tertiary* alkyl halide, e.g. t-butyl bromide, is treated with sodium ethoxide in ethanol:

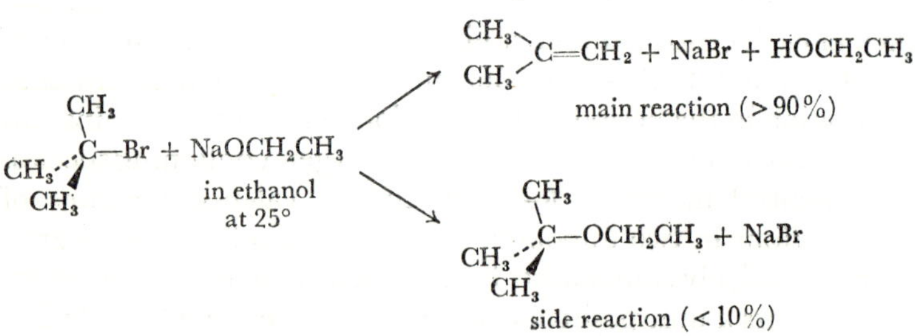

The products from the elimination reactions of alkyl halides contain carbon–carbon double bonds and are called *alkenes* (Chapter 5). Elimination reactions are discussed further on p. 120 and the behaviour of tertiary halides on p. 64.

QUESTIONS

1. Assign systematic names to the monochloro-compounds produced in the chlorination of 2-methylbutane (cf. question 6, Chapter 2). Classify each compound as a primary, secondary or tertiary halide.

2. Give names to the dichloro-compounds that can be formed in the chlorination of 1-chloro-2-methylbutane.

'One of these dichloro-compounds has a zero dipole moment'; is this statement right or wrong?

3. (*a*) Where appropriate, add the charge (magnitude and sign) to the following species, assuming that they are in their normal electronic states (eight electrons in the outer shell for elements other than hydrogen): O, CH_3O, $H_2OCH(CH_3)_2$, CH_2CH_3, ICl, NH_4, $NHCH_2CH_3$, $CH_3SCH_2CH_3$.

(*b*) Which two of the following species are unlikely to be able to act as nucleophiles: H_2S, $CH_3O^{\ominus}$, $(CH_3CH_2)_3N$, $(CH_3)_2CH^{\ominus}$, $CH_3CH_2\cdot$, $CH_3CH_2^{\oplus}$?

4. n-Butyl methyl ether can be prepared in the following way. Dissolve sodium (1 g.) in warm butan-1-ol (26 g.). Add methyl iodide (6·15 g.), and gently reflux the solution for 2 hours. Separate the ether product (b.p. 70–71°) from the excess of butan-1-ol (b.p. 118°) by distillation.

(*a*) What is the precipitate that separates during the 2 hour heating period, and how could you test your suggestion by experiment?

(*b*) Write a balanced equation for the main reaction.

(*c*) Give an electronic mechanism and draw an energy profile for the reaction (a second-order rate may be assumed).

(*d*) The weight of ether obtained was 2·7 g. What is its percentage yield based on methyl iodide?

(*e*) Calculate the atomic or mole ratios of sodium, butanol and methyl iodide that were put into the preparation. Can you suggest a reason for the fact that an excess of butanol was used?

5. (*a*) An apparatus for the continuous production of dichlorodifluoromethane can be readily designed because both reactants are liquid below 20°, and both products are gaseous at this temperature. Suggest a procedure for removing the hydrogen chloride from the required product, dichlorodifluoromethane.

(*b*) Radioactive bromine can be introduced into 1-bromobutane by the method given on p. 39. How may (i) the labelled alkyl bromide and (ii) the lithium bromide (still containing some $Li^{82}Br$ be isolated from the reaction mixture? Acetone (cf. p. 84) has a b.p. of 56°.

Sit down before fact as a little child, be prepared to give up every preconceived notion, follow humbly wherever and to whatever abysses nature leads, or you shall learn nothing.

THOMAS HUXLEY (1825–1895)

CHAPTER 4

ALCOHOLS AND RELATED COMPOUNDS

ALCOHOLS

Ethyl alcohol in the form of alcoholic beverages has been known from the dawn of human history. According to one legend it is a cure for all ailments, cf. whiskey (Gael. *uisge-beatha*, water of life).

Methyl alcohol, 'wood spirit', was discovered by Boyle in 1661 as a component of the distillate (pyroligneous acid) from the thermal decomposition of wood, although it was not until 1812 that its difference from ethyl alcohol was clearly recognised. Wood consists of cellulose and lignin, and the methyl alcohol is derived from the latter substance (p. 210). Esters of methyl alcohol also occur naturally, e.g. methyl salicylate (p. 210), 'oil of wintergreen'.

Cetyl alcohol, n-$C_{16}H_{33}OH$, is the alcoholic component of cetyl palmitate, the main component of spermaceti, a waxy solid obtained from sperm whale oil which, when refined, is used in emollients.

Structure and Nomenclature. The characteristic feature of the structure of alcohols is the presence of the group C—O—H. In other words, alcohols consist of alkyl and hydroxyl (OH) groups joined together, and the abbreviation, ROH, is often used.

The H—O—H bond angle in water and the C—O—H angle in alcohols are approximately 105°, which is close to the tetrahedral angle found in methane (p. 15). In water and alcohols, oxygen contains two unshared pairs of electrons, and the tetrahedral arrangement shown in diagram (1) (R=H or OH) minimises repulsion between bonded and unshared electron pairs.

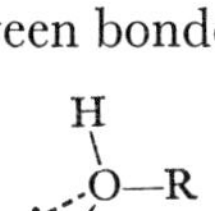

(1)

In systematic nomenclature, the presence of the hydroxyl group is denoted by the suffix *-ol*, which is attached to the name of the parent alkane (without the final *e*). The position of the hydroxyl group on the longest chain of carbons is indicated by a number placed before the suffix. There is also a semisystematic method, in which the term alcohol is preceded by the name of the alkyl group; this method is only suitable for the simpler alcohols.

	Systematic name (alkanol)	*Semisystematic name*
CH_3OH	methanol	methyl alcohol
CH_3CH_2OH	ethanol	ethyl alcohol
$CH_3CH_2CH_2OH$	propan-1-ol	(n-) propyl alcohol
$CH_3CH(OH)CH_3$	propan-2-ol	isopropyl alcohol
$CH_3(CH_2)_6CH_2OH$	octan-1-ol	(n-) octyl alcohol
$(CH_3)_2C(OH)CH_2CH_3$	2-methylbutan-2-ol	t-amyl alcohol
$HO\overset{1}{C}H_2\overset{2}{C}H(CH_3)\overset{3}{C}H_2\overset{4}{C}H_2Cl$	4-chloro-2-methyl butan-1-ol	—

The degree of substitution of the carbon atom bearing the hydroxyl group can be indicated, as with halides, by the terms primary ($-CH_2OH$), secondary ($>CHOH$) and tertiary ($\geqslant COH$). In the above list, propan-2-ol is a secondary alcohol, 2-methylbutan-2-ol is a tertiary alcohol and the remainder are primary alcohols.

The n-alkan-1-ols are liquid up to C_{12}. The higher compounds are solids.

CH_3OH	b.p. 65°	$CH_3(CH_2)_{10}CH_2OH$	m.p. 24°
CH_3CH_2OH	b.p. 78°	$CH_3(CH_2)_{14}CH_2OH$	m.p. 49°

Chain branching changes the physical properties; three of the butanols are liquids and one is a solid.

$CH_3CH_2CH_2CH_2OH$	b.p. 118°	$CH_3CH_2CH(OH)CH_3$	b.p. 108°
$(CH_3)_2CHCH_2OH$	b.p. 100°	$(CH_3)_3COH$	m.p. 25·5° (b.p. 83°)

The unexpectedly high boiling points of alcohols compared to alkyl halides or ethers of similar molecular weights, and of water compared, for example, to hydrogen sulphide is due to association of molecules by hydrogen bonding (p. 88).

General methods for the preparation of alcohols include the hydrolysis of alkyl halides (p. 33), the hydration of alkenes (p. 62), the reduction of aldehydes and ketones (p. 79), and the addition of Grignard reagents to compounds containing carbonyl groups (Chapter 10). Special methods for the preparation of certain alcohols are given later in this Chapter.

Alcohols can be recognised by their absorption bands near 3300 and 1100 cm.$^{-1}$ in the infrared region. Individual members of the series are identified by the formation of crystalline derivatives (e.g. 3,5-dinitrobenzoates, p. 223), which have well-defined melting points.

Formation of Metal Alkoxides. Sodium reacts vigorously with water to give hydrogen and sodium hydroxide. A similar, but less vigorous, reaction takes place when sodium is added to an alcohol; hydrogen and a *sodium alkoxide*, NaOR, are formed.

$$2Na + 2HOH \longrightarrow H_2 + 2NaOH(^{\oplus}Na\ ^{\ominus}OH)$$

$$2Na + 2CH_3CH_2OH \longrightarrow H_2 + \underset{\text{sodium ethoxide}}{2NaOCH_2CH_3}(^{\oplus}Na\ ^{\ominus}OCH_2CH_3)$$

Most of the electropositive metals react with alcohols to give alkoxides although anhydrous conditions are advantageous for those that form insoluble hydroxides.

$$2Mg + 2CH_3OH \longrightarrow H_2 + \underset{\text{magnesium methoxide}}{Mg(OCH_3)_2}$$

$$2Al + 6(CH_3)_3COH \longrightarrow 3H_2 + \underset{\text{aluminium t-butoxide}}{Al[OC(CH_3)_3]_3}$$

Replacement of Hydroxyl Group by Halogen. Hydrogen bromide dissolves in, and subsequently reacts with, butan-1-ol to give 1-bromobutane, but better yields are obtained when sulphuric acid is also present. The sulphuric acid combines with the water formed in the reaction:

$$CH_3CH_2CH_2CH_2OH + HBr \longrightarrow \underset{\text{1-bromobutane}}{CH_3CH_2CH_2CH_2Br} + H_2O$$

$$H_2O \xrightarrow{H_2SO_4} H_3O^{\oplus} + HSO_4^{\ominus}$$

The reaction proceeds by a two step mechanism. In the first step, a proton from one of the strong acids adds to the alcohol:

$$CH_3CH_2CH_2CH_2{-}\ddot{\underset{\cdot\cdot}{O}}{-}H + H^{\oplus} \rightleftharpoons CH_3CH_2CH_2CH_2{-}\overset{H}{\underset{\oplus}{\overset{|}{\ddot{O}}}}{-}H \quad (2)$$

This elementary process has a low activation energy, and is fast and reversible like the corresponding movement of protons between water molecules:

$$H_3O^{\oplus} + H_2O \rightleftharpoons H_2O + H_3O^{\oplus} \qquad E_A = 2{\cdot}6 \pm 0{\cdot}3 \text{ kcal./mol.}$$

$$HO^{\ominus} + H_2O \rightleftharpoons H_2O + HO^{\ominus} \qquad E_A = 4{\cdot}8 \pm 0{\cdot}5 \text{ kcal./mol.}$$

The positively charged *oxonium* cation (2) is similar electronically to the hydronium ion, $H_3O^{\oplus}$. In the second stage, the cation (2) reacts with a bromide anion (from HBr) by an S_N2 reaction, in which two neutral molecules are produced from the pair of oppositely charged ions.

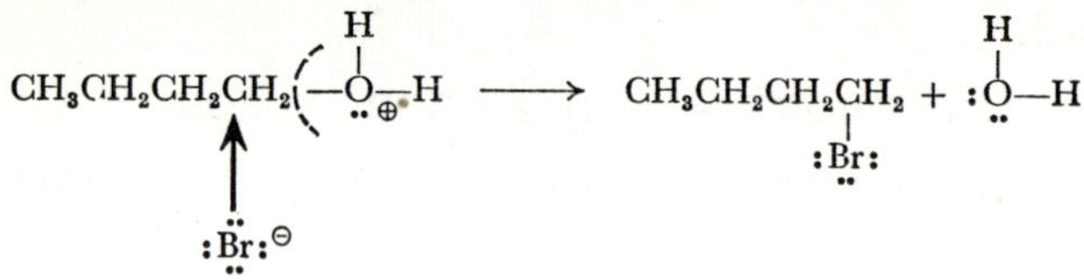

This is a fairly general reaction of alcohols, and other hydrogen halides can be used to prepare the corresponding alkyl halides. A typical procedure is given in question 5.

An alternative method of converting alcohols into halides consists in treating them with certain phosphorus or sulphur halides. Phosphorus tribromide and thionyl chloride can be used to make alkyl bromides and chlorides respectively.

$$3CH_3CH_2OH + PBr_3 \longrightarrow 3CH_3CH_2Br + H_3PO_3$$

$$CH_3CH(OH)CH_2CH_3 + SOCl_2 \longrightarrow CH_3CH(Cl)CH_2CH_3 + SO_2 + HCl$$

Dehydration. When alcohols are heated with inorganic oxyacids (sulphuric or, better, phosphoric acid), water is eliminated and *alkenes* (Chapter 5) are formed. A typical reaction is the dehydration of cyclohexanol to yield cyclohexene.

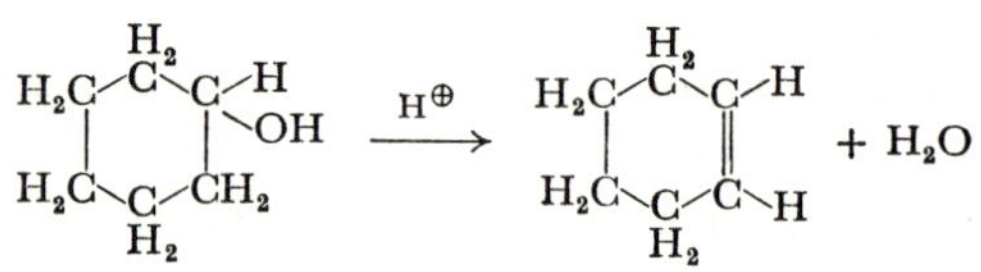

cyclohexanol, b.p. 161° cyclohexene, b.p. 83°

In this case a mixture of the alcohol and phosphoric acid is heated to 160°, and the lower-boiling cyclohexene distils out as it is formed. On a large scale, the alcohol (as vapour) can be passed over alumina at 350–450°, the reaction being performed as a continuous process.

The mechanism of the acid-catalysed alcohol dehydration reaction is discussed on p. 22.

Oxidation. Certain oxidising agents, including hexavalent chromium derivatives, oxidise primary and secondary alcohols to *aldehydes* and *ketones* respectively (Chapter 6):

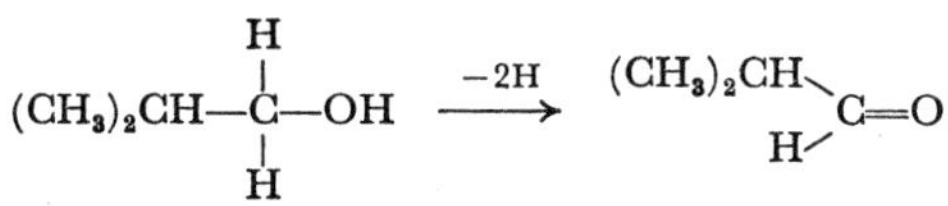

2-methylpropan-1-ol (isobutyl alcohol) → 2-methylpropanal (isobutyraldehyde)

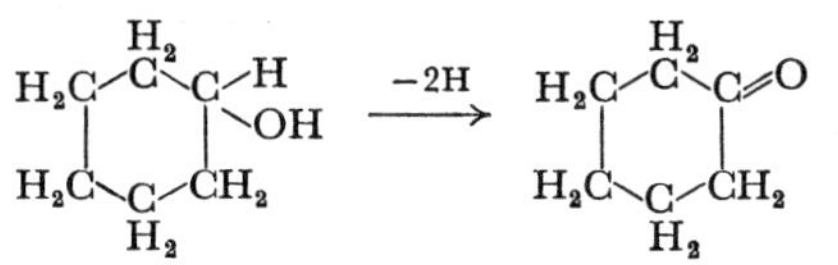

cyclohexanol, b.p. 161° → cyclohexanone, b.p. 156°

Chromic oxide, CrO_3, [or sodium dichromate, $Na_2Cr_2O_7$, and sulphuric acid] react with alcohols according to the equation:

$$3\ {>}CHOH + 2CrO_3 + 6H_3O^{\oplus} \longrightarrow 3\ {>}C{=}O + 2Cr^{3\oplus} + 12H_2O$$

The overall equation may be derived by first writing separate balanced equations for the organic and the inorganic starting components:

$$3[{>}CHOH \longrightarrow {>}C{=}O + 2H^{\oplus} + 2e]$$
$$2[Cr(VI) + 3e \longrightarrow Cr(III)]$$

In the preparation of aldehydes from primary alcohols an excess of the oxidising agent must not be used, since aldehydes can be oxidised further to carboxylic acids (cf. p. 82). Tertiary alcohols cannot be oxidised in this way because they contain no hydrogen attached to the carbon carrying the hydroxyl substituent.

Methanol. This alcohol can be separated from the mixture of products obtained by the decomposition of wood above 250°. Since the mid-1920's a cheaper method, in which carbon monoxide is reduced by hydrogen, has been used in most countries.

$$CO + 2H_2 \longrightarrow CH_3OH$$

The reactants are passed at a pressure of 200 atmospheres over a mixed metal oxide catalyst at 300–400°. Methanol is required in large quantities as a solvent and as a starting material for the preparation of other organic compounds, e.g. methyl esters and ethers, formaldehyde. Methanol is added to industrial ethyl alcohol to make it less suitable for drinking. Methanol has little inebriating effect but it produces acute acidosis due to its oxidation in the body to formic acid. It also has a potent effect on the cells of the retina and 15 ml. may suffice to produce blindness.

Ethanol is manufactured either by the hydration of ethylene (p. 62), or by the fermentation of sugars. The main sources of the appropriate sugars are molasses and starch (p. 179).

Starting from a C_6-sugar the overall equation is:

$$C_6H_{12}O_6 \xrightarrow{\text{yeast}} 2CH_3CH_2OH + 2CO_2$$

The carbon dioxide is a valuable by-product, which is converted into the solid form and sold as a cooling agent ('dry ice').

About half of the ethanol produced is converted, by dehydrogenation, into acetaldehyde. The remainder is used as a solvent and as a starting material for various syntheses.

Higher Alcohols. Isopropyl, sec. butyl, and t-butyl alcohols are obtained in the petroleum industry from propylene, butenes (1- or 2-ene), and isobutene respectively (p. 63).

n-Butyl alcohol (butan-1-ol) is the main product from reaction of propylene with carbon monoxide and hydrogen in the presence of cobalt carbonyl (oxo-reaction, 1943).

$$CH_3CH{=}CH_2 + CO + H_2 \xrightarrow{Co_2(CO)_8} CH_3CH_2CH_2CH_2OH$$

Of the eight isomeric pentanols, two have been known for a long time as by-products from the fermentation of crude potato starch. This starting material contains proteins (Chapter 13) as well as starch; during the fermentation process the proteins are broken down to a mixture of amino-acids, two of which are then converted into pentanols or 'amyl alcohols'.

$$\underset{\text{isoleucine}}{CH_3CH_2CH(CH_3)CH(NH_2)COOH} \longrightarrow \underset{\text{2-methylbutan-1-ol}}{CH_3CH_2CH(CH_3)CH_2OH}$$

$$\underset{\text{leucine}}{(CH_3)_2CHCH_2CH(NH_2)COOH} \longrightarrow \underset{\text{3-methylbutan-1-ol (isoamyl alcohol)}}{(CH_3)_2CHCH_2CH_2OH}$$

2-Methylbutan-1-ol is also called 'active amyl alcohol' because of its optical activity (Chapter 13). The preparation of longer chain alcohols is discussed on pp. 128–129.

THIOLS (THIOALCOHOLS)

The first thiol (ethanethiol) was obtained by heating a mixture of the barium salt of ethyl hydrogen sulphate with barium hydrosulphide. Thiols also occur in nature, one of the best known being the amino-acid, cysteine (see below).

Structure and Nomenclature. These compounds have the general formula, RSH; they are therefore the sulphur analogues of alcohols. The system of nomenclature is similar to that for alcohols, the suffix *thiol* being used to indicate the presence of the SH group.

	Systematic names	*Other names*
CH_3CH_2SH	ethanethiol	ethyl mercaptan
$CH_3CH_2CH(SH)CH_3$	butane-2-thiol	sec. butyl mercaptan

Other names for thiols are 'thioalcohols' or 'mercaptans', the latter name reflecting the fact that these compounds can be isolated as sparingly soluble mercuric salts, $(RS)_2Hg$ ('mercaptides') by treatment with aqueous mercuric acetate.

Most of the simpler thiols are liquids. Their boiling points are lower than those of the related alcohols, although the differences are not so pronounced as that between hydrogen sulphide (b.p. $-60°$) and water.

$$CH_3CH_2SH \quad \text{b.p. } 36° \qquad CH_3CH_2OH \quad \text{b.p. } 78°$$

The volatile thiols have unpleasant odours, which are detectable by the human nose at low concentrations, e.g. 1 part in 10^{10} parts of air.

Formation of Salts. Alkanethiols are more acidic than alcohols (cf. the similar difference between hydrogen sulphide and water), i.e. a proton is more easily removed from H—SR than from H—OR. One consequence of this difference in acidity is that metal salts of formula, $^{\oplus}$metal $^{\ominus}$SR, are formed when thiols are added to solutions of metal alkoxides in alcohols.

$$\text{H—SR} + {}^{\oplus}\text{Na}\,{}^{\ominus}\text{OR}' \rightleftharpoons {}^{\oplus}\text{Na}\,{}^{\ominus}\text{SR} + \text{H—OR}'$$

The dotted arrow in this general equation indicates that the components on the left hand side are present in the smaller concentrations at equilibrium (which is attained very rapidly). Such solutions can be used for the preparation of thioethers (p. 36). (Unwanted thiols in petroleum fractions can be removed by extraction with alkaline reagents.)

Thiols can be recognised generally by an absorption band near 2600 cm.$^{-1}$ in the infrared spectrum. Individual thiols may be identified by the preparation of mercuric derivatives which are more covalent than ionic in structure and often have sharp melting points. Thus the mercaptide of ethanethiol, $(CH_3CH_2S)_2Hg$, has m.p. 86°, and crystallises from ethanol.

Oxidation of Thiols. These reactions proceed very differently from those of alcohols. Mild oxidising agents (e.g. air) give *disulphides* by removal of hydrogen from pairs of thiol molecules:

$$2CH_3SH \underset{+2H}{\overset{-2H}{\rightleftharpoons}} CH_3\text{—S—S—}CH_3$$

dimethyl disulphide, b.p. 110°

Such disulphides are readily reduced, for example, by zinc and acid, to give back thiols.

Certain thiols and the related disulphides are present in living tissue where they are interconverted by oxidation-reduction reactions, e.g. the following pair of amino-acids:

$$2HS\text{—}CH_2CH(NH_2)COOH \underset{+2H}{\overset{-2H}{\rightleftharpoons}}$$

cysteine

$$HOOCCH(NH_2)CH_2\text{—S—S—}CH_2CH(NH_2)COOH$$

cystine

Treatment of thiols with more vigorous oxidising agents gives *sulphonic acids*:

$$CH_3CH_2{-}S{-}H \xrightarrow[\text{aq.}]{KMnO_4} CH_3CH_2{-}\overset{\overset{\large O}{\|}}{\underset{\underset{\large O}{\|}}{S}}{-}O{-}H$$ ethane sulphonic acid m.p. − 17°, hydrate, m.p. 54°

In this reaction, the carbon–sulphur bond is preserved and the product is different in structure from the alkyl esters of sulphuric acid (p. 102) in which carbon is linked to sulphur *via* oxygen. Sulphonic acids are considered in more detail in Chapter 17.

ETHERS AND THIOETHERS

Diethyl ether ('ether') was first obtained in 1540 by the action of vitriol (sulphuric acid) on spirits of wine (ethanol). Its effectiveness as an anaesthetic for dentistry was established (Morton) in 1846, and it continues to be used as a general anaesthetic. The amino-acid, methionine (p. 7) is an example of a naturally occurring thioether.

Structure and Nomenclature. An ether contains two alkyl groups linked together by an oxygen atom; a thioether contains sulphur in place of oxygen.

Discussion of nomenclature will be confined to simple compounds; in these, the two alkyl groups can be given as prefixes before the word ether or thioether;

$CH_3CH_2OCH_2CH_3$	diethyl ether, b.p. 35°
$CH_3CH_2CH_2CH_2OCH_3$	n-butyl methyl ether, b.p. 70°
$(CH_3)_2CHOCH(CH_3)_2$	diisopropyl ether, b.p. 67°
$CH_3CH_2SCH_2CH_3$	diethyl thioether, b.p. 91°
$CH_3SCH(CH_3)_2$	methyl isopropyl thioether, b.p. 94°

Diethyl and diisopropyl ethers are poorly soluble in water and are used for the extraction of organic compounds from their solutions or suspensions in water. Ethers have the disadvantage of being flammable and, for this reason, the less dangerous dichloromethane is employed nowadays for extractions.

Ethers and thioethers are prepared by the general methods discussed under S_N reactions (p. 36). Dialkyl sulphates (p. 103) may be used in place of more expensive alkyl halides.

Chemical Properties. Ethers and thioethers are fairly unreactive compounds. The following discussion is limited to aspects of the reactions of ethers with acids and of thioethers with oxidising agents.

Ethers dissolve in strong acids, e.g. concentrated sulphuric acid, with the immediate formation of oxonium salts, a property that differentiates them from alkanes.

$$CH_3CH_2{-}\ddot{\underset{\cdot\cdot}{O}}{-}CH_2CH_3 + H_2SO_4 \longrightarrow CH_3CH_2{-}\overset{H}{\underset{\oplus}{O}}{-}CH_2CH_3 + HSO_4^{\ominus}$$

$$\text{cf.}\quad H{-}\ddot{\underset{\cdot\cdot}{O}}{-}H + H_2SO_4 \longrightarrow H{-}\overset{H}{\underset{\oplus}{O}}{-}H + HSO_4^{\ominus}$$

Transference of a proton from sulphuric acid to the ether occurs, the process being analogous to the formation of the hydronium cation, $H_3O^{\oplus}$, from water and sulphuric acid. When the solution of the ether in sulphuric acid is diluted with water, a proton is transferred from the oxonium ion to a water molecule, and the ether is regenerated:

$$CH_3CH_2{-}\overset{H}{\underset{\oplus}{O}}{-}CH_2CH_3 + HSO_4^{\ominus} + H_2O \longrightarrow$$

$$CH_3CH_2{-}\ddot{\underset{\cdot\cdot}{O}}{-}CH_2CH_3 + H_3O^{\oplus} + HSO_4^{\ominus}$$

Whereas ethers are attacked by oxidising agents at C—H bonds next to the oxygen atom, reactions of thioethers with appropriate oxidising agents leads to the addition of oxygen to the sulphur atom; for example, treatment of diethyl thioether with hydrogen peroxide in acetic acid yields a sulphone:

$$\underset{\text{diethyl thioether}}{CH_3CH_2{-}\ddot{\underset{\cdot\cdot}{S}}{-}CH_2CH_3} + 2H_2O_2 \xrightarrow[\text{acetic acid}]{\text{in}} \underset{\text{diethyl sulphone, m.p. 72°}}{CH_3CH_2{-}\overset{O}{\underset{O}{\overset{\|}{\underset{\|}{S}}}}{-}CH_2CH_3} + 2H_2O$$

QUESTIONS

1. (*a*) Sketch an apparatus that should be suitable for purifying t-butyl alcohol by sublimation, and discuss the temperature control that could be necessary during the operation.

(*b*) t-Butyl alcohol has a lower boiling point than its isomers.

How might this be caused by the methyl groups making the hydroxyl group less accessible for hydrogen bonding (p. 88)?

(*c*) Aluminium t-butoxide sublimes when heated under reduced pressure (at *ca.* 160° at 0·05 mm.). Is this fact more in keeping with an ionic or with a covalent structure for the compound?

(*d*) Suggest a reason for the need to exclude water in order that the reaction of aluminium with t-butyl alcohol will proceed satisfactorily.

2. (*a*) Give structural formulae and systematic names to (i) the isomeric monobromopentanes, and (ii) the compounds that may be produced from each of them by replacement of bromine by hydroxyl.

(*b*) When this replacement is carried out in practice (using aqueous potassium hydroxide), the yields of alcohols from four of the bromoalkanes are high, whereas one of the yields is exceptionally low. Comment on these results.

3. (*a*) Explain how each of the following transformations may be carried out in two stages.

$$CH_3CH_2OH \longrightarrow \quad \xrightarrow{X} CH_3CH_2CN$$

$$CH_3CH(OH)CH_3 \longrightarrow \quad \xrightarrow{X} CH_3CH(OCH_3)CH_3$$

(*b*) Write an electronic mechanism for each of the reactions marked X.

4. (*a*) Analysis of a compound gave 60% carbon and 13·3% hydrogen; the molecular weight was 60. Write the structural formulae of all compounds that would give this result. Which of the compounds would react (i) with sodium and (ii) with thionyl chloride?

(*b*) Inductive reasoning is the process of arriving at a generalisation from particular instances (or facts). Suggest a general rule that may be adduced from the following refractive index values (n_D at 20°):

X	$X(CH_2)_4CH_3$	$X(CH_2)_5CH_3$	$X(CH_2)_6CH_3$	$X(CH_2)_7CH_3$
Br	1·4444	1·4478	1·4505	1·4527
HO	1·4101	1·4178	1·4241	1·4293
H	1·3579	1·3749	1·3876	1·3975

5. 1-Bromopentane can be made by heating a mixture of pentan-1-ol (22 g.), a 48% solution (53 g.) of hydrogen bromide in water, and concentrated sulphuric acid (17·5 g.). As the reaction proceeds the bromoalkane separates as an upper layer. The yield of 1-bromopentane is in the range of 30 to 35 g.

(*a*) Calculate the number of moles of each of the three reactants. (*b*) Write a balanced equation for the reaction. (*c*) Calculate the percentage yield (range) of the 1-bromopentane. (*d*) Give an electronic mechanism for the reaction. (*e*) What would be the best procedure at the end of the reaction for separating and purifying the bromoalkane?

6. Which one of the following reactions is 'impossible' in a single operation? (Reagents are not indicated.)

(*a*) chloroform $\longrightarrow$ chlorodifluoromethane
(*b*) diethyl disulphide $\longrightarrow$ ethane thiol
(*c*) ethanol $\longrightarrow$ ethyl cyanide
(*d*) but-1-ene $\longrightarrow$ pentan-1-ol
(*e*) neopentane $\longrightarrow$ 1-chloro-2,2-dimethylpropane

... those sciences are vain and full of errors which are not born from experiment, the mother of all certainty ...

LEONARDO DA VINCI (1452–1519)

CHAPTER 5

ALKENES AND ALKYNES

ALKENES

The first member of this group of compounds, ethylene, C_2H_4, was obtained in 1795 by the action of sulphuric acid on ethyl alcohol. It was called 'olefiant gas' because, on admixture with chlorine, an oil (ethylene dichloride) was formed. Although 'olefiant gas' became replaced by the word 'ethylene', the older term gave rise to 'olefin', a general name for ethylene and compounds related to it in structure.

Ethylene was first used as an anaesthetic for clinical surgery in 1923 but, although very safe from the medical viewpoint, it has the practical disadvantage of forming easily exploded mixtures with oxygen.

Structure and Nomenclature. An alkene contains a carbon–carbon 'double bond'. Various ways of expressing this structural feature in the simplest alkene, ethylene, are:

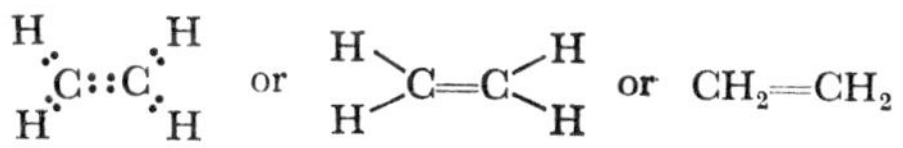

The two carbon atoms share the four electrons of the double bond so that each atom has an octet of electrons around it. Ethylene contains two hydrogens less than ethane, and this is a general relation between alkenes and alkanes containing the same number of carbons. Alkenes therefore have the general formula C_nH_{2n}.

Alkenes are named systematically as derivatives of the related alkanes, the suffix *-ene* being used to denote the presence of a double bond.

Of the three isomeric butenes, two are related to n-butane and one to isobutane. The name, but-1-ene, signifies that the double bond begins at carbon atom numbered one in the chain of four carbons, and extends to carbon atom two (which is not numbered). In but-2-ene, the double bond is located between carbon atoms two and three. In the last compound of the list, the stem name hex- refers to the longest chain of carbon atoms; the double bond extends from position two to three, and the ethyl and methyl groups are regarded as substituents in the main chain.

	Systematic names	*Other names*
$CH_2{=}CH_2$	ethene	ethylene
$CH_3CH{=}CH_2$	propene	propylene
$\overset{4}{C}H_3\overset{3}{C}H_2\overset{2}{C}H{=}\overset{1}{C}H_2$	but-1-ene	1-butene
$\overset{4}{C}H_3\overset{3}{C}H{=}\overset{2}{C}H\overset{1}{C}H_3$	but-2-ene	2-butene
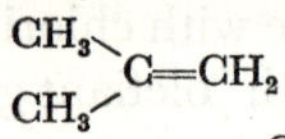	2-methylpropene	isobutene
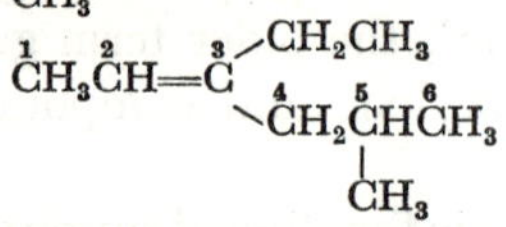	3-ethyl-5-methylhex-2-ene	

An alternative general name for an alkene is *olefin*. Carbon–carbon double bonds are referred to as *olefinic* or *ethylenic* bonds.

Boiling points are changed only to a small extent on introduction of a double bond into an alkane, and alkenes containing one to four carbon atoms are gases.

ethane	b.p. −127·5°	ethylene	b.p. −105°
pentane	b.p. 36°	pent-1-ene	b.p. 30°
octane	b.p. 258°	oct-1-ene	b.p. 250°
		oct-2-ene	b.p. 257°

Reference has already been made to the preparation of alkenes from alkanes (p. 27) and from alcohols (p. 46). Other general methods include the partial hydrogenation of alkynes (p. 68), and the elimination reactions described on pp. 40, 120.

Electronic states. The relation between tetrahedral structure and sp^3 hybridisation in compounds containing four-valent carbon has been considered (p. 17). When carbon has fewer substituents, bonding orbitals can develop by hybridisation of either three atomic orbitals (sp^2) or two atomic orbitals (sp) (p. 66). At the same time, one or more electrons may be left in dumb-bell shaped *p*-orbitals; these orbitals are shown unhatched in the diagrams below (and on p. 66). The unbonded electrons in *p*-orbitals in species (1) and (2) are indicated on the diagrams by single dots, but the electron density is distributed equally between both lobes of each orbital. Removal of an electron from an uncharged methyl radical (1) (p. 20) leaves a positively charged methyl *carbonium ion*, $CH_3^{\oplus}$. The planar arrangements

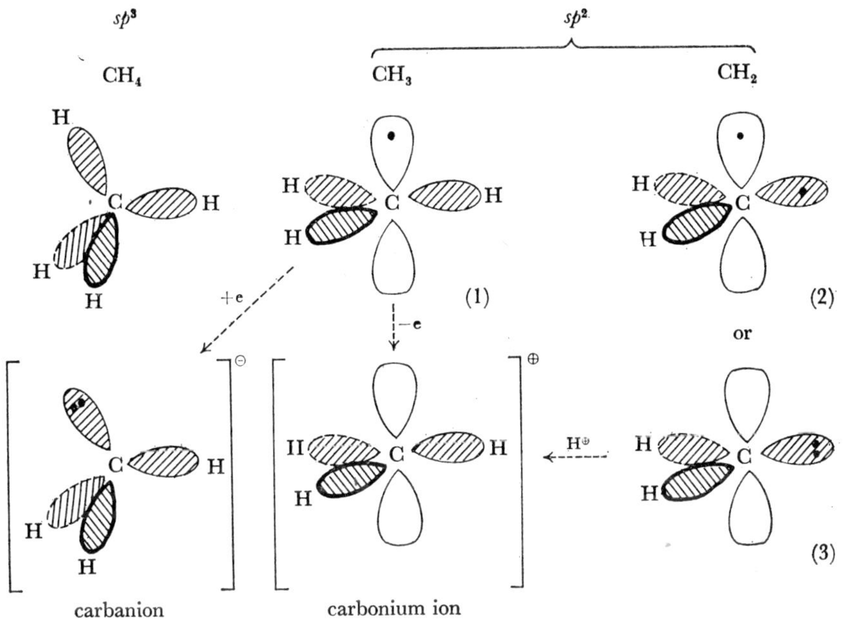

(1) (2) (3)

carbanion carbonium ion

of sp^2 bonds in these species allows the orbitals to be as far apart as possible. Addition of an electron to a methyl radical gives a methyl *carbanion*, $:CH_3^{\ominus}$, which contains the normal complement of eight electrons in the valence shell and has an sp^3 hybrid configuration.

A methylene (CH_2) species can adopt alternative electronic states, (2) and (3), both of which are sp^2 hybridised. In either form the *p*-orbital is perpendicular to the plane of the sp^2 orbitals, including the two C—H bonds. This spatial arrangement is preserved in ethylene (and other alkenes) where two of these units are joined together. In ethylene therefore, the four hydrogens lie in the same plane as the carbons, and the molecule has a thick, oval shape. Bond lengths and angles are given on the accompanying diagram.

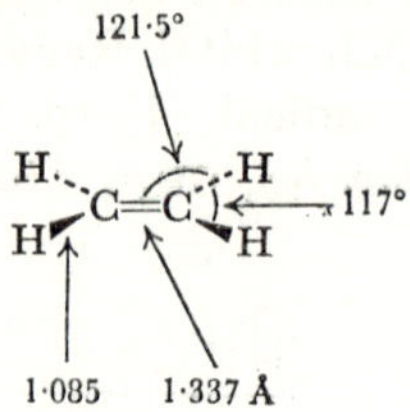

Geometrical Isomers (Diastereomers). The electronic structure of an olefinic bond is firm enough to prevent rotation about the axis of the bond. For rotation to occur it is necessary to convert the double bond into a single bond, and the energy required (up to 60 kcal./mole) is too high to be produced by molecular collisions alone at room temperature. This rigidity has important stereochemical consequences. 1,2-Disubstituted alkenes, for example, exist as two stereoisomers (*geometrical isomers* or *diastereomers*). The two isomers of but-2-ene are depicted in diagrams (4) and (5). Compound (4) in which similar atoms

(4) 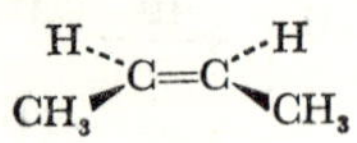

cis-

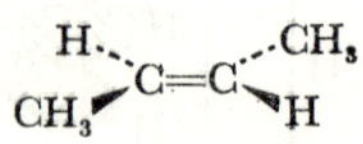 (5)

trans-

are on the same side of the molecule is described as the *cis*-isomer; in the *trans*-isomer (5) similar atoms are on opposite sides of the molecule. The boiling points of *cis*- and *trans*-but-2-enes are 3° and 1° respectively; a larger difference is encountered in the melting points of *cis*- and *trans*-crotonic acids:

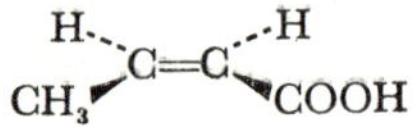

cis-crotonic acid, m.p. 15°

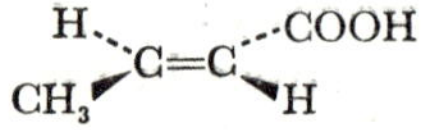

trans-crotonic acid, m.p. 72°

Differences in physical and chemical properties can assist in the assignment of a *cis*- or a *trans*-structure to a particular isomer. The geometrical isomers of 1,2-dichloroethylene, for example, have dipole moments of zero and 1·85 D. The compounds contain polar C—Cl bonds in which electronic charge is displaced towards chlorine (p. 32). The C—Cl bond dipoles of the *trans*-

isomer (7) are in opposite directions and exactly compensate, resulting in a zero dipole moment for the molecule. An example

cis- (6)

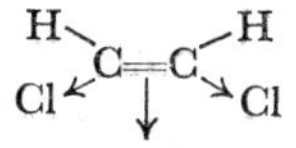

resultant dipole moment of 1·85D

trans- (7)

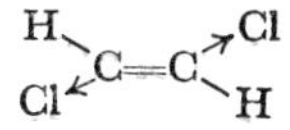

dipole moment, zero

of the use of a chemical method to distinguish *cis-* from *trans-* isomers is described later (p. 152). In the molecule of *cis*-1,2-dichloroethylene (6), the two C—Cl bond moments are in the same general direction and this isomer is clearly the one with a molecular dipole moment.

The presence of a double bond is not the only condition for restricted rotation, and many saturated cyclic compounds, for example, exist as geometrical isomers (p. 174).

ADDITION REACTIONS

The most characteristic feature of the chemistry of alkenes, and a major difference from that of alkanes, is that many reagents add to carbon–carbon double bonds. A general equation for these reactions is:

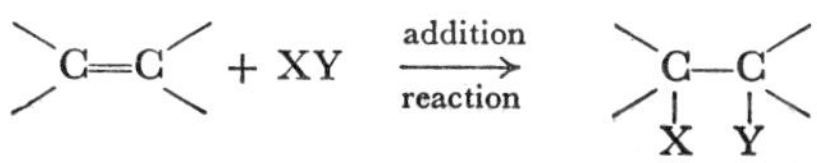

addenda adduct

Alkanes are often described as 'saturated' hydrocarbons because they do not undergo addition reactions. By contrast, alkenes are 'unsaturated'; addition of the reagent XY is said to 'saturate' the olefinic bond. Reagents do not add to alkenes unless the addenda can form covalent bonds to the two carbons.

Hydrogen. Although the reaction is thermodynamically very favourable, hydrogen does not add to an alkene unless a catalyst is present, e.g., finely divided nickel, palladium or platinum. An alkane is then produced by *hydrogenation* of the double bond.

$$CH_2{=}CHCH_2C(CH_3)_3 + H_2 \longrightarrow CH_3CH_2CH_2C(CH_3)_3$$

4,4-dimethylpent-1-ene 2,2-dimethylpentane

The hydrogen and the alkene are adsorbed on to the metal, the former giving metal–hydride groups on the surface. Reaction occurs on the surface between the adsorbed alkene and the metal–hydride, and the product formed is desorbed into the surrounding liquid or gas phase (*heterogeneous catalysis*).

Hydrogenation of alkenes proceed exothermically at ordinary temperatures, the reaction heats being in the range of 26–33 kcal./mole. These results, taken with those obtained from the heats of combustion of alkenes and alkanes, have helped to establish bond energy values for carbon to carbon double and single bonds. Typical figures are:

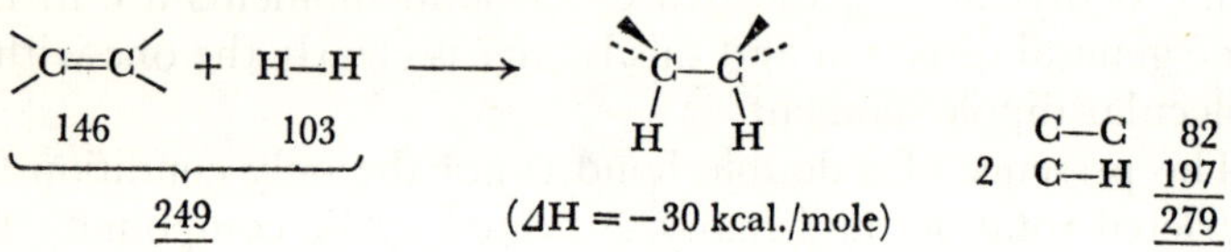

These figures show that alkanes are considerably more stable than alkenes, but the converse becomes increasingly apparent as the temperature is raised. Useful proportions of alkenes can then be generated from alkanes by *dehydrogenation*:

$$\text{n-butane } (+\text{ steam}) \xrightarrow[\text{containing chromia}]{600°;\ \text{alumina}} \text{butene} + \text{butadiene (p. 138)} + \text{hydrogen } (+\text{ steam} + \text{butane})$$

Halogens. In contrast with alkanes, alkenes react rapidly with chlorine or bromine at ordinary temperatures in the absence of light. The decolourisation of a solution of bromine in carbon tetrachloride or acetic acid provides a convenient method for the rapid differentiation of alkanes and alkenes. The reactions are exothermic, for example, a rise in temperature of *ca.* 20° occurs when a solution of bromine (0·25 ml.) in carbon tetrachloride (5 ml.) is added to cyclohexene (0·5 ml.) dissolved in the same solvent (10 ml.). Other examples of the general reaction are:

$$BrCH_2CH_2Br \xleftarrow{Br_2} CH_2{=}CH_2 \xrightarrow{Cl_2} ClCH_2CH_2Cl$$

1,2-dibromoethane, b.p. 112° (ethylene dibromide) — 1,2-dichloroethane, b.p. 84° (ethylene dichloride)

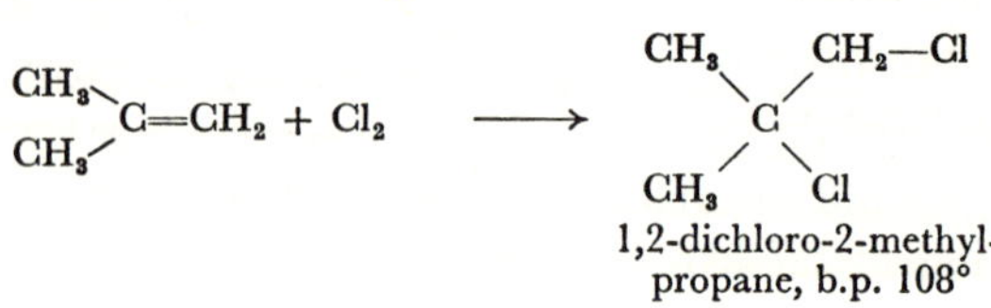

1,2-dichloro-2-methylpropane, b.p. 108°

The mechanism of these addition reactions is visualised as proceeding in two steps. In step A, the electrons of the double bond react with one end of a bromine molecule which cleaves unsymmetrically to give a negatively charged bromide ion. The olefin is thereby converted into a bromonium ion in which a divalent (and therefore positively charged) bromine is linked to the two carbon atoms of the original olefinic bond:

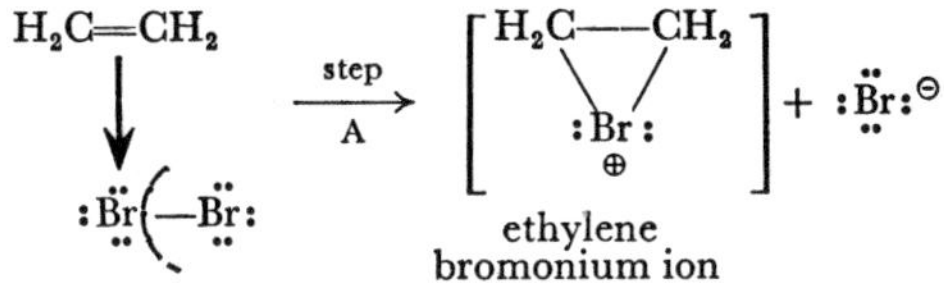

These ions only have a transitory existence before the bromide anion reacts with the bromonium cation to give the final product; step B can be represented:

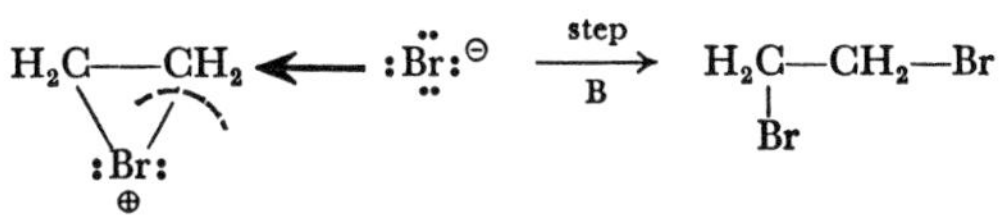

Each step of the reaction is an S_N process, the double bond and the bromide ion being the respective nucleophiles. In each stage the bromine that acquires a pair of electrons acts as the electrophile.

The degree of unsaturation of a substance can be determined by treating it with a known amount of iodine chloride (or bromide) in acetic acid, followed by titration of the unused halogen reagent (p. 107).

Hydrogen Halides. Hydrogen halides also add to olefinic bonds although less readily than chlorine and bromine. At 100°, ethylene and hydrogen bromide react at a convenient rate to give bromoethane:

$$CH_2{=}CH_2 + HBr \longrightarrow CH_3CH_2Br$$

Hydrogen chloride and hydrogen iodide also add to ethylene to give chloroethane and iodoethane respectively.

Additions of hydrogen halides to unsymmetrically substituted alkenes can in theory result in the formation of a mixture of two products, but in practice only one compound may be formed.

Thus addition of hydrogen iodide to propylene yields 2-iodo-propane, not 1-iodopropane:

$$CH_3{-}CH{=}CH_2 + HI \longrightarrow CH_3{-}\underset{\displaystyle I}{\underset{|}{CH}}{-}CH_3 \quad (\text{not } CH_3{-}CH_2{-}\underset{\displaystyle I}{\underset{|}{CH_2}})$$

Several observations of this kind led Markovnikov to propose (in 1870) a rule for the direction of addition of hydrogen halides to unsymmetrically alkylated olefins: the negative part of the addendum adds on to the carbon atom that is joined to the least number of hydrogen atoms. Further examples of reactions that conform to the Markovnikov rule are:

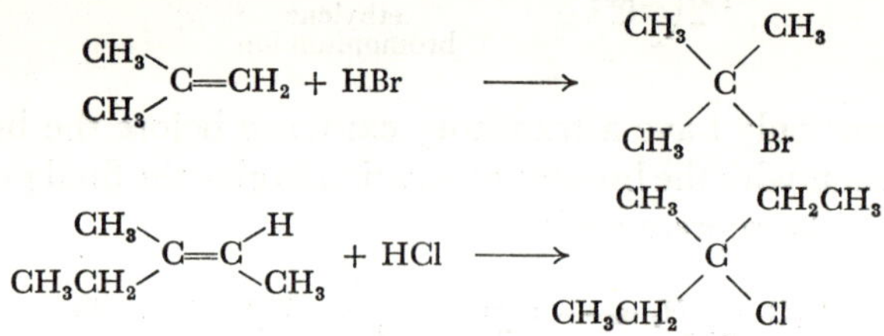

The mechanism of the addition of hydrogen bromide to iso-butene is considered in the next section.

Hydration. Carbonium Ions. When ethylene is passed into concentrated sulphuric acid at 80°, the mono-ethyl derivative of sulphuric acid is formed, and this can be isolated as its crystalline mono-potassium salt:

$$CH_2{=}CH_2 + H_2SO_4 \longrightarrow CH_3{-}CH_2{-}O{-}\overset{\displaystyle O}{\overset{\|}{\underset{\displaystyle O}{\underset{\|}{S}}}}{-}OH$$

ethyl hydrogen sulphate

Ethyl hydrogen sulphate undergoes hydrolysis in hot aqueous solution to give ethanol and sulphuric acid:

$$CH_3CH_2{-}O{-}\overset{\displaystyle O}{\overset{\|}{\underset{\displaystyle O}{\underset{\|}{S}}}}{-}OH + 2H_2O \longrightarrow CH_3CH_2{-}OH + H_3O^{\oplus} + HSO_4^{\ominus}$$

As a result of these two reactions, ethylene has been *hydrated* to ethanol, a process that cannot be performed using water alone. However, hydration of the olefin can be effected with the aid of a catalytic amount of acid if higher temperatures and pressures are employed; the use of increased pressure illustrates Le Chatelier's principle.

$$\text{ethylene} + \text{water} \xrightarrow[\text{phosphoric acid}]{300°/70 \text{ atmos.}} \underset{(4\%)}{\text{ethanol}} + \underset{(96\%)}{\text{ethylene}} + \text{water}$$

Optimum conditions are reached when *ca.* 4% of the ethylene is hydrated during its passage through the reaction vessel. After cooling, the aqueous ethanol and unchanged ethylene are easily separated. The alkene is recycled, and the process is designed to run as a continuous operation. A longer reaction time would enable more of the ethylene to become hydrated, but this would be offset by conversion of part of the ethanol into diethyl ether (p. 51).

Propylene is hydrated under similar conditions (270°) to propan-2-ol. The direction of hydration conforms to the Markovnikov rule, the hydroxyl group being the negative part of the addendum.

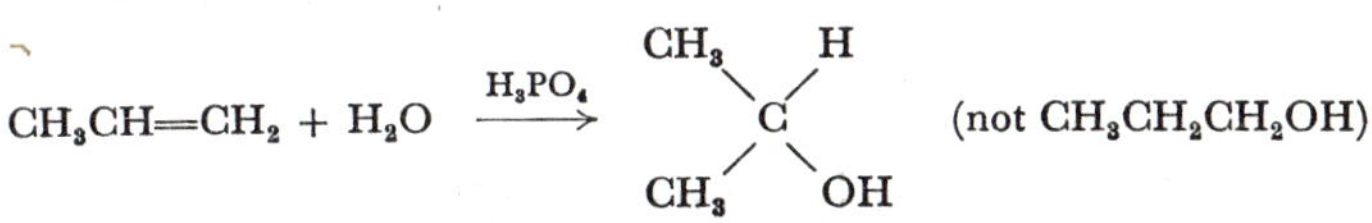

propan-2-ol
(isopropyl alcohol)

Hydration occurs surprisingly easily when two alkyl groups are attached to one of the olefinic carbons: for example, isobutene is converted into t-butyl alcohol, $(CH_3)_3COH$, at an appreciable rate when brought into contact with 1M aqueous mineral acid at 20°. This phenomenon is due to the exceptional ease of formation of a *tertiary* carbonium ion as a reaction intermediate:

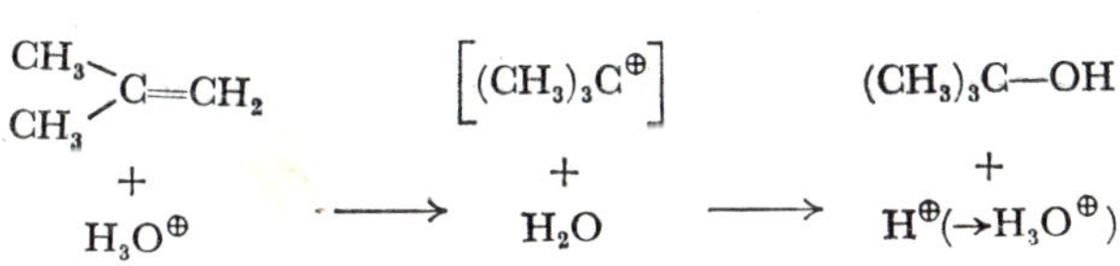

The addition of hydrogen bromide to isobutene may be represented similarly:

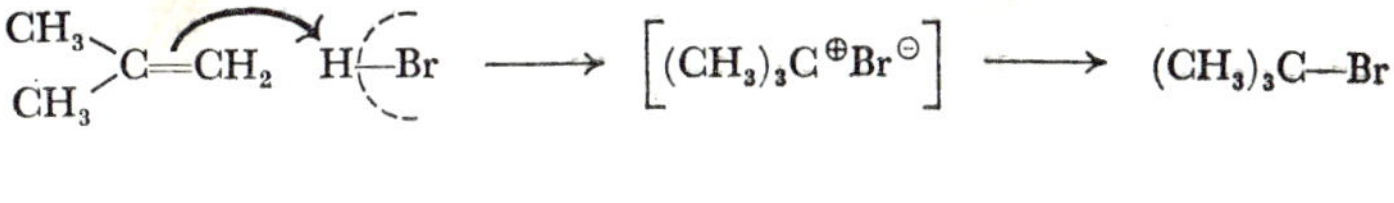

The tendency for carbonium ion stability to increase in the order primary < secondary < tertiary provides a basis for the Markovnikov rule. In the above reactions of isobutene, for example, the alternative products, $(CH_3)_2CHCH_2OH$ and $(CH_3)_2CHCH_2Br$, are not present, since their formation would involve the primary carbonium ion, $(CH_3)_2CHCH_2^{\oplus}$, which is *ca.* 35 kcal./mole less stable than the tertiary ion, $(CH_3)_3C^{\oplus}$.

The relative stability of a tertiary carbonium ion also accounts for the difference in the kinetic behaviour of tertiary and primary alkyl halides towards nucleophilic reagents in solution. Whereas primary halides react by second order kinetics (p. 33), the concentration of a tertiary halide diminishes at a rate that is *independent* of the concentration of added nucleophile

$$\text{rate} = k[(CH_3)_3CBr]$$

The result shows that the reaction does not proceed by an S_N2 mechanism, but it is consistent with the following picture in which the slower (rate-controlling step) is the production of the t-butyl carbonium ion:

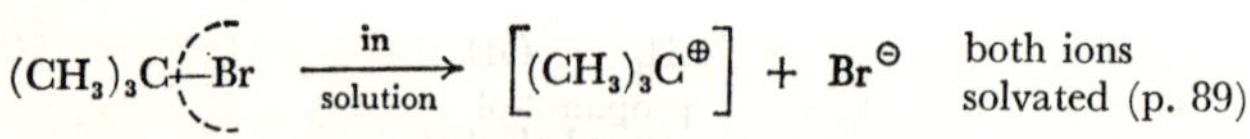

The fate of the very reactive carbonium ion will depend on its external environment; competing reactions include the loss of a proton to give isobutene and, in water (with or without hydroxide ion present), the formation of t-butyl alcohol:

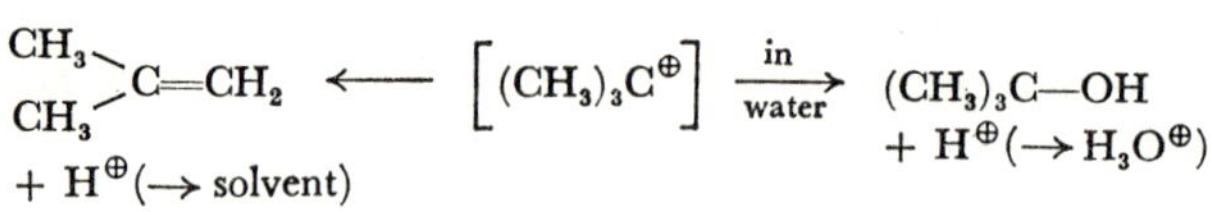

The decomposition of a tertiary halide in solution by first order kinetics is designated as an S_N1 reaction.

Oxidative Cleavage. Carbon–carbon double bonds can be broken by reagents that do not react with single bonds. This difference is of value for the determination of the position of a double bond in large molecules, for the smaller fragments produced are usually easier to identify.

In one method, disubstituted alkenes are oxidised to carboxylic acids in 95–98% yields:

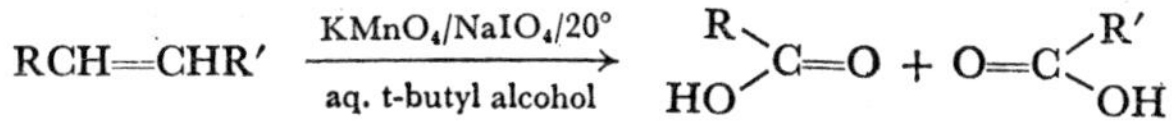

Another approach involves the very ready reaction between an alkene and ozone. The vigour of such reactions is moderated by using ozone diluted with oxygen (a mixture of approximate composition 1:19 is obtained by passing the diatomic gas through an electric discharge), and by using a solution of the olefin in an inert solvent such as dichloromethane or acetic acid. Addition of ozone to an olefinic bond produces an ozonide, in which the carbons are now linked only through oxygen atoms.

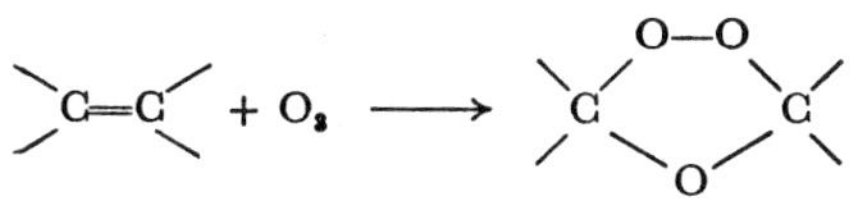

ozonide

In practice ozonides are not usually isolated but are decomposed by treating the reaction solution with a reducing or an oxidising agent. The products depend on the structure of the initial alkene, for example, on the degree of substitution of the double bond. Reaction of the disubstituted olefin, pent-2-ene with ozone followed by treatment of the resultant solution with the reducing agent sodium tetrahydroborate ($NaBH_4$) affords a mixture of two primary alcohols, each in over 90% yield:

$$CH_3CH_2CH{=}CHCH_3 \xrightarrow[\text{NaBH}_4\text{ in aq. }CH_3OH]{O_3\text{ then}} \underset{\text{propan-1-ol}}{CH_3CH_2CH_2OH} + \underset{\text{ethanol}}{HOCH_2CH_3}$$

If the olefin-ozone reaction solution is treated with zinc/acetic acid, two aldehydes are produced:

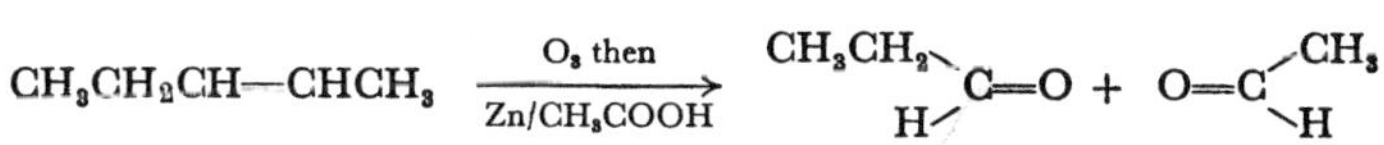

propionaldehyde acetaldehyde

The isomeric pent-l-ene affords a different pair of aldehydes:

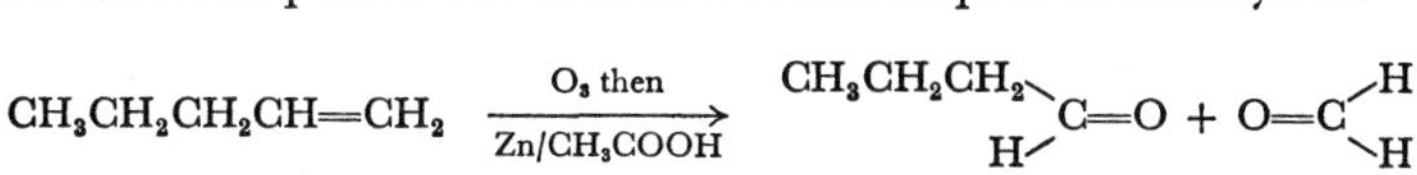

n-butyraldehyde formaldehyde

Yields of aldehydes by this method are often moderate. Better results (75–90% yields) are obtained by performing the initial ozonolysis (Gk. *lysis*, loosening) reaction in methanol at $-60°$, followed by addition of dimethyl thioether to the solution:

$$RCH{=}CHR' + O_3[+ (CH_3)_2S] \xrightarrow[\text{methanol}]{\text{in}} RCHO + OCHR'[+ (CH_3)_2SO]$$

Chemical evidence regarding molecular structure is desirably based on reactions that proceed in high yields, for the isolation and identification of a product formed in low yield may represent an unexpected side reaction.

ALKYNES

Hydrocarbons of the general formula C_nH_{2n-2} that contain a carbon–carbon 'triple bond' are called acetylenes or, more systematically, *alkynes*. An acetylenic bond is formed by the sharing of six electrons between two carbons as exemplified in the structure of the parent compound, *acetylene*:

$$H{:}C{:}{:}{:}C{:}H \quad \text{or} \quad H{-}C{\equiv}C{-}H \quad \text{or} \quad C_2H_2$$

The triple bond of acetylene is derived from two CH units (8). Each unit contains two linear *sp* hybrid orbitals and two non-bonded electrons distributed in two dumb-bell shaped *p*-orbitals perpendicular to each other and to the axis of the *sp*-orbitals. In the composite structure of acetylene (9), overlap between the

(8)

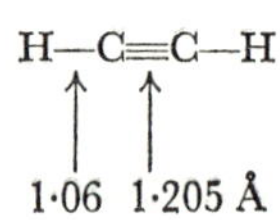

(9)

p-orbitals gives a cylindrical distribution of electrons around the carbon–carbon axis. The $-C{\equiv}C-$ groups of other alkynes also have a linear structure, and there is thus no possibility of geometrical isomerism.

The term *bond order* refers to the number of electron pairs shared by the atoms concerned: the bonds C—Br, C═C and C≡N have orders of 1, 2 and 3 respectively. In general it is found that an increase of bond order causes a decrease in bond length and an increase in bond strength:

	Bond length, Å	*Bond strength, kcal./mole*
$H_3C—CH_3$	1·540	83
$H_2C{=}CH_2$	1·337	100
$HC{\equiv}CH$	1·205	123

Nomenclature. The suffix *-yne* is used in systematic nomenclature to denote the presence of an acetylenic bond. The systematic name for acetylene is *ethyne*, although the former name, proposed by Berthelot in 1862, is still in general use. Many of the simpler alkynes can also be named as alkyl or dialkyl derivatives of acetylene:

	Systematic name	*Semisystematic name*
$CH_3C{\equiv}CH$	propyne	methylacetylene
$\overset{1}{C}H_3\overset{2}{C}{\equiv}\overset{3}{C}\overset{4}{C}H_2\overset{5}{C}H_2\overset{6}{C}H_3$	hex-2-yne (or hex-4-yne)	methyl n-propylacetylene
$(\overset{5}{C}H_3)(CH_3)\overset{4}{C}H\overset{3}{C}{\equiv}\overset{2}{C}\overset{1}{C}H_2Cl$	1-chloro-4-methylpent-2-yne (or 5-chloro-2-methylpent-3-yne)	

In the last two compounds, the term 2-yne indicates that the triple bond extends from C2 to C3 in the chain. The two alternative names (given in parentheses) also define the structures of these molecules but lower numbers are preferred.

The simpler alkynes are gases but but-2-yne and the pentynes are liquids (at 20°). The boiling points of acetylenes are in general slightly higher than those of related alkenes and alkanes.

pentane, b.p. 36°	pent-1-ene, b.p. 30°	pent-1-yne, b.p. 39·5°
	pent-2-ene, b.p. 36°	pent-2-yne, b.p. 55°

METAL DERIVATIVES OF ALK-1-YNES

The triple bond confers acidic character on a hydrogen atom that is linked directly to it, and metal derivatives of general formula, metal$C{\equiv}CR$ (including metal$C{\equiv}CH$) can be prepared. This property of alk-1-ynes (including acetylene) serves to differentiate them from compounds in which the triple bond is in a non-terminal position. Thus, insoluble copper and silver derivatives separate when a terminal acetylene is mixed with an ammoniacal solution of cuprous chloride or silver nitrate.

Sodium salts of acetylene and monoalkyl acetylenes are conveniently prepared by adding the organic compound to a solution of the strong base, sodamide, dissolved in liquid ammonia:

$$HC{\equiv}CH + {}^{\oplus}Na\ {}^{\ominus}{:}\ddot{N}H_2 \longrightarrow HC{\equiv}\overset{\ominus}{C}{:}\ \overset{\oplus}{Na} + {:}NH_3$$

ethynylsodium
(sodium acetylide)

$$CH_3C{\equiv}CH + {}^{\oplus}Na\ {}^{\ominus}{:}\ddot{N}H_2 \longrightarrow CH_3C{\equiv}\overset{\ominus}{C}{:}\ \overset{\oplus}{Na} + {:}NH_3$$

propynylsodium

There is a structural similarity between these anions and the cyanide ion, and all behave as nucleophilic reagents towards alkyl bromides. Such replacement reactions may be performed by adding the alkyl halide, e.g. 1-bromobutane, to the liquid ammonia solution of the alkynylsodium prepared by the above method.

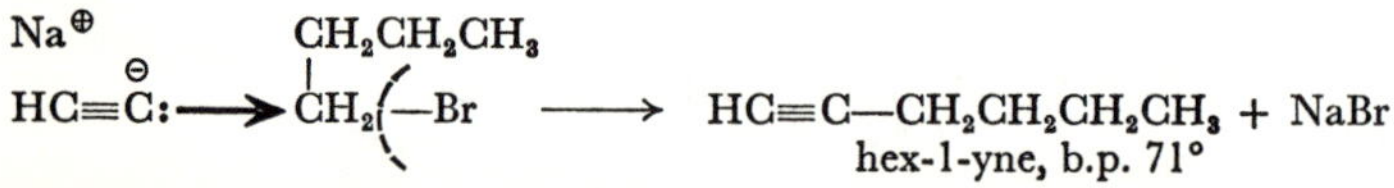

Fractional distillation separates the product from the reaction solvent (b.p. $-33{\cdot}5°$). Alternatively, another alkyl group can be introduced by adding sodamide to the liquid ammonia solution of the hexyne (to form its sodium derivative), followed by an alkyl halide, e.g. ethyl bromide:

$$HC{\equiv}C(CH_2)_3CH_3 \xrightarrow[\text{liq. } NH_3]{NaNH_2} \overset{\oplus}{Na}\ {:}\overset{\ominus}{C}{\equiv}C(CH_2)_3CH_3$$

$$\Big\downarrow\ CH_3CH_2Br$$

$$CH_3CH_2C{\equiv}C(CH_2)_3CH_3$$

oct-3-yne, b.p. 130°

Acetylene can be converted into a variety of mono- and dialkylacetylenes by these reactions.

ADDITION REACTIONS OF ALKYNES

Hydrogen. Alkynes resemble alkenes in undergoing addition reactions. They are hydrogenated over metal catalysts, the heat of reaction ΔH, for the initial stage, alkyne + H_2 $\rightarrow$ alkene, being *ca.* —40 kcal./mole. This stage usually proceeds more rapidly than the subsequent step, alkene + H_2 → alkane. Consequently, the reduction of an alkyne slows down after the absorption of one molecule of hydrogen, and if the reaction is terminated at this point the olefin is obtained:

$$CH_3CH_2CH_2C{\equiv}CH \xrightarrow{2H} CH_3CH_2CH_2CH{=}CH_2$$

Disubstituted alkynes, on similar partial hydrogenation, give *cis*-alkenes, and this reaction is a valuable general method for preparing such geometrical isomers.

Polymerisation of ethylene (p. 142) requires a pure starting material; the small amount of acetylene (0·1–1 %) present in the ethylene obtained by cracking is removed by partial selective hydrogenation.

Hydrogen Halides. One or two equivalents of a hydrogen halide may add to an acetylene:

$$HC{\equiv}CH + HBr \longrightarrow \underset{\text{bromoethene (vinyl bromide)}}{H_2C{=}CHBr} \xrightarrow{HBr} \underset{\text{1,1-dibromoethane}}{CH_3CHBr_2}$$

Addition of the second molecule of hydrogen bromide is in accord with the Markovnikov rule. Vinyl chloride, $H_2C{=}CHCl$, is manufactured by the addition of hydrogen chloride to acetylene using a mercuric salt as catalyst (see p. 144).

Hydration. When acetylene is passed into aqueous sulphuric acid containing mercuric sulphate, addition of water to the triple bond occurs and acetaldehyde is formed:

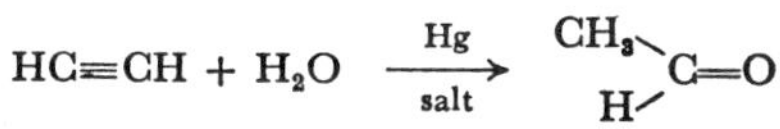

acetaldehyde, b.p. 20°

Acetaldehyde is made from acetylene on an industrial scale by this method. When mono-alkyl acetylenes are hydrated in this way, methyl ketones (p. 74) are formed, not aldehydes:

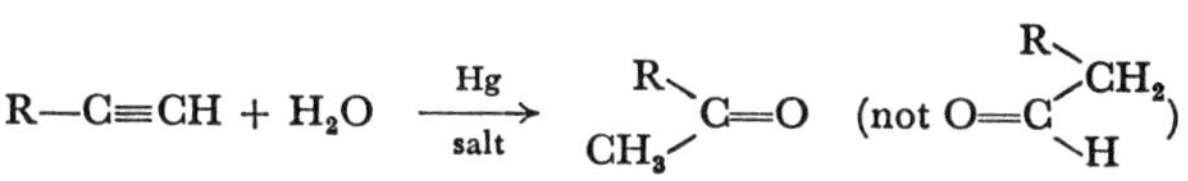

This reaction is similar to the hydration of alkenes in that the oxygen-bearing group becomes bonded to the carbon carrying the lesser number of hydrogens. A mixture of the two possible ketones is produced by the hydration of an unsymmetrical dialkyl acetylene:

$$R{-}C{\equiv}C{-}R' + H_2O \xrightarrow[\text{salt}]{\text{Hg}} O{=}C\begin{matrix}\nearrow R \\ \searrow CH_2R'\end{matrix} + O{=}C\begin{matrix}\nearrow CH_2R \\ \searrow R'\end{matrix}$$

Acetylene. E. Davy (1836) found that a product obtained by heating carbon with the potassium salt of tartaric acid gave a reactive gas on treatment with water. The intermediate product probably contained potassium acetylide, C_2K_2, from which acetylene is liberated by hydrolysis. During the present century acetylene has been chiefly made by hydrolysis of calcium acetylide (calcium 'carbide'), which is prepared by heating lime and coke together in an electric furnace.

$$CaO + 3C \xrightarrow{2500^\circ} CO + \overset{\oplus\oplus}{Ca}\ :\overset{\ominus}{C}{\equiv}\overset{\ominus}{C}:$$

calcium acetylide

$$\overset{\oplus\oplus}{Ca}\ :\overset{\ominus}{C}{\equiv}\overset{\ominus}{C}: + 2H_2O \xrightarrow{20^\circ} Ca(OH)_2 + C_2H_2$$

Advantages of the method are that the carbide can be transported to where it is required and that the hydrolysis stage is easily controlled. The cost of electricity in the first stage is a disadvantage.

The thermochemical feasibility of producing acetylene from alkanes at a suitably high temperature has been realised for some time. However at the temperatures required (above 1000°) acetylene is unstable relative to carbon and hydrogen, and very short reaction times become necessary, e.g. 0·05 sec. The following diagram summarises the operations required for the process, alkanes → acetylene + hydrogen.

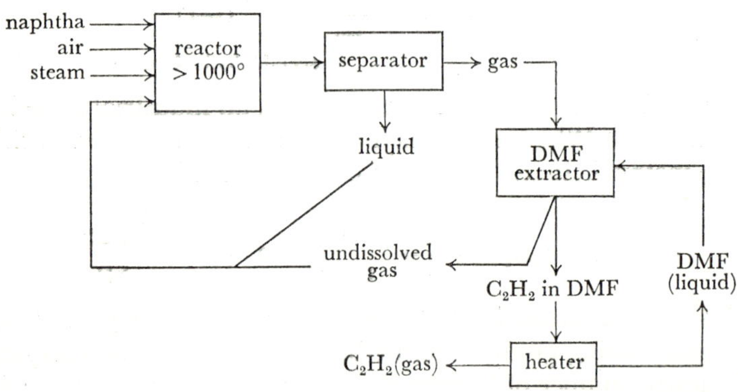

The effluent hydrocarbons from the reactor contain up to 10% of acetylene. After separation from liquids, including unchanged naphtha, the acetylene is selectively removed from the gaseous products by extraction with dimethylformamide (DMF, p. 94). It is later recovered by heating the solution. The capital cost of this method of producing acetylene is high, but it can be run as a continuous process on a large scale.

Acetylene and ethylene often compete with one another as sources of other C_2-compounds, e.g. vinyl chloride and acetaldehyde. Chloroprene (p. 145) is one product that is most readily obtained starting from acetylene:

$$2C_2H_2 \xrightarrow[\text{in aq. acid}]{(NH_4)_2CuCl_3} \underset{\text{vinylacetylene}}{CH_2{=}CH{-}C{\equiv}CH} \xrightarrow{HCl} \underset{\text{chloroprene}}{CH_2{=}CHCCl{=}CH_2}$$

QUESTIONS

1. Give the structures and names of the isomeric olefins (including stereoisomers) of molecular formula C_5H_{10}.

2. Give the products and mechanisms of the following reactions:

hex-3-ene and bromine;
2-methylpent-1-ene and hydrogen chloride.

3. (*a*) A plant wax was separated into alkanes and alkenes, the latter ranging from $C_{15}H_{30}$ up to $C_{33}H_{66}$. One of the alkenes (A) had a molecular weight of 462, and showed bands in the infrared spectrum at 911 and 933 cm.$^{-1}$ characteristic of a vinyl group ($-CH{=}CH_2$.) Reduction of (A) (suggest a method) gave $CH_3(CH_2)_{31}CH_3$. What is the structure of (A)?

An isomeric olefin (B) also gave $CH_3(CH_2)_{31}CH_3$ on reduction. Work out the structure of (B) from the additional fact that on ozonolysis, followed by reduction of the ozonide with sodium tetrahydroborate, a mixture of decan-1-ol and $CH_3(CH_2)_{21}CH_2OH$ was obtained.

(*b*) Filtered air was passed for several days over a culture of the fungus, *Fomes annosus*, growing at 23°. The effluent gas was passed through a trap (−73°) where a material collected, which was subsequently dried with phosphorus pentoxide at a low temperature. The dried material (A) formed colourless crystals which, in bulk, decomposed explosively at room temperature. A gave an *m*/*e* ratio of 74 by mass spectrometry, and it could also be hydrogenated to n-hexane. What is the structure of A?

4. (*a*) But-2-ene is an olefin that is sufficiently stable to be converted into the *cis*: *trans* equilibrium mixture by heat alone. At 390°, a 47·2:52·8 *cis*:*trans* ratio is obtained. Calculate ΔG for the equilibrium. What technique might be suitable for determining the ratio?

(*b*) When a mixture of X, C_4H_8, and hydrogen sulphide is passed over a silica-alumina catalyst at 100°, 1,1-dimethylethane thiol is produced. What is X?

5. (*a*) One gram of a hydrocarbon, A, absorbs 400 ml. of hydrogen (at N.T.P.) over a platinum catalyst. Ethanol was the only product from ozonolysis (then $NaBH_4$ reduction) of the hydrocarbon. What is the structure of A?

(*b*) What will be formed by the action of ozone (followed by $NaBH_4$ reduction) upon (i) 2-methylbut-2-ene, (ii) natural rubber

$$(\text{-}CH_2\overset{\overset{\large CH_3}{|}}{C}{=}CHCH_2\text{-})_n \quad ? \quad \text{(p. 145)}$$

6. Draw structures for the following compounds to illustrate the arrangement in space of the component atoms:

dichloromethane, bromoethene, propyne, dimethyl ether.

7. What elements, apart from carbon, can pair to form multiple bonds?

One result can prove me wrong.

ALBERT EINSTEIN (1879–1955)

CHAPTER 6

ALDEHYDES AND KETONES

The name *aldehyde* is derived from the words '*al*cohol *dehyd*rogenation', as acetaldehyde, the first aldehyde to be discovered (in 1774), was obtained later by the dehydrogenation of ethanol. Formaldehyde, although it is the simplest member of the series, was not prepared until 1867. The term *ketone* is derived from *acetone*, which was prepared in 1831 by heating calcium acetate obtained from 'pyroligneous acid' (p. 43).

Structure and Nomenclature. Aldehydes and ketones have the following general structures:

aldehydes $\begin{matrix} R \diagdown \\ \quad C{=}O \\ H \diagup \end{matrix}$ or RCHO (R is an alkyl group or a hydrogen atom)

ketones $\begin{matrix} R \diagdown \\ \quad C{=}O \\ R' \diagup \end{matrix}$ or RCOR′ (R and R′ are alkyl groups)

Both groups of compounds contain the *carbonyl group* ($>C{=}O$ or $>C::\ddot{\underset{\cdot\cdot}{O}}$), and hence have many chemical properties in common. Differences in properties usually arise from the fact that aldehydes, unlike ketones, contain a hydrogen linked directly to the carbon of the carbonyl group.

Two systems for naming *aldehydes* are in general use, and are understood by considering the following successive oxidations involving typical aldehydes.

alkan-1-ol		*alkanal or aldehyde*		*carboxylic acid*
CH_3OH methanol	→	HCHO methanal or formaldehyde	→	HCOOH formic acid
CH_3CH_2OH ethanol	→	CH_3CHO ethanal or acetaldehyde	→	CH_3COOH acetic acid
$CH_3CH_2CH_2CH_2OH$ butan-1-ol	→	$CH_3CH_2CH_2CHO$ butanal or n-butyraldehyde	→	$CH_3CH_2CH_2COOH$ n-butyric acid

In the *systematic* method, the name of the aldehyde is derived by adding *-al* to the name of the corresponding alkane (without the terminal -*e*). Note that the corresponding hydrocarbon is that containing the same number of carbon atoms. When there is a branched alkyl chain, a numbering system is employed (as with halides and alcohols) and the longest chain containing the aldehyde group is numbered. The aldehyde carbon is regarded as C–1, but this is not usually numbered, as an aldehyde group must always be at the end of a chain.

$$CH_3CH_2CH(CH_3)CHO \quad \text{or} \quad \overset{4}{C}H_3\overset{3}{C}H_2\overset{2}{C}H(CH_3)\overset{1}{C}HO \quad \text{2-methylbutanal}$$

$$(CH_3CH_2)_2CHCH_2CHO \quad \text{or} \quad \overset{5}{C}H_3\overset{4}{C}H_2\overset{3}{C}H(CH_2CH_3)\overset{2}{C}H_2\overset{1}{C}HO \quad \text{3-ethylpentanal}$$

Alternatively, the word *aldehyde* is added to the root of the specific name of the corresponding carboxylic acid. This method, which is also illustrated in the oxidative sequence above, is used only for simple aldehydes.

Ketones are named systematically by adding -*one* to the name of the corresponding alkane. The method can be appreciated by considering the following compounds:

alkane	*alkanol*	*alkanone or ketone*
$CH_3CH_2CH_2CH_3$	$CH_3CH_2CH(OH)CH_3$	$CH_3CH_2COCH_3$
butane	butan-2-ol	butanone
$CH_3CH(CH_3)CH_2CH_2CH_3$	$CH_3CH(CH_3)CH(OH)CH_2CH_3$	$CH_3CH(CH_3)COCH_2CH_3$
2-methylpentane	2-methylpentan-3-ol	2-methylpentan-3-one

Usually the position of the carbonyl group must be indicated by a number, but in the case of butanone such a number is unnecessary since, although there are two n-butanols, there is only one butanone.

An alternative system of nomenclature, which can be cumbersome when applied to higher members of the series, is to specify the alkyl groups attached to the ketone group:

$$CH_3CH_2COCH_3 \quad \text{ethyl methyl ketone}$$

Further examples of the use of these methods are:

$CH_3CH_2COCH_2CH_3$	pentan-3-one or diethyl ketone, b.p. 102°
$(CH_3)_2CHCOCH_2CH_2CH_2CH_3$	2-methylheptan-3-one or n-butyl isopropyl ketone, b.p. 160°
$CH_3(CH_2)_3CH(CH_3)COCH_2C(CH_3)_3$	2,2,5-trimethylnonan-4-one

Aldehydes and ketones are usually prepared by oxidation of alcohols (p. 47). Other methods include ozonolysis of alkenes (p. 65), oxidation of α-glycols (p. 133) and hydration of acetylenes (p. 69).

The Carbonyl Bond. X-ray studies show that the carbonyl group and its two substituents lie in one plane, the bond angles being *ca.* 120°, as in olefins. Another similarity with alkenes (p. 67) is that carbon–oxygen double bonds are shorter (*ca.* 1·22 Å) than carbon–oxygen single bonds (*ca.* 1·43 Å). In contrast to olefins, aldehydes and ketones have appreciable dipole moments (2·25–2·8 D), indicating that the carbonyl group is polarised as represented in diagram (1).

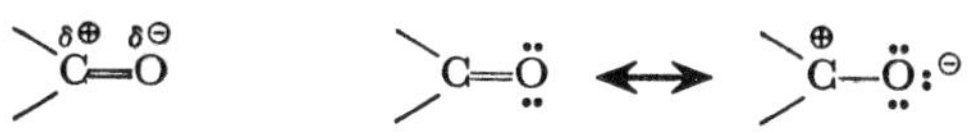

(1) (2) (3)

Alternatively, the structure of the carbonyl group may be said to be a hybrid of the two extreme forms (2) and (3) (cf. p. 91). The effect arises because oxygen has a greater power to attract electrons (is more electronegative) than carbon.

ADDITION REACTIONS

Whether a chemical change will occur or not depends on the colliding molecules being able to reach the energy level of a transition state. The character of the ground state of either reactant is therefore not a reliable guide to reactivity. Nevertheless, its dipolar double bond structure makes a carbonyl group conducive to the attack of nucleophiles at the electron-deficient carbon and of electrophiles (e.g. protons) at the oxygen atom.

In the addition reactions of aldehydes and ketones discussed below, equilibrium signs are used in the equations because either the forward or reverse reactions can be promoted by adjusting the experimental conditions (e.g. pH). The reactions are often performed in hydroxylic solvents, and solvation effects are energetically important especially when ionic species are present.

Water. Unlike the behaviour encountered with alkenes, addition occurs when water is mixed with an aldehyde or ketone, e.g.,

$$CH_3CHO + H_2O \rightleftharpoons CH_3CH(OH)_2 \quad \text{acetaldehyde hydrate}$$

The extent of *hydration* varies enormously with individual compounds. In water at 20° conversion into hydrate is nearly complete for formaldehyde, is 58% for acetaldehyde, and is almost negligible for acetone. Nevertheless acetone, in contrast with a hydrocarbon of similar molecular weight, is miscible with water in all proportions. This is due to intermolecular* hydrogen bonding (p. 88): $(CH_3)_2C{=}O\ldots HOH$. Most hydrates cannot be isolated, although chloral (trichloroacetaldehyde), when exposed to moisture is converted into its stable crystalline hydrate:

$$\underset{\text{chloral, b.p. 98°}}{Cl_3CCHO} + H_2O \rightleftharpoons \underset{\text{chloral hydrate, m.p. 52°}}{Cl_3CCH(OH)_2}$$

The sedative and hypnotic properties of chloral have been recognised for almost a hundred years, but it suffers the disadvantage of having an unpleasant taste. This has been surmounted (1960) by preparation of the inoffensive, but equally effective, mono-trichloroethyl phosphate, $Cl_3CCH_2OP(O)(OH)_2$ (p. 102).

The following scheme depicts the mechanism of hydration. Water acts as an uncharged nucleophile (p. 36), and attacks the carbon of the carbonyl group. The second stage of the reaction involves the transference of a proton from one oxygen to the other. Each of the species is likely to be hydrogen bonded to other water molecules.

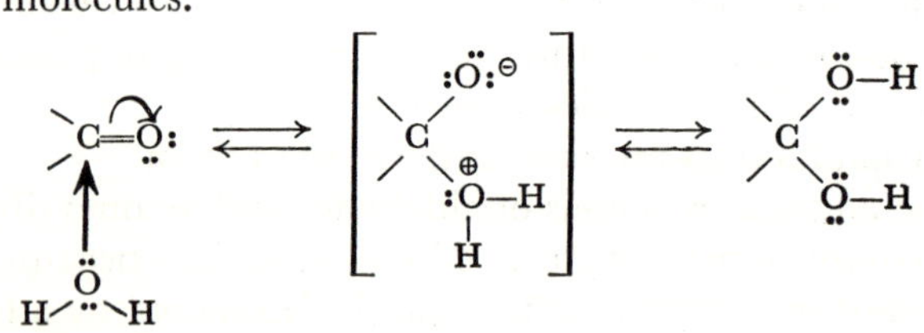

* Intermolecular = *between* molecules; intramolecular (p.83) = *within* a molecule.

Hydrogen Cyanide. The reaction of aldehydes and some ketones with hydrogen cyanide and a trace of potassium cyanide gives *cyanohydrins*. The mechanism can be expressed in the following simplified form:

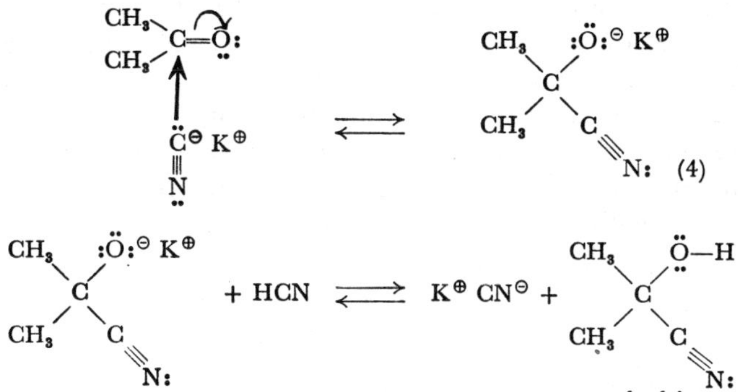

(4)

acetone cyanohydrin

The reaction is initiated by attack of the nucleophilic cyanide ion on the electron-deficient carbonyl carbon atom. The intermediate (4) extracts a proton from hydrogen cyanide, giving the cyanohydrin and regenerating cyanide ion catalyst. Typically, 0·01–0·1 mole of cyanide salt is used per mole of each of the two addenda. Hydrolysis of cyanohydrins gives α-hydroxy-acids (p. 149).

Amino-compounds. Aldehydes and ketones react with a variety of reagents containing basic amino-groups (p. 115), e.g.

$$CH_3CHO + NH_2CH_2CH_3 \rightleftharpoons CH_3CH{=}NCH_2CH_3 + H_2O$$

$$(CH_3)_2CO + NH_2NH_2 \rightleftharpoons (CH_3)_2C{=}NNH_2 + H_2O$$

hydrazine — acetone hydrazone

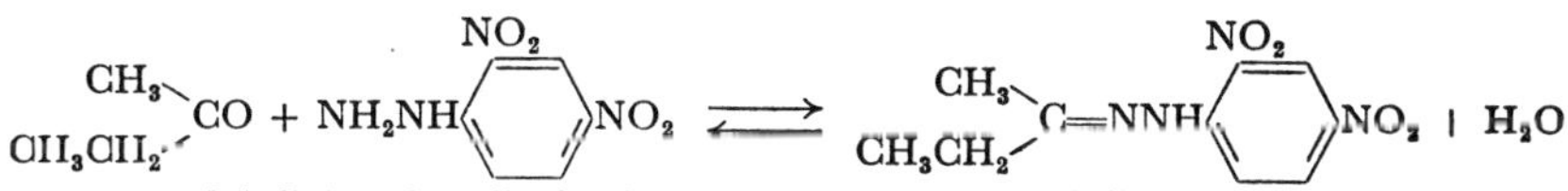

2,4-dinitrophenylhydrazine — butanone 2,4-dinitrophenylhydrazone, m.p. 115°

$$(CH_3)_2CHCHO + NH_2OH \rightleftharpoons (CH_3)_2CHCH{=}NOH + H_2O$$

hydroxylamine — 2-methylpropanal oxime, m.p. 139°

$$CH_3CH_2CHO + NH_2NHCONH_2 \rightleftharpoons CH_3CH_2CH{=}NNHCONH_2 + H_2O$$

semicarbazide — propanal semicarbazone, m.p. 89°

Aldehydes and ketones are distinguished from other organic compounds by the formation of yellow, orange or red 2,4-dinitrophenylhydrazones, and by the strong infrared absorption of the carbonyl group at 1700–1725 cm.$^{-1}$. Many aldehydes and ketones are liquids which can be identified individually by the preparation of crystalline derivatives. 2,4-Dinitrophenylhydrazones, oximes and semicarbazones are suitable for such a purpose, as shown by the melting points of representative compounds given above. Under appropriate conditions these compounds are formed in high yield, and the reverse reactions, especially with oximes and semicarbazones, can be carried out by heating with aqueous acid. Aldehydes and ketones can therefore be purified by the successive stages of formation, crystallisation, and hydrolysis of these derivatives.

A solution of 2,4-dinitrophenylhydrazine (0·30 g.) in concentrated sulphuric acid (1 ml.) and methanol (15 ml.) was added to butan-2-one (0·10 g.), followed by water (3 ml.). After 10 min. the precipitate was filtered off, washed with 6 ml. of a 1 : 1 mixture of cold methanol and water, and dried (0·31 g.), m.p. 110–115°. Crystallisation from methanol gave orange needles (0·21 g.), m.p. 114–116°.

Taking hydrazine as an example, the following mechanism for these reactions may be written:

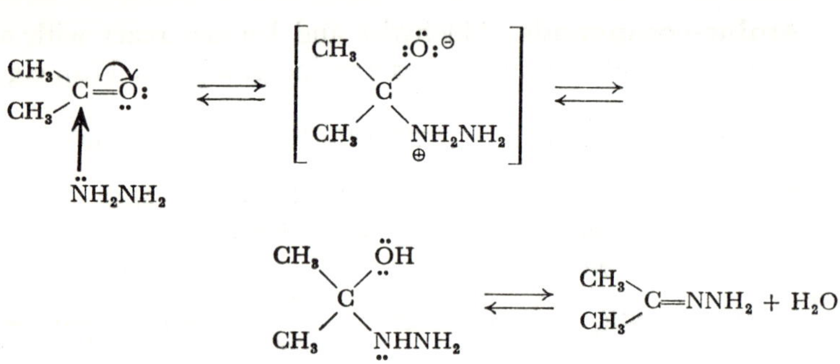

This scheme, which is analogous to those given for other carbonyl addition reactions, involves the addition of a nucleophilic amino-group, followed by the transfer of a proton from nitrogen to oxygen. However, the final elimination of water gives a new product, whereas in hydration (p. 76) loss of water from the product leads to reversal of the reaction and formation of the carbonyl compound. The elimination of water cannot occur in the hydrogen cyanide reaction, as the incoming -CN group does not contain a hydrogen atom.

In the experiment described above, the sulphuric acid has two functions: it keeps the weakly basic dinitrophenylhydrazine in solution and it protonates the ketone, $RR'C{=}\overset{\oplus}{O}H$, thereby making it more susceptible to nucleophilic attack.

Reduction. Aldehydes are reduced to primary alcohols and ketones to secondary alcohols by hydrogenation in the presence of a metal catalyst, e.g. nickel or platinum:

$$CH_3CH_2CH_2CHO \xrightarrow{2H} CH_3CH_2CH_2CH_2OH$$

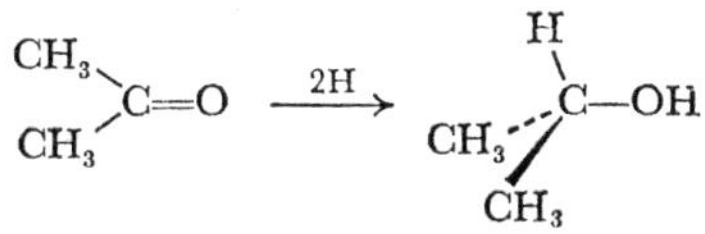

These hydrogenations are less exothermic (by 11–19 kcal./mole) than the corresponding reduction of carbon–carbon double bonds. Convenient reagents for reduction of aldehydes and ketones on a small scale are lithium tetrahydroaluminate ($LiAlH_4$) in dry ether solution and sodium tetrahydroborate ($NaBH_4$) in water or aqueous methanol. The stoichiometries of these reactions are the same:

$$4R_2CO + LiAlH_4 \longrightarrow Li^{\oplus}[Al(OCHR_2)_4]^{\ominus} \xrightarrow[HCl]{aq.} 4HOCHR_2$$

$$4R_2CO + NaBH_4 \longrightarrow Na^{\oplus}[B(OCHR_2)_4]^{\ominus} \xrightarrow[HCl]{aq.} 4HOCHR_2$$

Reactions involving α-C—H Bonds. Certain reactions of aldehydes and ketones depend primarily on the reactivity of C—H bonds of alkyl groups adjacent (i.e. α) to the carbonyl group. As explained below, these reactions are catalysed by acids and by bases.

(*a*) The *aldol reaction* occurs readily with aldehydes in the presence of aqueous alkali:

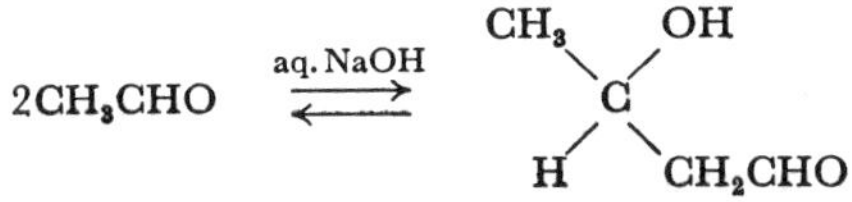

aldol, b.p. 77° at 16 mm. pressure
(3-hydroxybutanal)

Ketones react in the same way, but less readily:

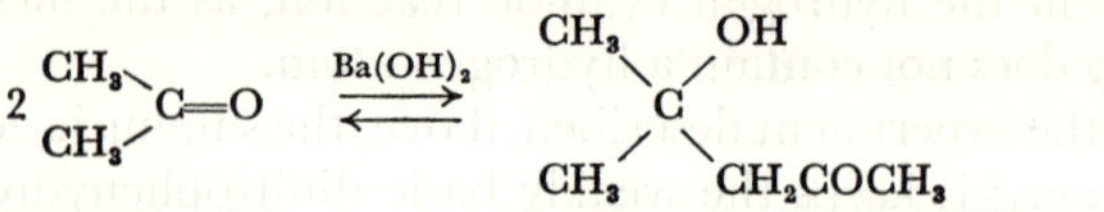

acetone, b.p. 56° — diacetone alcohol, b.p. 164° (4-hydroxy-4-methylpentan-2-one)

In the preparation of diacetone alcohol the equilibrium is so far to the left that, in order to obtain a good yield of product, only the acetone must be in contact with the alkali. This is achieved by heating acetone under reflux so that the ketone is returned to the boiling solution *via* a container holding solid barium hydroxide. Diacetone alcohol accumulates in the boiling solution, as its rate of dissociation into acetone is slow in the absence of barium hydroxide.

The aldol reaction, as it involves carbon–carbon bond formation, is a very useful and general method of synthesis, e.g. of unsaturated aldehydes and ketones, which can be obtained by dehydration of the hydroxy-carbonyl compounds. The mechanism of the reaction of, for example, acetaldehyde is as follows:

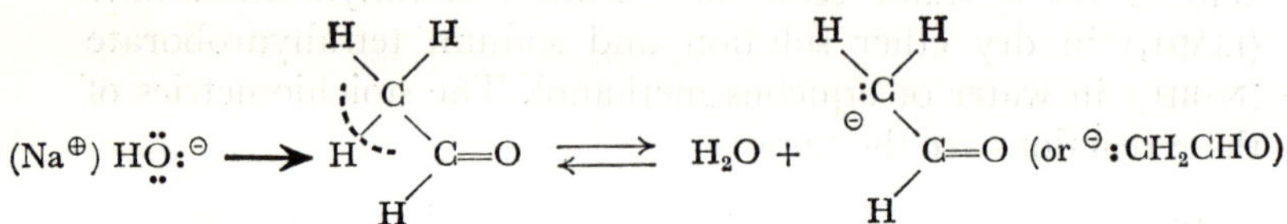

(5)

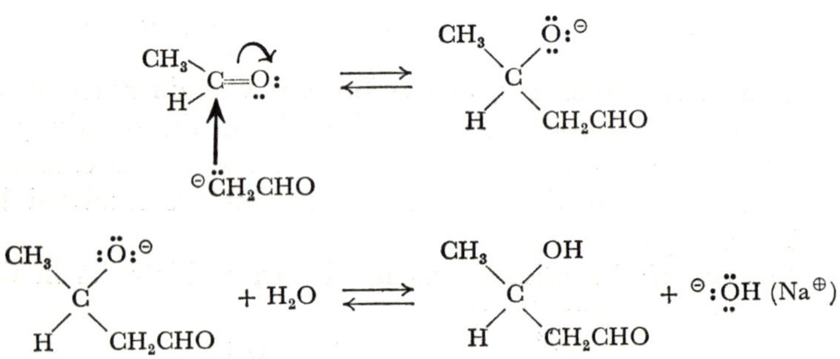

Initially the hydroxide ion removes a proton from the α-methyl group, leaving a new species (5), which retains the pair of electrons originally shared in the C—H bond. This reaction occurs because of the special stability of (5), which is discussed later (p. 91). It must be emphasised that only α-C—H bonds are reactive; propionaldehyde, for example, gives only one anion:

$$\overset{\beta}{C}H_3\overset{\alpha}{C}H_2CHO \longrightarrow CH_3\overset{\ominus}{\ddot{C}}HCHO \quad (\textit{not}\ \overset{\ominus}{\ddot{C}}H_2CH_2CHO,\ CH_3CH_2\overset{\ominus}{\ddot{C}}O)$$

The later stages of the aldol condensation are very similar to the addition reactions already described. In this case the intermediate anion (5) is the nucleophile that attacks the carbonyl carbon atom of a second molecule of acetaldehyde. The adduct so formed extracts a proton from a water molecule, thus regenerating the catalytic hydroxide ion.

Compounds such as 2,2-dimethylpropanal, $(CH_3)_3CCHO$, and formaldehyde, HCHO, do not contain an α-hydrogen and hence do not undergo the aldol reaction.

(*b*) *Halogenation* of aldehydes and ketones only occurs readily in the presence of a base or a trace of mineral acid. Hydrogens on α-carbon atoms are substituted by halogen:

$$CH_3COCH_3 + Br_2 \xrightarrow{\text{trace HBr}} CH_3COCH_2Br + HBr$$

Lapworth found (in 1904) that bromination of acetone was a second order reaction, dependent on the concentration of ketone and *acid* (not bromine). The function of the acid is to promote the formation of the *enol* form of acetone:

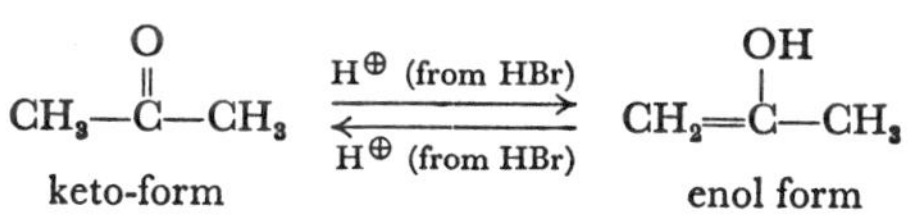

This process, involves the transfer of a hydrogen atom from carbon to oxygen with a consequent shift of the double bond. In the subsequent more rapid step the olefinic bond of the enol reacts with bromine affording a bromonium ion, which loses a proton to give the product:

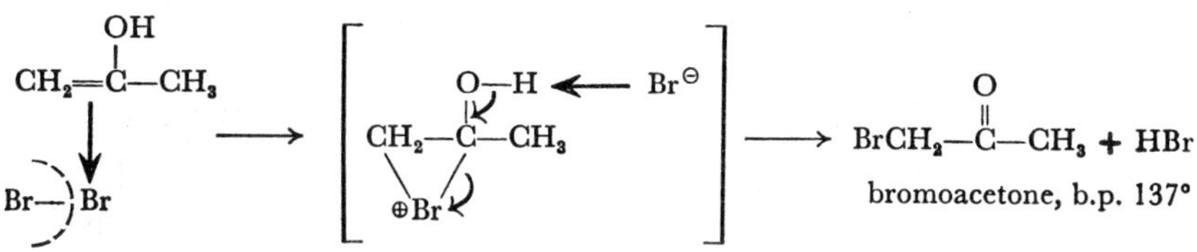

Even in the absence of acid, carbonyl compounds that possess an α-hydrogen are a mixture of keto- and enol forms in equilibrium, but simple aldehydes and ketones contain only a trace of enol isomer (at 25°, 0·00025% for acetone, 0·02% for cyclohexanone). The keto-form is more stable, mainly because of the greater strength of a carbonyl compared to an olefinic bond, as the following calculation shows:

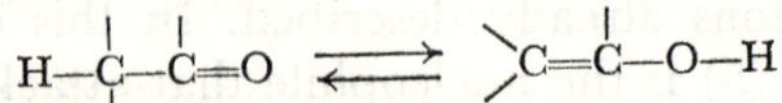

bond energies 98·5 82·5 177·5 — 146 85·5 110·5 kcal./mole

$$\Delta H = 358{\cdot}5 - 342 = +\ 16{\cdot}5 \text{ kcal./mole}$$

Bromoacetone is not necessarily the final product of the bromination of acetone, as further α-hydrogen atoms can be replaced if more bromine is employed. The chlorination of acetaldehyde proceeds similarly:

$$CH_3CHO + Cl_2 \xrightarrow{H^{\oplus}} ClCH_2CHO + HCl$$
chloroacetaldehyde

$$ClCH_2CHO + Cl_2 \xrightarrow{H^{\oplus}} Cl_2CHCHO + HCl$$
dichloroacetaldehyde

$$Cl_2CHCHO + Cl_2 \xrightarrow{H^{\oplus}} Cl_3CCHO + HCl$$
trichloroacetaldehyde (chloral)

Oxidation. Aldehydes are oxidised easily to carboxylic acids containing the same number of carbon atoms. The reaction involves the C—H bond that is part of the aldehyde group, and can be expressed in general terms as:

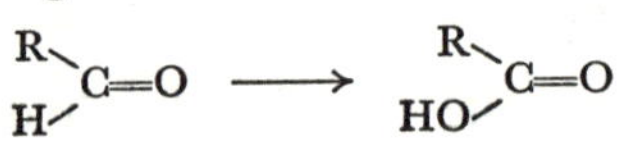

As ketones do not contain such a C—H bond, they are not oxidised in this way, and this difference is the basis of two laboratory tests used to distinguish between aldehydes and ketones. Aldehydes are oxidised by Tollens reagent (an aqueous solution of silver nitrate, ammonia and sodium hydroxide) to give a silver mirror:

$$RCHO + 2[Ag(NH_3)_2]^{\oplus}\,OH^{\ominus} \longrightarrow RCOO^{\ominus}\,NH_4^{\oplus} + 3NH_3 + H_2O + 2Ag\downarrow$$

Oxidation with cupric salts (often employed as Fehling's solution) gives a red precipitate of cuprous oxide, Cu_2O. This test is often used to detect the presence of 'reducing sugars' (i.e. carbohydrates containing potential aldehyde groups, p. 172) in the urine of diabetic patients.

Ketones can be oxidised by strong oxidising agents, e.g. potassium permanganate in acid solution, but the carboxylic acids formed must contain less carbon atoms than the original ketone. The reaction proceeds through the enol form which undergoes oxidative cleavage at the double bond:

$$CH_3CH_2COCH_2CH_3 \overset{H^{\oplus}}{\rightleftharpoons} CH_3CH{=}\overset{\overset{OH}{|}}{C}CH_2CH_3 \xrightarrow{KMnO_4} CH_3COOH + HOOCCH_2CH_3$$

Formaldehyde, a gas (b.p. −21°) is manufactured by passing a mixture of methanol and air over a copper or silver catalyst at 250–300°:

$$2CH_3OH + O_2 \longrightarrow 2HCHO + 2H_2O$$

The products are absorbed in water to give 'formalin', a 35–40% solution of formaldehyde. The presence of methanol in formalin inhibits the polymerisation of the aldehyde to solid paraformaldehyde $(\text{-}CH_2\text{—}O\text{-})_n$ which, however, is obtained by evaporation of the solution. Gaseous formaldehyde is obtained when paraformaldehyde is heated.

Formaldehyde is very reactive and is used extensively in the synthesis of organic compounds including the phenol-formaldehyde (bakelite) and urea–formaldehyde plastics. Biological materials are often preserved in formalin; the aldehyde by reacting with protein molecules prevents disintegration of the tissue, and also serves as an anti-bacterial agent.

Acetaldehyde, b.p. 21°, is prepared by catalytic oxidation (dehydrogenation) of ethanol (cf. production of formaldehyde), and by hydration of acetylene (p. 69). A recent process (1959) takes advantage of the availability of ethylene; the olefin is oxidised by aqueous palladium chloride in the presence of oxygen and copper salts:

$$CH_2{=}CH_2 \xrightarrow[CuCl_2,\ O_2]{PdCl_2,\ H_2O} CH_3CHO$$

Deuterium is not incorporated into the acetaldehyde when the reaction is performed in deuterium oxide (^2H_2O) instead of water. The experiment shows that intramolecular (p. 76) migration of hydrogen occurs during the oxidation.

Acetone. Propylene from petroleum is now the best starting material for the preparation of acetone. The conversion is accomplished either by hydration of the olefin to propan-2-ol followed by catalytic oxidation, or by oxidation by the palladium chloride method:

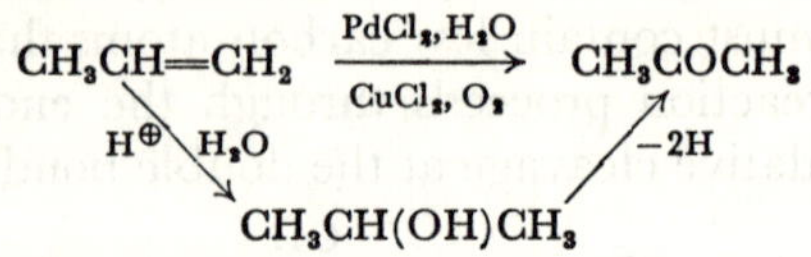

Acetone, b.p. 56°, is miscible with water, and is an important solvent.

QUESTIONS

1. (*a*) Write unambiguous structural formulae for the following compounds:

pentan-2-one; n-octanal; 4,4-dimethylhexan-3-one; chloro-ethanal.

(*b*) Give systematic names for the following molecules:

$CH_3CH_2CH_2CH_2CHO$ $BrCH_2CH_2COCH_3$ $(CH_3)_3CCOCH_2CH_2CH_3$

2. Aqueous sodium carbonate (70 ml. of 10% solution) is added to a stirred mixture of cyclohexanone (9·8 g.) and hydroxylamine hydrochloride (8·4 g.) in water (20 ml.). A precipitate forms which after 10 min. is collected and crystallised from hot water; it then has m.p. 88°.

(i) Give the structure and the name of the product (analysis: C, 63·7%; H, 9·8%; N, 12·4%).
(ii) Write a mechanism for its formation.
(iii) What is the function of the sodium carbonate?
(iv) Suggest a reason for the fact that a 1:1 ratio of ketone: hydroxylamine hydrochloride is not used for the reaction.

3. (*a*) Acetaldehyde labelled with ^{18}O is formed when water containing $H_2{}^{18}O$ is mixed at 20° with unlabelled acetaldehyde. Explain, *via* the mechanism of the reaction how this occurs, and suggest a practical method of recovering a (dry) sample of the labelled acetaldehyde.

(*b*) How is water isotopically enriched with ^{18}O obtained?

(*c*) Interpret the mass spectrum of propanal which shows two main peaks at 58 and 29 (cf. p. 7). How would the mass spectrum of propanal labelled with ^{18}O be helpful for testing your explanation?

4. Discuss with the aid of energy profiles the mechanisms of hydration of (*a*) acetaldehyde and (*b*) isobutene.

5. (*a*) Outline a practical synthesis of (i) 2-methylpropanal from 1-bromo-2-methylpropane, (ii) butanone from but-1-ene, and (iii) the cyanohydrin of butanone.

(*b*) The sedative, 3-methylpent-1-yn-3-ol, is made by reacting butanone with sodium acetylide, followed by treatment of the product with acid (1 equivalent). Discuss the chemistry involved in the preparation.

6. (*a*) A preparation described in a book published in 1900 refers to the necessity of removing acetone from commercial methyl alcohol. What would be the best procedure for determining the presence of acetone in methanol (i) in 1900 and (ii) at the present time? Why would the methanol available in 1900 be more likely to contain acetone than present-day material? How might a large quantity of a mixture of acetone and methanol be separated?

(*b*) Estimation of the amount of aldehyde or ketone in a dilute aqueous solution can be carried out by adding an excess of 2,4-dinitrophenylhydrazine in 2N hydrochloric acid, when the sparingly soluble derivative separates in almost 100% yield. A saturated solution (1 ml.) of cyclohexanone in water at 20° gave 153 mg. of derivative. What is the solubility of cyclohexanone in water (100 g.) at 20°?

The secret of all who make discoveries is looking on nothing as impossible.

J. VON LIEBIG (1803–1873)

CHAPTER 7

CARBOXYLIC ACIDS AND THEIR DERIVATIVES

Acetic acid, in the form of an aqueous solution (vinegar) is produced by the 'souring' of wine, and its name is derived from *acetum*, the Latin word for vinegar. Formic acid was first obtained by heating red ants (Latin, *formica*) with water.

Carboxylic acids are sometimes called 'fatty acids' because certain of the higher members of the series are obtained in the hydrolysis of natural fats (Chapter 8).

Structure. Carboxylic acids and their derivatives contain the carbonyl group and have the general structure

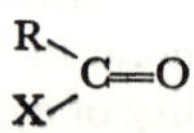

where R is an alkyl group or hydrogen and X contains an element from Group V, VI or VII linked directly to the carbonyl carbon. The presence of X causes the chemical properties of these compounds to be different in many respects from those of aldehydes and ketones, which also contain the carbonyl group. The general structures of the carboxylic acids and their principal derivatives are given below in full and also in abbreviated form. The structures of aldehydes and ketones are included for comparison.

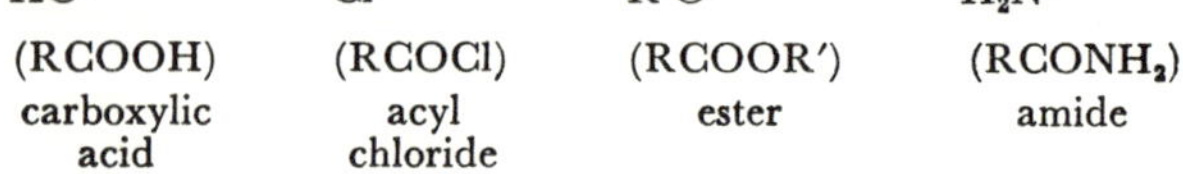

(RCOOH) carboxylic acid — (RCOCl) acyl chloride — (RCOOR′) ester — ($RCONH_2$) amide

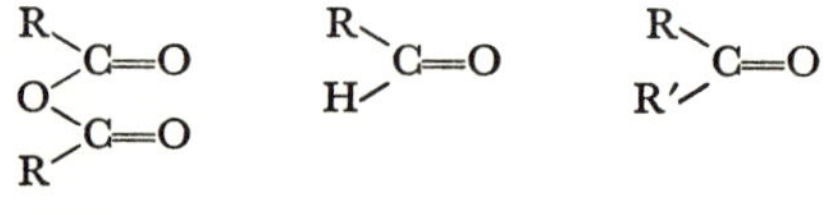

[$(RCO)_2O$] acid anhydride — (RCHO) aldehyde — (RCOR′) ketone

The group RCO– is known as the *acyl* group, and acid derivatives RCOX in which X is halogen are described as acyl halides.

Nomenclature. Carboxylic acids are often known by specific names. Alternatively, they can be named systematically by adding *-oic acid* to the name of the corresponding alkane (omitting the terminal *-e*). The carbon atom of the *carboxyl* (i.e. –COOH) group is part of the chain, and is regarded as the first carbon atom when a numbering system is used:

$(CH_3)_2CH(CH_2)_3CH(Cl)COOH$ or 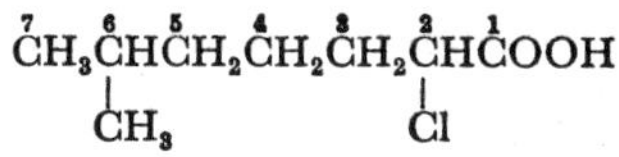

2-chloro-6-methylheptanoic acid

Examples of both methods of nomenclature are shown in the following table:

	Systematic name	*Specific name*	
HCOOH	methanoic acid	formic acid	b.p. 101°
CH_3COOH	ethanoic acid	acetic acid	b.p. 118°
CH_3CH_2COOH	propanoic acid	propionic acid	b.p. 141°
$CH_3(CH_2)_2COOH$	butanoic acid	n-butyric acid	b.p. 163°
$(CH_3)_2CHCOOH$	2-methylpropanoic acid	isobutyric acid	b.p. 155°
$CH_3(CH_2)_3COOH$	pentanoic acid	n-valeric acid	b.p. 186°
$(CH_3)_2CHCH_2COOH$	3-methylbutanoic acid	isovaleric acid	b.p. 176°
$(CH_3)_3CCOOH$	2,2-dimethylpropanoic acid	pivalic acid	m.p. 35°
$CH_3(CH_2)_{14}COOH$	hexadecanoic acid	palmitic acid	m.p. 63°
$CH_3(CH_2)_{16}COOH$	octadecanoic acid	stearic acid	m.p. 71°

The C_1–C_3 acids are pungent liquids, soluble in water. Straight-chain (n-alkanoic) acids containing 4 to 9 carbon atoms are rancid liquids, partially soluble in water, whereas the higher acids are solids, nearly insoluble in water. The presence of a carboxyl group increases the solubility of a compound in water. This is particularly obvious in the earlier members of the series, in which the carboxyl group represents the major portion of the molecule. A hydrocarbon chain has the opposite effect, and so when this feature becomes predominant as the series is ascended, solubility in water decreases.

Carboxylic acids are formed by oxidation of alkenes (p. 65), alcohols (p. 47) and aldehydes (p. 82), and by the hydrolysis of acid derivatives, the nitriles being particularly useful (cf. p. 102) for the synthesis of acids. Another method of preparation involves Grignard reagents (p. 129).

The Hydrogen Bond and Solvation. As with alcohols (p. 44), the boiling points of carboxylic acids are higher than those of isomeric substances not containing a hydroxyl group:

CH_3CH_2OH	b.p. 78°	CH_3CH_2COOH	b.p. 141°
CH_3OCH_3	b.p. −23·5°	CH_3COOCH_3	b.p. 57°

The general phenomenon indicates that hydroxylic molecules can associate in some way. This occurs *via* the formation of *hydrogen bonds* which link together oxygen atoms of different molecules.

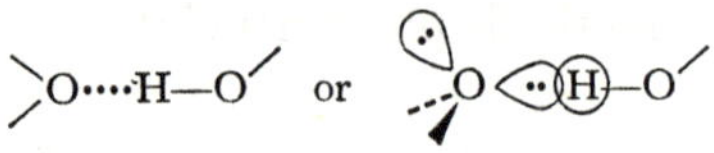

(1)

The probability of hydrogen bonding is greatest with water, and this high boiling compound (mol. wt. only 18) consists of a mobile, three dimensional network of hydrogen bonded molecules. The similar intermolecular hydrogen bonds in alcohols and carboxylic acids may be depicted:

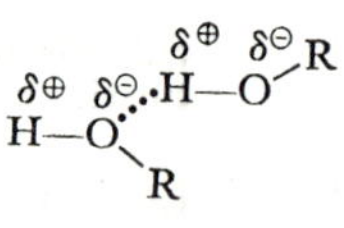

(2)

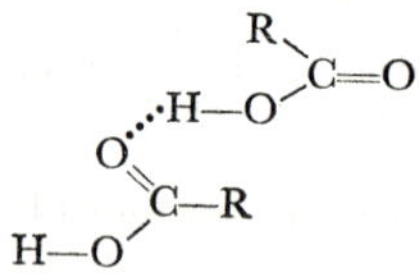

Hydrogen bonds are generally intermediate in strength (3–10 kcal./mole) between covalent bonds (usually 50–226 kcal./mole) and intermolecular non-bonding attractive forces. They develop when the hydrogen is able to bridge two electronegative centres, e.g. oxygen, nitrogen, halogens. Despite the amount of work on the topic, it is not simple to express the origin of hydrogen bonding. Orbital overlap between the two electron-rich atoms and the 2*s* hydrogen shell may be involved cf. (1). An alternative view of the situation is that bonding is due to electrostatic attraction between adjacent atoms carrying unlike charges [see diagram (2)].

The fact that alcohols and carboxylic acids are more soluble in water than related non-hydroxylic compounds is also attributed to hydrogen bonding, in this case between different molecules (compound and water). Such solute–solvent interactions are described by the general term *solvation* (in water, *hydration*). Much greater energy changes may be entailed in the solvation of ions.

Acidity and Resonance. The most characteristic property of carboxylic acids is their acidity, defined as the tendency to lose a proton to a base, thereby forming a salt. Thus, an exothermic reaction occurs when acetic acid is added to aqueous sodium hydroxide solution; sodium acetate is formed, and this colourless salt is obtained in *ca.* 100% yield by evaporation of the solution. The negatively charged anion in this salt is an example of a *carboxylate ion* (or anion).

That salt formation is essentially complete is also shown by the fact that the heat of reaction of acetic acid with sodium hydroxide (13·35 kcal./mole) is very similar to that of the neutralisation of a strong acid and a strong base. [Carboxylic acids are stronger acids than carbonic acid, so that carbon dioxide is evolved when sodium carbonate or sodium bicarbonate is added, for example, to a solution of acetic acid. This evolution of carbon dioxide provides one test for an organic acid.]

A solution of acetic acid in water alone gives an acid reaction, but measurement of the *dissociation constant*, K_a, reveals that only a very small fraction of the acetic acid is present as hydrated ions:

$$CH_3COOH(aq.) + H_2O \rightleftharpoons CH_3COO^{\ominus}(aq.) + H_3O^{\oplus}(aq.)$$

$$K_a = \frac{[H_3O^{\oplus}][CH_3COO^{\ominus}]}{[CH_3COOH]} = \underset{(\text{at } 25°)}{1{\cdot}85 \times 10^{-5}} \ \text{(concentration of water taken as unity)}$$

[Alternatively the strength of an acid can be expressed as its pK_a value, where $pK_a = -\log K_a$. For acetic acid, $pK_a = 4{\cdot}76$; stronger acids give lower pK_a values.] An unfavourable entropy change is mainly responsible for the low value of K_a for the equilibrium.

Acetic acid, like most carboxylic acids, is much weaker than a mineral acid. Consequently the organic acid can be recovered from one of its salts by appropriate treatment with a mineral acid.

Although both ethanol and acetic acid contain a hydroxyl group, the former compound is by far the weaker acid: $K_a = 7{\cdot}3 \times 10^{-20}$. Acetic acid, unlike ethanol, has the hydroxyl group attached to a carbonyl group, and it is this feature which must be responsible for the considerable difference in their acid strengths. Bond lengths in organic molecules can be measured by electron diffraction methods, and it is found that double bonds are shorter than single bonds (cf. p. 67). For example, the carbon–oxygen bond lengths in the carboxyl group of acetic acid are 1·24 Å and 1·43 Å. On the basis of structure (3) the carboxylate ion might also be expected to have carbon–oxygen bonds of different length, but in fact it is found that these bonds are *equal* in length (1·27 Å). The carboxylate ion is symmetrical, and half the negative charge resides on each oxygen atom. The ion can then be regarded as a *hybrid* of the two extreme forms (3) and (4); the actual structure is not identical with either, but is intermediate between the two.

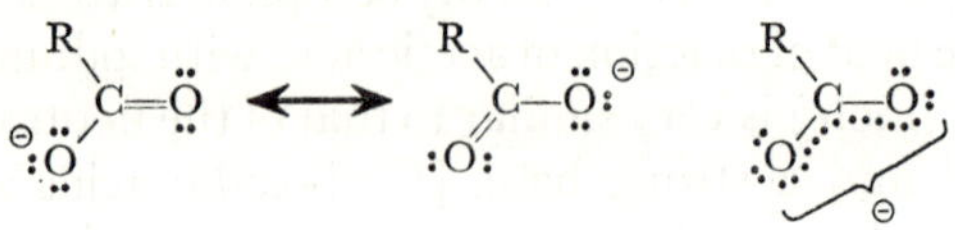

(3) (4) (5)

Alternatively, and more conveniently, the hybrid structure can be represented as (5). The dotted line on this diagram represents the delocalisation of four electrons (two electrons from the carbonyl bond and an unbonded pair from oxygen). The concept involved is known as *resonance* (or *mesomerism*), and the molecule is described as a *resonance hybrid.* (In diagrams the two contributing forms are linked by a double-headed arrow.) It must be emphasised that the carboxylate ion has only one structure, and is *not a mixture* of the forms (3) and (4). Forms contributing to a resonance hybrid differ *only* in electron distribution. The hybrid anion (5) has a lower energy content ('is more stable') than either of the contributing forms.

Ionisation proceeds to equilibrium, and the acidity of an alcohol or a carboxylic acid (the dissociation constant for ionisation) is determined by the relative stability of reactants and products (ΔG). The reason why acetic acid is much more acidic than ethanol is that loss of a proton gives an anion *stabilised by resonance*; the ethoxide ion, $^{\ominus}OCH_2CH_3$, derived from ethanol, is not stabilised in this way. These points are illustrated by the energy profiles; the lower (less positive) value of $\Delta G_{(acid)}$ compared to $\Delta G_{(alcohol)}$ arises from the greater resonance stabilisation in the carboxylate ion.

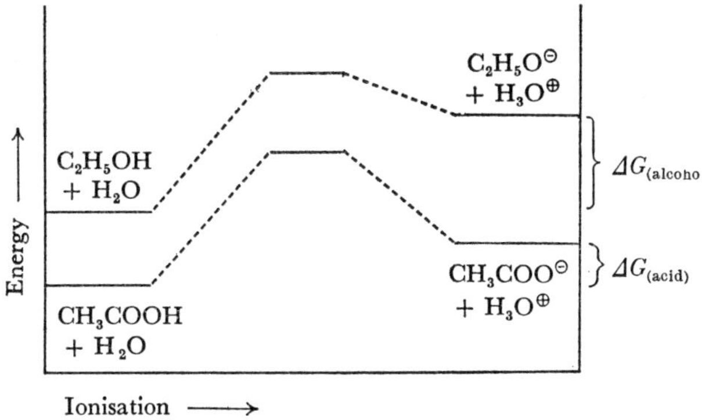

The contributing forms of the carboxylate ion are equivalent, but this is not an essential condition for resonance. For example, the carbonyl group of aldehydes and ketones is a resonance hybrid (6), in which two non-equivalent forms are involved. Another example is the anion (5, p. 80) derived from acet-

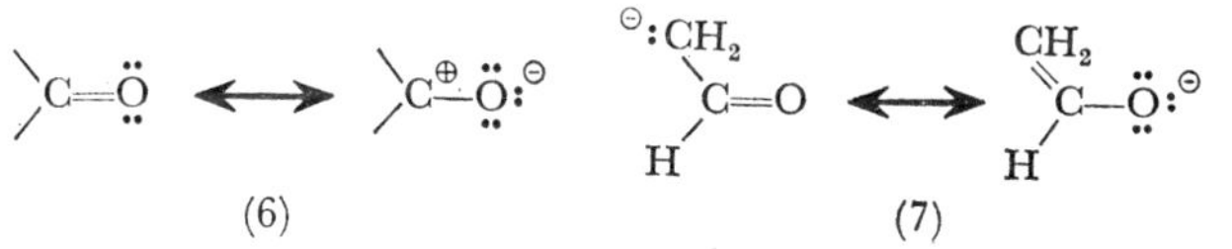

(6) (7)

aldehyde which is a resonance hybrid (7). The stability of the anion encourages the loss of a proton from acetaldehyde when this compound is treated with alkali.

Information concerning intramolecular effects can be obtained from study of the strengths of various acids. Introduction of an electronegative element, e.g. chlorine, into acetic acid greatly increases the acid strength:

pK_a values:
CH_3COOH, 4·76 $ClCH_2COOH$, 2·86 CH_3CH_2COOH, 4·88

On the other hand, the extra methyl group in propionic acid slightly reduces the degree of ionisation.

Conversion of Carboxylic Acids into their Derivatives. (*a*) The hydroxyl group of carboxylic acids can be replaced by halogen to give, for example, *acyl chlorides* (RCOCl) or *acyl bromides* (RCOBr). The following reactions occur with phosphorus halides:

$$3CH_3COOH + PBr_3 \longrightarrow \underset{\text{acetyl bromide, b.p. 81°}}{3CH_3COBr} + H_3PO_3$$

$$CH_3CH_2CH_2COOH + PCl_5 \longrightarrow \underset{\text{n-butyryl chloride, b.p. 102°}}{CH_3CH_2CH_2COCl} + POCl_3 + HCl$$

The by-products (e.g. $POCl_3$, b.p. 107°) are sometimes difficult to remove from the acyl halides and an alternative procedure for chlorides is to heat the acid with thionyl chloride, when the other products are gaseous:

$$(CH_3)_2CHCOOH + SOCl_2 \longrightarrow \underset{\text{isobutyryl chloride, b.p. 92°}}{(CH_3)_2CHCOCl} + SO_2 + HCl$$

The acyl halides can be named by adding *-yl halide* to the root of the specific name of the corresponding carboxylic acid, or by adding *-oyl halide* to the name of the corresponding alkane:

CH_3COCl	acetyl chloride or ethanoyl chloride
CH_3CH_2COCl	propionyl chloride or propanoyl chloride
$CH_3(CH_2)_2COBr$	n-butyryl bromide or butanoyl bromide
$\overset{3}{C}H_3\overset{2}{C}H(CH_3)\overset{1}{C}OCl$	isobutyryl chloride or 2-methylpropanoyl chloride

(*b*) When the salts of acids are heated with acyl chlorides, *acid anhydrides* are formed:

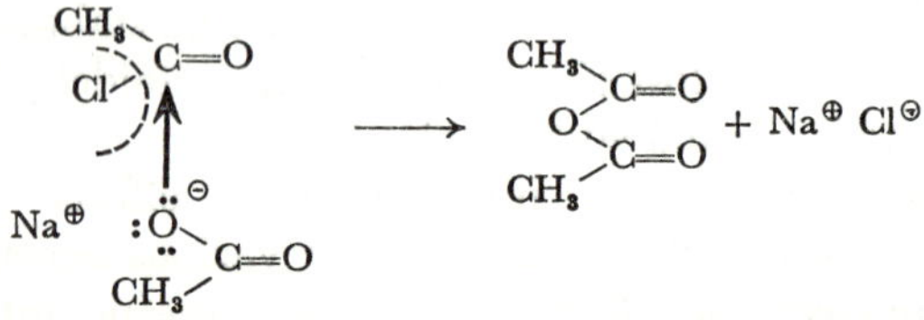

Note that this is the reaction of an acyl chloride with a nucleophilic carboxylate anion, and does not involve the replacement of the hydroxyl group of a carboxylic acid.

The names given to acid anhydrides are often derived from the specific names of the acids:

$(CH_3CO)_2O$	acetic anhydride, b.p. 140°
$(CH_3CH_2CO)_2O$	propionic anhydride, b.p. 167°

(*c*) Heating an acid with an alcohol slowly gives an *ester*, by replacement of the OH group by O-alkyl:

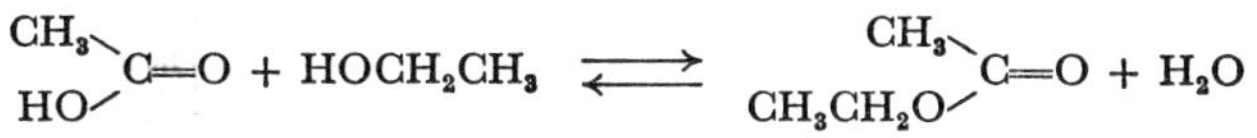

The following *K* values show that the ester and water are the major components at equilibrium; also that the equilibrium value is only approximately constant in this system.

Mole ratio of ethanol : acetic acid	*K*	*Proportion of ester*
3:1	2·45	86%
1:1	3·8	89%
1:3	4·75	92%

A small amount (e.g. 1 %) of hydrogen chloride or sulphuric acid increases the yield of ester by converting the water formed into hydronium ions, $H_3O^{\oplus}$. The strong acid also has the useful practical effect of accelerating (catalysing) the rate of ester formation from the two starting materials (see p. 98).

The lower members of the ester series are pleasant-smelling volatile liquids which are useful as solvents. Some of the higher boiling esters serve as plasticisers and in stationary phases for gas chromatography. Two systems of nomenclature are in common use, and correspond to the two methods of naming acids. In each case the name of the alkyl group (R′ in RCOOR′). precedes the acid name in which the suffix -*ic* has been replaced by -*ate*:

$HCOOCH_3$	methyl formate or methyl methanoate	b.p. 31·5°
$CH_3COOCH_2CH_3$	ethyl acetate or ethyl ethanoate	b.p. 77°
$CH_3CH_2COOCH_3$	methyl propionate or methyl propanoate	b.p. 80°
$(CH_3)_2CHCH_2COOCH_2CH_2CH_3$	n-propyl 3-methylbutanoate	b.p. 156°

Similar methods are used for naming carboxylic acid salts:

$(CH_3)_2CHCOO^{\ominus}\ K^{\oplus}$	potassium isobutyrate or potassium 2-methylpropanoate

(*d*) When carboxylic acids are heated strongly with ammonia or amines, the hydroxyl group is replaced by $-NH_2$, $-NHR$, or NR_2 to give *amides*. As the reagents are basic, carboxylate salts are intermediates:

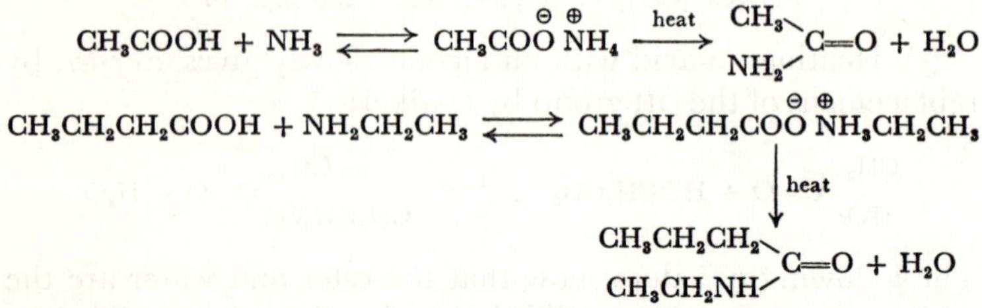

This reaction does not occur easily; alternative methods of preparing amides are discussed later (pp. 102, 119).

The classification of amides into primary, secondary and tertiary depends on the degree of substitution on the nitrogen; in the following structures, R, R′ and R″ are alkyl groups:

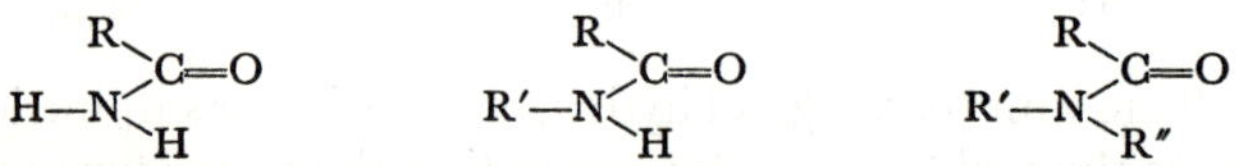

primary   secondary   tertiary

The primary amides are named by adding *-amide* to the root of the specific name of the corresponding acid, or by adding *-amide* to the name of the corresponding alkane (without the terminal *-e*):

$HCONH_2$	formamide or methanamide	b.p. 195°
$HCON(CH_3)_2$	dimethylformamide	b.p. 153°
CH_3CONH_2	acetamide or ethanamide	m.p. 82°
$CH_3CH_2CONH_2$	propionamide or propanamide	m.p. 81°
$\overset{5}{C}H_3\overset{4}{C}H(CH_3)\overset{3}{C}H_2\overset{2}{C}H_2\overset{1}{C}ONH_2$	4-methylpentanamide	m.p. 120°

Formamide and dimethylformamide are liquids, but most other amides are crystalline solids, the lower members of the series being soluble in water. Although amides contain amino-groups, they are very much weaker bases than amines (p. 115).

Reactions of Carboxylic Acid Derivatives. Most of the reactions of acid derivatives involve replacement of X in the general structure

It is therefore convenient to consider the chemistry of this group together, and to refer their reactions to this general structure.

(*a*) Hydrolysis by water, aqueous acid or aqueous alkali replaces X by OH to give *carboxylic acids* (or their salts if alkali is used).

$$\underset{\text{acetyl chloride}}{CH_3COCl} + H_2O \longrightarrow CH_3COOH + HCl$$

$$\underset{\text{propionic anhydride}}{(CH_3CH_2CO)_2O} + H_2O \longrightarrow 2CH_3CH_2COOH$$

$$\underset{\text{ethyl acetate}}{CH_3COOCH_2CH_3} + H_2O \xrightarrow{H^{\oplus}} CH_3COOH + HOCH_2CH_3$$

$$\underset{\text{ethyl acetate}}{CH_3COOCH_2CH_3} + H_2O \xrightarrow{HO^{\ominus}\ Na^{\oplus}} CH_3COO^{\ominus}\ Na^{\oplus} + HOCH_2CH_3$$

$$CH_3CONH_2 + H_2O + HCl \longrightarrow CH_3COOH + NH_4Cl$$

Acyl chlorides and anhydrides are usually hydrolysed fairly quickly by water alone at 20°. Esters and amides are less reactive; acidic or basic reagents are used in practice (e.g. hydrolysis of fats, p. 106) to accelerate their hydrolysis.

The hydrolysis of esters could proceed in more than one way:

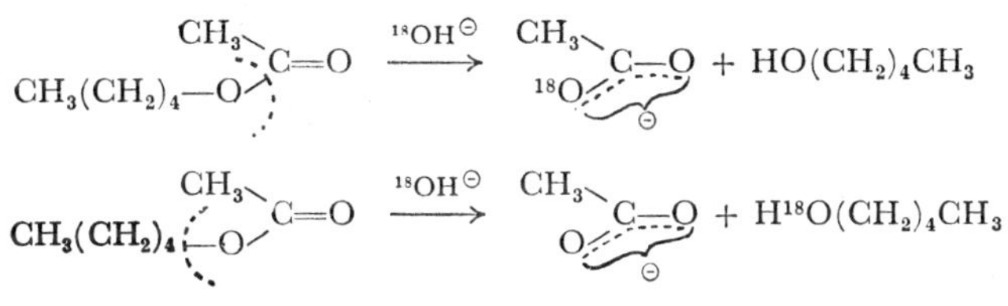

As indicated by the dotted lines in the ester structures, either the carbonyl–oxygen bond or the oxygen–alkyl bond could break, to give the same products. This question was decided, as shown in the diagram, by hydrolysis with hydroxide ion containing the ^{18}O isotope of oxygen (Polanyi and Szabo, 1934). Analysis of the products (by mass spectrometry) showed that only the carboxylate ion contained ^{18}O, thus supporting the first mechanism for the hydrolysis of the ester. It is this mechanism (replacement of OR′ in RCOOR′) (cf. p. 97) which conforms to the general pattern of acid derivative reactions.

(*b*) Reaction of acyl halides and anhydrides with alcohols gives *esters* by replacement of X by OR.

$$(CH_3)_2CH(Cl)C{=}O + HOCH_2CH_3 \xrightarrow{20^\circ} (CH_3)_2CH(CH_3CH_2O)C{=}O + HCl$$

$$(CH_3C{=}O)_2O + HOCH_2CH_2CH_3 \xrightarrow{20^\circ} CH_3(CH_3CH_2CH_2O)C{=}O + CH_3COOH$$

These reactions (in which the alcohol is *acylated* by the acid derivative) often proceed in high yields at room temperature, and provide useful methods for preparing esters.

Thiols react with acyl halides to give thiol esters:

$$CH_3(Cl)C{=}O + HSCH_2CH_3 \longrightarrow CH_3(CH_3CH_2S)C{=}O + HCl$$

b.p. 116°
(*cf.* ethyl acetate, b.p. 77°)

(*c*) Reaction of acid derivatives with ammonia or amines is a good method of preparing *amides*.

$$CH_3(Cl)C{=}O + 2NH_3 \xrightarrow{20^\circ} CH_3(NH_2)C{=}O + NH_4Cl$$

The reaction with esters involves the replacement of the O-alkyl substituent by the nitrogen-containing group.

(*d*) Acyl chlorides, acid anhydrides and esters are reduced to *primary alcohols* by various reducing agents, but most conveniently, by lithium tetrahydroaluminate, e.g.,

$$2\ CH_3CH_2CH_2(CH_3O)C{=}O + LiAlH_4 \xrightarrow[20^\circ]{\text{in ether}} [(CH_3CH_2CH_2CH_2O)_2Al(OCH_3)_2]^{\ominus}\,Li^{\oplus}$$

$$\downarrow \text{aq. acid}$$

$$2\ CH_3CH_2CH_2CH_2OH + 2CH_3OH$$

All derivatives of one carboxylic acid are reduced to the same alcohol, e.g. acetyl chloride, acetic anhydride and any acetate are all reduced to ethanol.

Amides behave differently, and give *amines* on reduction with lithium tetrahydroaluminate:

$$CH_3CONH_2 \longrightarrow CH_3CH_2NH_2$$
ethylamine, b.p. 19°

$$CH_3CH_2CONHCH_3 \longrightarrow CH_3CH_2CH_2NHCH_3$$
methylpropylamine, b.p. 62°

Mechanisms of Replacement Reactions. Acid derivatives, like aldehydes and ketones, contain a carbonyl group and react with nucleophiles. This similarity is reflected in the mechanisms of their reactions. Consider the general scheme:

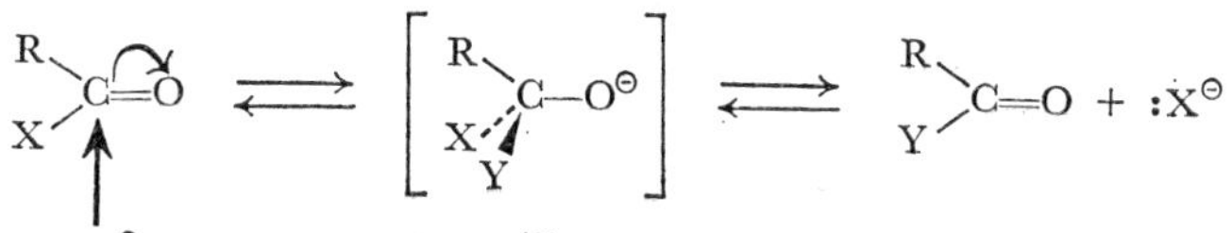

(8)

The first stage (attack of nucleophile $^{\ominus}$:Y) is an addition process, and occurs with both groups of compounds (cf. pp. 75–79). In the resultant intermediate (8) from aldehydes and ketones, the C—X bond (C—H and C—C respectively) is stable; with acid derivatives loss of X$^{\ominus}$ by cleavage of the C—X bond of the intermediate results in the overall replacement of X by Y.

This general mechanism possibly applies to all replacement reactions of acid derivatives, e.g. alkaline hydrolysis of an ester (cf. p. 95):

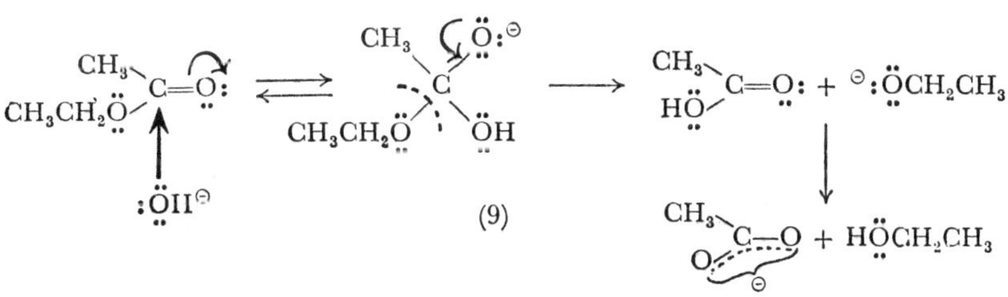

(9)

This picture is consistent with the isotopic labelling experiment (p. 95) that showed that fission of the oxygen–carbonyl bond occurred during the reaction, and with the kinetic behaviour:

rate $= k$[ester] [HO$^{\ominus}$]. The three elementary processes in the reaction are illustrated in the energy profile

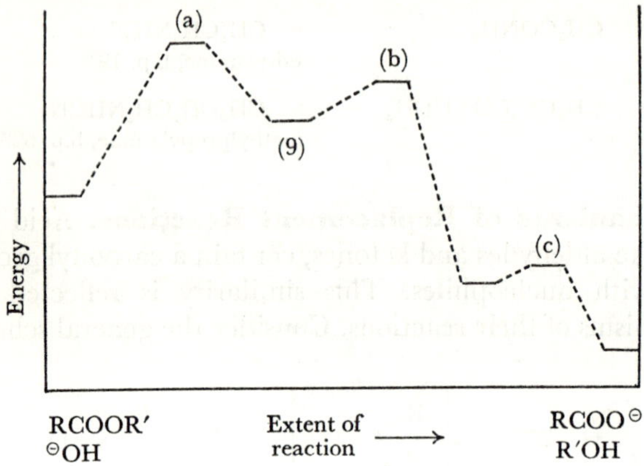

Point (a) represents the transition state in which the intermediate adduct (9) is formed by attack of the nucleophilic hydroxide ion on the ester, and (b) represents the transition state for the decomposition of (9) into acetic acid and ethoxide ion. [Note that the energy difference between (a) and (b) may be sufficiently small to permit the temporary reversal of the initial addition step.] The final step, proton transference, proceeds *via* transition state (c), and requires only a small activation energy.

In carboxylic acids the addition properties of the carbonyl group are repressed in basic solution because the stable carboxylate ion is formed preferentially. This does not apply in acid solution. The principal function of the acid catalyst used in the esterification of carboxylic acids (p. 93) is to protonate the oxygen of the carbonyl group, thus facilitating attack of the nucleophile, R—$\ddot{\underset{\cdot\cdot}{O}}$—H, at the carbon of the carbonyl group (cf. p. 79). The mechanism of the formation of ethyl acetate can be expressed as shown on p. 99.

Rapid exchange of a proton between the oxygen atom of intermediate (10) gives (11). Subsequent stages are similar in principle to those involved in the hydrolysis of esters by base (p. 97), except that all stages are reversible. The mechanism of the reverse reaction (acid-catalysed hydrolysis of an ester, p. 95) must necessarily involve the same stages as the esterification process, but in the reverse sequence.

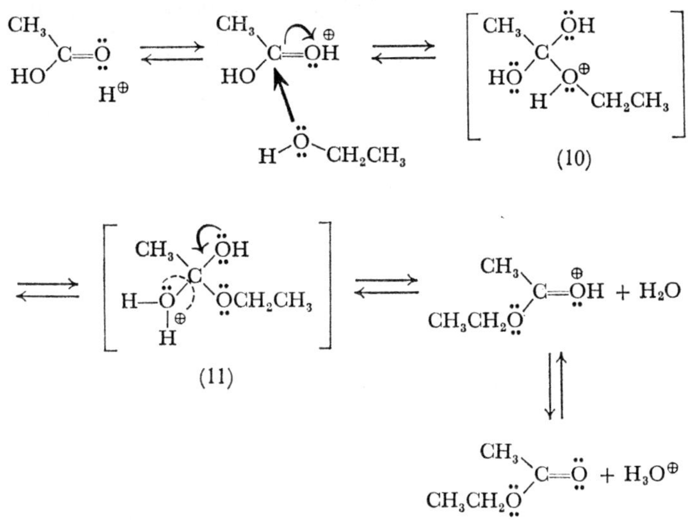

Electrolysis of Carboxylic Salts. Electrolysis of concentrated solutions of sodium or potassium salts of carboxylic acids using a platinum anode gives alkanes:

$$2CH_3COO^{\ominus}\,K^{\oplus} + 2H_2O \longrightarrow CH_3CH_3 + 2KHCO_3 + H_2$$

$$2CH_3(CH_2)_8COO^{\ominus}\,Na^{\oplus} + 2H_2O \longrightarrow \underset{\text{octadecane}}{CH_3(CH_2)_{16}CH_3} + 2NaHCO_3 + H_2$$

This is a valuable method of preparing higher hydrocarbons, as it involves the coupling of alkyl radicals to give a carbon chain of twice the length. The carboxylate ion loses an electron at the anode to give a radical, which decomposes into carbon dioxide and an alkyl radical. Two alkyl radicals combine to form an alkane:

$$CH_3{-}CO_2{:}^{\ominus} \longrightarrow CH_3{-}CO_2\cdot + e$$

$$CH_3{-}CO_2\cdot \longrightarrow \cdot CH_3 + CO_2$$

$$2CH_3\cdot \longrightarrow CH_3CH_3$$

The following synthesis of methyl stearate (p. 107) illustrates how an unsymmetrical product was obtained by electrolysis of a mixture of two salts.

$$CH_3(CH_2)_8COOH + HOOC(CH_2)_8COOCH_3 \xrightarrow[\text{alkaline solution}]{\text{electrolyse in}} CH_3(CH_2)_{16}COOCH_3$$

A solution of decanoic acid (5·2 g., 3 mol. (p. 36)), methyl hydrogen octane-1,8-dicarboxylate (8·6 g., 4 mol.) and sodium hydroxide (0·02 mol.) in dry methanol (30 ml.) was added to a glass vessel containing two platinum foil electrodes 2 mm. apart. A current of 1·5–2 amps was passed through the solution until it gave an alkaline reaction, the temperature being kept below 50° by external cooling. Water and light petroleum were then added. The solvent was removed from the latter extract by evaporation and the residue was fractionally distilled under reduced pressure to give octadecane (4·4 g.), b.p. 100°/0·15 mm., m.p. 28°, and methyl stearate (2·6 g.), b.p. 136°/0·3 mm., which by alkaline hydrolysis gave stearic acid (2 g.), m.p. 68·8°.

Infrared Spectra of Carbonyl Compounds. Carbonyl derivatives display intense infrared absorption in the frequency range 1820–1620 cm.$^{-1}$ due to stretching of the carbon–oxygen double bond. Different types of carbonyl compound can often be distinguished by their characteristic absorption within the quoted range. The following table gives representative data for aldehydes, ketones, and acid derivatives of general structure C_6H_5COX (C_6H_5 = phenyl, p. 184) measured in hexane solution:

X	=	Br	OCH_3	H	CH_3	$N(CH_3)_2$	
ν(C=O)	=	1760	1735	1714	1697	1660	cm.$^{-1}$

The fall in frequency in this series is a consequence of increasing single bond character in the carbonyl groups due to increasing contribution of the ionic form to the hybrid structure (p. 184):

There is the expected correlation with bond lengths of the carbonyl group which, for example, are longer in amides (1·24 Å) than in acid halides (1·17 Å).

Formic Acid, HCOOH, is obtained either by the hydration of carbon monoxide under high pressure over a metal oxide catalyst at 200–300°, or as a by-product in the large scale aerial oxidation of alkanes to acetic acid.

Like aldehydes, formic acid contains a hydrogen joined directly to a carbonyl group, and a chemical similarity is that formic acid (in contrast with other carboxylic acids) also reduces Tollens reagent (p. 82) to silver:

$$\begin{matrix}H\diagdown \\ C{=}O \\ HO\diagup\end{matrix} + 2\overset{\oplus}{Ag}(NH_3)_2\ {}^{\ominus}OH \longrightarrow \begin{matrix}{}^{\oplus}NH_4\ {}^{\ominus}O\diagdown \\ C{=}O \\ {}^{\oplus}NH_4\ {}^{\ominus}O\diagup\end{matrix} + H_2O + 2NH_3 + 2Ag\downarrow$$

Acetic Acid, CH_3COOH, was for many years isolated from 'pyroligneous acid' (p. 43) *via* the calcium salt, but is now generally obtained either by oxidation of acetaldehyde by air;

$$2CH_3CHO + O_2 \longrightarrow 2CH_3COOH$$

or by vigorous oxidation of lower alkanes from petroleum.

Malt vinegar, which contains 4–6% acetic acid, is prepared from a mixture of barley and either rice or maize; the aqueous ethanol obtained initially by fermentation is oxidised by specific bacteria (e.g. *B. aceti*) first to acetaldehyde and then to acetic acid. The bacterial oxidation is carried out at 32–44° in the presence of air. Much commercial 'vinegar' is produced artificially from synthetic acetic acid flavoured with caromel.

Acetic acid and sodium acetate have many industrial applications, for example, the production of acetic anhydride, used to prepare cellulose acetate, acetylsalicylic acid (aspirin) and diacetyl morphine (heroin). In living organisms acetic acid occupies a key position in biosynthesis and metabolism (pp. 110, 111).

Nitriles. Nitriles (or alkyl cyanides) have the general structure $R{-}C{\equiv}N$. They are prepared either from alkyl halides (p. 36) or by the dehydration of primary amides, e.g. with phosphorus pentoxide:

$$CH_3CONH_2 \xrightarrow{P_2O_5} CH_3{-}C{\equiv}N \xleftarrow{KCN} Br{-}CH_3$$

This dual origin is reflected in two of the systems of nomenclature; either they can be named as alkyl cyanides, or *-onitrile* is added to the root of the specific name of the carboxylic acid which contains the same number of carbon atoms. In a systematic method, the group CN (known as *cyano-*) is regarded as the substituent in a hydrocarbon chain:

CH_3CN	cyanomethane	methyl cyanide	acetonitrile	b.p. 82°
CH_3CH_2CN	cyanoethane	ethyl cyanide	propionitrile	b.p. 98°
$(CH_3)_2CHCN$	2-cyanopropane	isopropyl cyanide	isobutyronitrile	b.p. 108°

$$\overset{1}{C}H_3\overset{2}{C}H_2\underset{\displaystyle CN}{\underset{|}{\overset{3}{C}}}H(CH_2)_3CH_3$$ 3-cyanoheptane

The cyano-group, like the carbonyl group, is polarised so that the carbon atom is electron deficient (e.g. the dipole moment of acetonitrile is 3·5 D). Reactions of nitriles involve addition of nucleophiles to this carbon atom, as with acid derivatives, aldehydes and ketones. The hydrolysis of nitriles to carboxylic acids is an important example of this type of reaction:

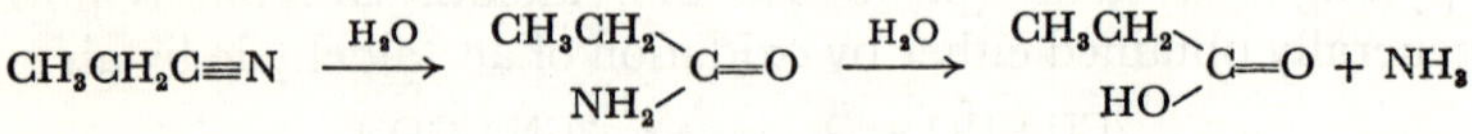

In practice water alone is inadequate for hydrolysis and prolonged boiling with aqueous acid or alkali is necessary. As the above equation shows, amides are intermediates, and can be isolated under appropriate conditions.

Reduction of nitriles, either by catalytic hydrogenation or with lithium tetrahydroaluminate, involves addition of two molecules of hydrogen to the cyano-group to give *primary amines*:

$$CH_3(CH_2)_3CN \xrightarrow{4H} CH_3(CH_2)_3CH_2NH_2$$

When nitriles are prepared from alkyl halides, an extra carbon atom is introduced; this is the key step in an important method for ascending the series of carboxylic acids (or primary alcohols):

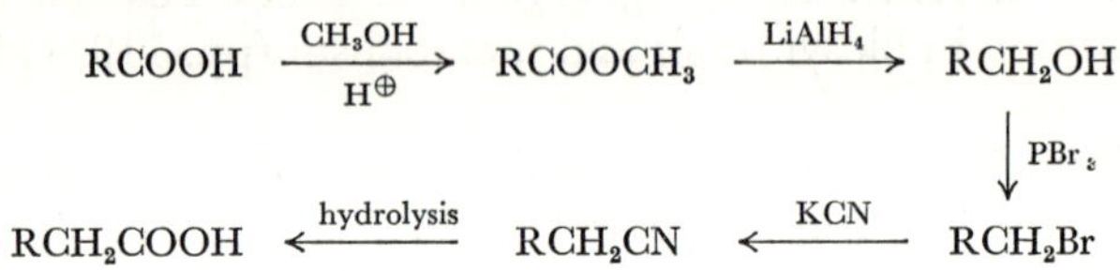

Esters of Inorganic Acids. The dibasic sulphuric acid, H_2SO_4, and the tribasic phosphoric acid, H_3PO_4, form esters, known as dialkyl sulphates and trialkyl phosphates respectively. There is a structural similarity between these esters and carboxylic esters:

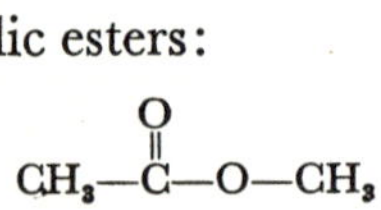

methyl acetate, b.p. 57°

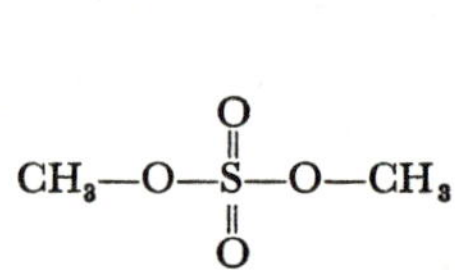

dimethyl sulphate, b.p. 188°

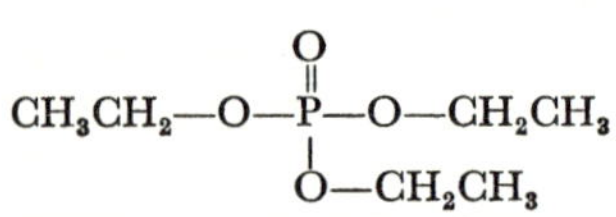

triethyl phosphate, b.p. 216°

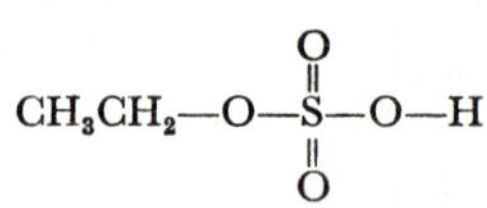

ethyl hydrogen sulphate, b.p. 280°

Dimethyl sulphate and triethyl phosphate are examples of esters in which all the acidic hydrogen atoms of the acids are replaced by alkyl groups. Intermediate compounds are known in which one hydrogen atom of sulphuric acid is replaced by an alkyl group, e.g. ethyl hydrogen sulphate (p. 62), or in which only one or two hydrogen atoms of phosphoric acid are replaced (pp. 76, 109, 177).

Sulphates and phosphates can be prepared from acid halides, e.g. from phosphorus oxychloride, the acid chloride of phosphoric acid:

$$POCl_3 + 3CH_3CH_2OH \longrightarrow \underset{\text{triethyl phosphate}}{(CH_3CH_2O)_3PO} + 3HCl$$

This method is analogous to that used for the formation of carboxylic esters from acyl halides. The preparation of a nitrate is referred to on p. 136.

Alkyl sulphates and phosphates resemble alkyl halides in undergoing replacement reactions with nucleophiles. Thus, ethers are formed from alkyl sulphates and alkoxides:

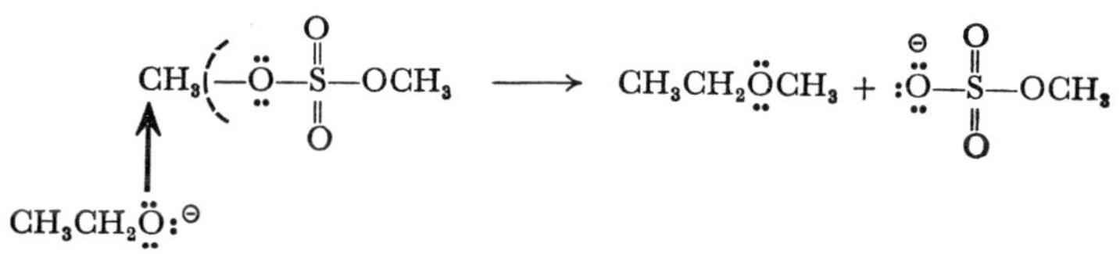

The nucleophilic ethoxide ion attacks the electron-deficient carbon atom of the C—O bond, displacing the sulphate residue.

Sulphate and phosphate esters occur in biological tissues. One of the constituents of skin and cartilage is a salt formed from a basic protein (collagen) and the monosulphate of a carbohydrate. Adenosine triphosphate (ATP), the storehouse of energy at the molecular level in cells, is generated by oxidative phosphorylation, the immediate precursor being adenosine diphosphate (ADP). The complete oxidation of one mole of glucose (pp. 172–180) may give rise to more than 30 moles of ATP from a similar amount of ADP:

$$\text{glucose} + \text{oxygen} + \text{ADP} + \text{phosphate} \longrightarrow \text{ATP} + \text{carbon dioxide} + \text{water}$$

Phosphate structures are also present in phospholipids (p. 109), in nucleic acids, in vitamin B_{12}, and in many co-enzymes, particularly those concerned with energy-transfer.

QUESTIONS

1. Write the structural formula of each of the following compounds:

formic acid, acetyl fluoride, 4-cyanoheptane, hexanoic acid, dimethylacetamide, sodium propionate, ethyl butanoate, n-butyric anhydride.

2. (*a*) Describe what happens (i) visually, and (ii) in molecular terms, when octanoic acid (1 g.) is added to a solution of sodium carbonate (1 g.) in water (20 ml.).

(*b*) Calculate the percentage of acetic acid in an ionised form in 1M aqueous solution from the K value given on p. 89.

3. (*a*) 'Normal' oxygen to nitrogen single and double bond lengths are 1·48 and 1·10 Å respectively. Discuss the finding that, in nitric acid, the bond lengths are 1·44 (HO—N bond) and 1·20 Å (either of the other two O to N bonds).

(*b*) Comment on the bond lengths in propanal and methyl formate:

CH_3—CH_2—CH═O
↑ ↑ ↑
1·54 1·50 1·22

CH_3—O—CH═O
↑ ↑ ↑
1·44 1·37 1·22 Å

(*c*) Draw a possible conformation for methyl formate that shows the positions of all the atoms in space.

(*d*) Which species would you expect to contain the longer oxygen to carbon bond: carbonate$^{2\ominus}$ or carbon dioxide?

4. Suggest methods for converting propionic acid into (*a*) propionyl chloride, (*b*) propionamide, (*c*) propionic anhydride, (*d*) ethyl propionate, (*e*) propionitrile and (*f*) butane.

5. Write the structures of the products of the following reactions:

$(CH_3)_2CHCOCH(CH_3)_2 + Br_2 + HBr$ (trace)
$CH_3CH(CH_3)CH_2CH_2COOCH(CH_3)_2 + LiAlH_4$
$CH_3CH_2CH(CH_2CH_3)CH_2CH(CH_3)CH_2CONH_2 + LiAlH_4$

Give systematic names for the three organic starting compounds, and indicate how they could be distinguished from each other by infrared spectroscopy.

6. Give balanced equations, electronic mechanisms and energy profiles for the reactions of n-propylamine (i) with acetic anhydride, and (ii) with acetyl chloride. Both reactions proceed readily. How could you tell which one occurs more quickly? How could you isolate a pure sample of a product from either reaction?

7. With reference to the electrolysis described on p. 100, (i) why should only a small proportion of sodium hydroxide be used for the reaction, and (ii) why was octadecane another product of the process?

8. (*a*) Discuss the following observation.
No precipitate is formed when aqueous solutions of barium chloride and potassium ethyl sulphate are mixed, but when the mixture is heated a precipitate slowly separates.

(*b*) How could the presence of phosphorus in a sample of ATP be detected?

9. Comment on the statement: as ethyl butanoate and methyl pentanoate are isomeric esters, they will have the same b.p., density, refractive index and infrared absorption spectrum.

10. Why is it an advantage to use a high pressure for the synthesis of formic acid from carbon monoxide and steam?

11. (*a*) Comment on the fact that ethyl methyl thioether has a higher b.p. (66°) than ethyl methyl ether (10°), whereas ethanethiol has a lower b.p. than ethanol (36° and 78·3° respectively).

(*b*) Hexanoic acid and water are partly soluble in each other. Suggest procedures for determining the solubility of each compound in the other phase.

(*c*) Which compound might you expect to have the higher viscosity, ethyl acetate or butanoic acid? How could you test your suggestion by experiment?

Vision is the art of seeing things invisible.

JONATHAN SWIFT (1667–1745)

CHAPTER 8

LIPIDS

Lipids are naturally-occurring esters in which at least one of the acid components is a long-chain carboxylic acid (*fatty acid*). Depot fats (a collective term applied to animal fats and vegetable oils) are lipids that are stored by living organisms and are used as fuel, their chemical breakdown eventually yielding carbon dioxide, water and energy. Lipids must also have more specific biological functions, as indicated, for example, by the occurrence of certain types of lipids, e.g. *cerebrosides*, *phospholipids*, in highly specialised tissue such as brain, spinal chord and egg yolk.

Triglycerides. Depot fats, of general structure (1), are *triglycerides*, which are triesters derived from glycerol (2) (p. 139) and fatty acids.

Fat from a particular plant or animal source is not a single compound, but is composed of esters of glycerol with a number of different fatty acids, probably in a variety of combinations. The structures of fats are shown by alkaline hydrolysis (a typical reaction of esters) to glycerol and the salts of fatty acids:

$$\begin{array}{lcll} CH_2OCOR & & CH_2OH & {}^{\oplus}K\ {}^{\ominus}OCOR \\ | & & | & \\ CHOCOR' + 3KOH & \longrightarrow & CHOH + & {}^{\oplus}K\ {}^{\ominus}OCOR' \\ | & & | & \\ CH_2OCOR'' & & CH_2OH & {}^{\oplus}K\ {}^{\ominus}OCOR'' \\ \quad (1) & & \quad (2) & \end{array}$$

The average molecular weight of the fat is derived experimentally from the *saponification value* (the number of milligrams of potassium hydroxide required to hydrolyse 1 g. fat).

The *waxes* are esters of fatty acids and alcohols other than glycerol, e.g. the wax of the sperm whale is derived from cetyl alcohol, $CH_3(CH_2)_{14}CH_2OH$.

Fatty Acids. Some of the acids obtained from fats are shown below:

$CH_3(CH_2)_{10}COOH$
lauric acid, m.p. 44°
$CH_3(CH_2)_{12}COOH$
myristic acid, m.p. 58°
$CH_3(CH_2)_{14}COOH$
palmitic acid or hexadecanoic acid, m.p. 63°
$CH_3(CH_2)_{16}COOH$
stearic acid or octadecanoic acid, m.p. 70°
$CH_3(CH_2)_7CH{=}CH(CH_2)_7COOH$
oleic acid or octadec-9-enoic acid, m.p. 13°
$CH_3(CH_2)_5CH(OH)CH_2CH{=}CH(CH_2)_7COOH$
ricinoleic acid, m.p. 50°
$CH_3(CH_2)_4CH{=}CHCH_2CH{=}CH(CH_2)_7COOH$
linoleic acid or octadeca-9,12-dienoic acid, m.p. −5°
$CH_3CH_2CH{=}CHCH_2CH{=}CHCH_2CH{=}CH(CH_2)_7COOH$
linolenic acid or octadeca-9,12,15-trienoic acid, m.p. −11°
$CH_3(CH_2)_7CH{=}CH(CH_2)_{11}COOH$
erucic acid, m.p. 34°

It should be noted that the fatty acids in this list contain an even number of carbon atoms joined in a straight chain.

The acids that occur most frequently in depot fats are the saturated palmitic and stearic acids and the C_{18} unsaturated acid, oleic acid. Vegetable oils contain more unsaturated fatty acid components than animal fats, as indicated by the following figures:

olive oil: palmitic + stearic acids, 10%; oleic acid, 83%; linoleic acid, 7%.
beef fat: palmitic + stearic acids, 47%; oleic acid, 48%; linoleic acid, 3%.

This difference in composition is responsible for the lower melting points of vegetable oils, because the unsaturated fatty acids and their esters (e.g. methyl stearate, m.p. 39°; methyl oleate, m.p. −20°) melt lower than their saturated analogues. The degree of unsaturation in a fat, given by the *iodine value*, is determined by the use of iodine monochloride, ICl, which adds to olefinic bonds.

As well as oleic acid, many other unsaturated fatty acids occur as glycerides in lipids. For example, linseed oil contains a high proportion of the diene, linoleic acid, and the triene, linolenic acid; castor oil contains 88% ricinoleic acid (a hydroxy-derivative of oleic acid); and the C_{22}-unsaturated acid, erucic acid, is found in the seed oil of mustard and nasturtium. All the acids mentioned have *cis*-double bonds (cf. Chapter 5), but the *trans*-arrangement occurs in a few fatty acids.

Unusual fatty acids are found in bacterial fats. Thus, the bacilli responsible for tuberculosis and leprosy contain tuberculostearic acid which is a methyl derivative of stearic acid, and therefore has an odd number of carbon atoms. Furthermore, the molecule is asymmetric, and the acid occurs as the optically-active (−)-enantiomer (cf. Chapter 13).

$$CH_3(CH_2)_7\underset{\displaystyle CH_3}{\underset{|}{C}}H(CH_2)_8COOH$$

tuberculostearic acid, m.p. 11°
(10-methylstearic acid)

Acetylenic acids sometimes occur in plant lipids, but more frequently in fungi. For example, the antibiotic, mycomycin, has the remarkable structure (3) containing two acetylenic groups, an allene system (p. 138), and both a *cis*- and a *trans*- double bond:

$$HC{\equiv}C{-}C{\equiv}C{-}CH{=}C{=}CH{-}CH{=}CH{-}CH{=}CH{-}CH_2COOH$$

(3) allene *cis* *trans*

Industrial Applications. (*a*) In the *hardening of oils*, available vegetable oils, e.g. peanut or soyabean oil, are heated at 150–200° with hydrogen and a nickel catalyst when some of the olefinic bonds are reduced, e.g. oleate → stearate:

$$\begin{array}{l} CH_2OCO(CH_2)_7CH{=}CH(CH_2)_7CH_3 \\ | \\ CHOCOR \\ | \\ CH_2OCOR' \end{array} \xrightarrow{2H} \begin{array}{l} CH_2OCO(CH_2)_{16}CH_3 \\ | \\ CHOCOR \\ | \\ CH_2OCOR' \end{array}$$

an oleate a stearate

This produces a solid fat (used, for example, in lard substitutes), which has a higher melting point because it contains a higher proportion of saturated ester groups. Partial hydrogenation of whale oil eliminates the fishy smell, and the product can then be used in salad oil or in cosmetics.

(*b*) *Soap* is a mixture of the alkali–metal salts of fatty acids. Ordinary soap (sodium salts) is often manufactured by hydrolysing animal fats or hardened vegetable oils with boiling aqueous sodium hydroxide (cf. equation on p. 106). Addition of sodium chloride precipitates the sodium salts, which are separated, boiled with water and allowed to set. The useful glycerol is recovered by evaporation of the aqueous solution followed by vacuum distillation. The potassium salts of fatty acids (*soft soaps*), are lower melting and more soluble in water, and are used, for example, in shaving creams.

Soap has the disadvantage that in hard water the insoluble calcium or magnesium salts are precipitated. Synthetic surfactants (surface active agents) designed to overcome this problem contain groups, e.g. $-OSO_2O^{\ominus}Na^{\oplus}$ which increase solubility in water. Surfactants (detergents) containing *unbranched* n-alkyl chains, e.g. n-$C_{16}H_{33}OSO_2ONa$ and the salt (4), are degraded by micro-organisms in rivers, thus avoiding the formation of persistent foams.

$CH_2OCO(CH_2)_{11}CH_3$
|
CHOH (4)
|
$CH_2OSO_2O^{\ominus}Na^{\oplus}$

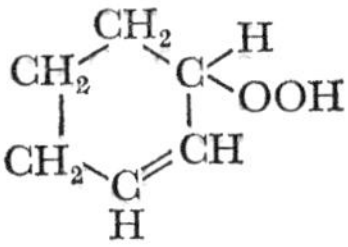
(5)

(*c*) *Drying oils*, which are used in the preparation of paints and varnishes, harden when exposed to air. This property is associated with a high degree of unsaturation in the fatty acid components, linseed oil, for example, containing 10–25% linoleic acid and 45–65% linolenic acid (as esters).

The chemistry of the hardening process is complicated, but reactions with oxygen ('autoxidation') and cross linking of chains are involved. Simpler alkenes react with oxygen to give *hydroperoxides*, ROOH, the product from cyclohexene (p. 46) having the structure (5).

Phospholipids and Cerebrosides. *Phospholipids* occur in brain and spinal chord and in many other tissues; an example (6) of this class of compound is shown below:

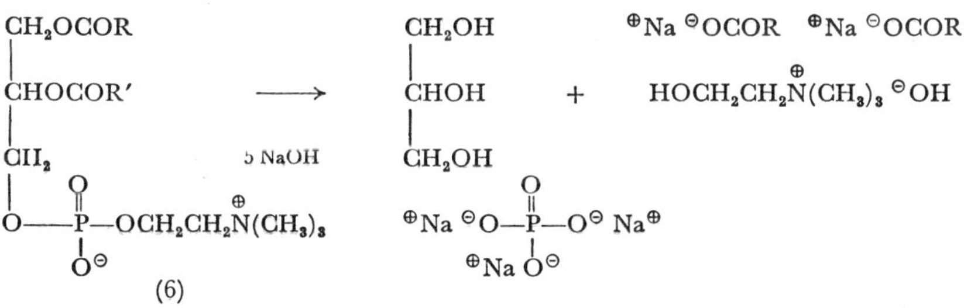
(6)

This phospholipid is an ester of glycerol with typical fatty acids (RCOOH, R′COOH) as two of the acid components. The compound is also a dialkylphosphate (p. 102), phosphoric acid being esterified by the third hydroxyl group of glycerol, and also by the quaternary ammonium alcohol, choline (p. 135). The anion associated with the quaternary group of choline is derived from the third hydroxyl group of phosphoric acid by removal of a proton. The phospholipid contains therefore four ester links, and hydrolysis with sodium hydroxide (as shown) gives a mixture of glycerol, choline, the sodium salts of the fatty acids and sodium phosphate. The identification of such hydrolysis products is the first stage in the elucidation of the structure of an unknown phospholipid. In this group there is considerable variation in the amino-component. For example, α-cephalin from brain tissue contains the $-OCH_2CH_2NH_2$ group in place of choline, and on hydrolysis yields the amino-alcohol, $HOCH_2CH_2NH_2$ (p. 135).

Recent work indicates that the phospholipids present in the walls of cells are involved in the mechanism by which inorganic ions are transported through the membrane.

Cerebrosides have the partial structure (7) in which the amide link is derived from a fatty acid, RCOOH, and the amino-group of a complex amino-alcohol. The fatty acid constituents of cerebrosides are C_{24} compounds, for example, nervonic acid (8).

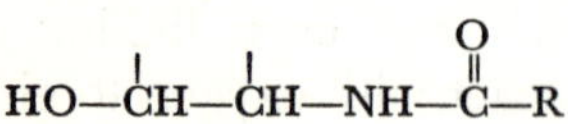

(7)

$$CH_3(CH_2)_7CH{=}CH(CH_2)_{13}COOH$$

(8)

Biosynthesis of Fatty Acids. The most abundant fatty acids are straight-chain compounds containing an even number of carbon atoms. This suggests that the biological 'building unit' is a C_2 compound, and as early as 1905 acetic acid was proposed as the biological precursor of the fatty acids. One possible scheme is illustrated for hexanoic acid:

$$CH_3COOH + CH_3COOH \longrightarrow CH_3COCH_2COOH + H_2O$$

$$CH_3COCH_2COOH + CH_3COOH \longrightarrow CH_3COCH_2COCH_2COOH + H_2O$$

$$CH_3COCH_2COCH_2COOH \xrightarrow{\text{reduction}} CH_3CH_2CH_2CH_2CH_2COOH$$

The methyl group of one molecule of acetic acid reacts with the carboxyl group of another molecule to produce a keto-acid containing four carbon atoms. This process can continue indefinitely, thus building up acids two carbon atoms at a time.

Reduction of the ketonic carbonyl groups must occur, but not necessarily at the end of the synthesis.

This theory did not gain much credence, partly because acetic acid is a rather unreactive molecule, and also because no experimental proof was available. During the last twenty years the general theory has been vindicated by radioactive tracer experiments. For example, feeding acetic acid, $CH_3{}^{14}COOH$ (in which only the carboxyl–carbon is labelled with the radioactive isotope) to a suitable organism and isolating the fats gave, after hydrolysis, fatty acids that were radioactive. Furthermore, a chemical breakdown of the fatty acids, one carbon atom at a time, showed that the carboxyl group and *alternate* carbon atoms contained ^{14}C; other carbon atoms were unlabelled:

i.e. $$CH_3{}^{14}COOH \longrightarrow RCH_2{}^{14}CH_2CH_2{}^{14}CH_2CH_2{}^{14}COOH$$

This result is in complete accord with the theory outlined above.

Many of the enzymes required for each step in the synthesis and in the breakdown of fatty acids have been isolated, and it is now known that C_2 units are supplied by acetic acid or by malonic acid (p. 149) in the form of thiol esters (p. 96).

Metabolism. Metabolism refers to the energetic and molecular courses involved in the chemistry of living cells. The overall metabolism of plants is energy-demanding and depends on the reductive conversion of carbon dioxide into organic molecules by photosynthesis. The following diagram summarises the metabolic relation between lipids, carbohydrates and proteins, three of the major constituents of cells. The cycle of compounds that includes citric acid is summarised later (p. 168).

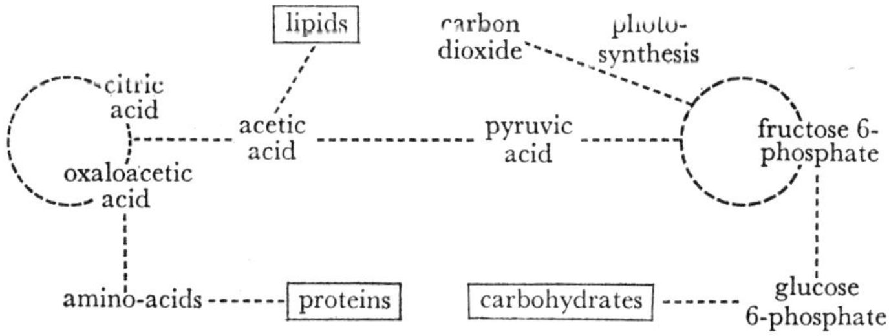

QUESTIONS

1. Give structures and names for the products from the alkaline hydrolysis of the following substances:

(*a*) $CH_3(CH_2)_{16}COOCH_2(CH_2)_{14}CH_3$

(*b*) $CH_2OCO(CH_2)_7CH{=}CH(CH_2)_7CH_3$
$|$
$CHOCO(CH_2)_6CH_3$
$|$
$CH_2OCO(CH_2)_7CH{=}CHCH_2CH{=}CH(CH_2)_4CH_3$

(*c*) $CH_2OCO(CH_2)_{16}CH_3$
$|$
$CHOCO(CH_2)_{14}CH_3$
$|$
$CH_2\quad O$
$|\qquad \|$
$O{-}{-}P{-}OCH_2CH_2\overset{\oplus}{N}(CH_3)_3$
$\qquad |$
$\qquad O^{\ominus}$

2. Calculate (i) the saponification value (p. 106), and (ii) the iodine value of methyl linolenate. [The iodine value (p. 107) is defined as the number of grams of I_2 equivalent to 100 g. of fat.]

3. Suggest reagents for the steps numbered (1) to (10) in the synthesis of vaccenic acid outlined below:

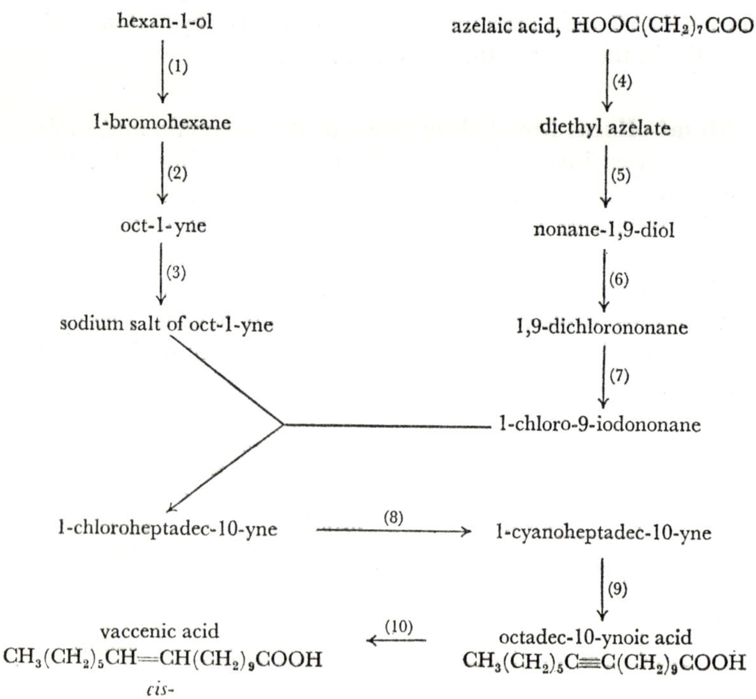

Suggest electronic mechanisms for reactions (1), (2), (3), (7) and (8).

4. (*a*) Magnesium stearate is a frequent component of cosmetic powders. How could you find out if this substance was present in a sample of face powder?

(*b*) Suggest a practical procedure for determining the relative effectiveness of two different samples of linseed oil as 'drying oils'.

(*c*) What structural feature is present in all surfactants used for washing fabrics? How would you envisage a molecule of a surfactant, e.g. sodium stearate, to arrange itself at the interface between an aqueous and a lipid phase?

5. (*a*) Suggest a method for preparing labelled acetic acid ($CH_3{}^{14}COOH$) from sodium cyanide ($Na^{14}CN$) and any necessary organic compounds. Would it be necessary to take safety precautions during the preparation?

(*b*) How many hydrogens in each molecule of ricinoleic acid will be replaced by deuterium if the compound is treated with deuterium oxide (^2H_2O) at 20°?

Ricinoleic acid is sparingly soluble in water; what solvent could be added to this mixture to produce a homogeneous solution?

The kernel of the scientific approach is the refusal to regard our own desires, tastes and interests as affording a key to the understanding of the world.

BERTRAND RUSSELL (1872–)

CHAPTER 9

AMINES

Although aromatic amines, heterocyclic amines* and alkaloids (p. 122) were isolated in the early part of the nineteenth century, the first primary alkyl amine, ethylamine, b.p. 18°, was discovered by Wurtz in 1849, and initially mistaken for ammonia.

Structure and Nomenclature. Amines are derivatives of ammonia, in which one or more hydrogen atoms are replaced by alkyl groups. Primary amines have one alkyl group, and thus contain the $-NH_2$ (*amino*) group; secondary amines have two alkyl groups and contain the >NH group; tertiary amines are trialkyl derivatives of ammonia and have no hydrogen atom attached to the nitrogen. The general structures are:

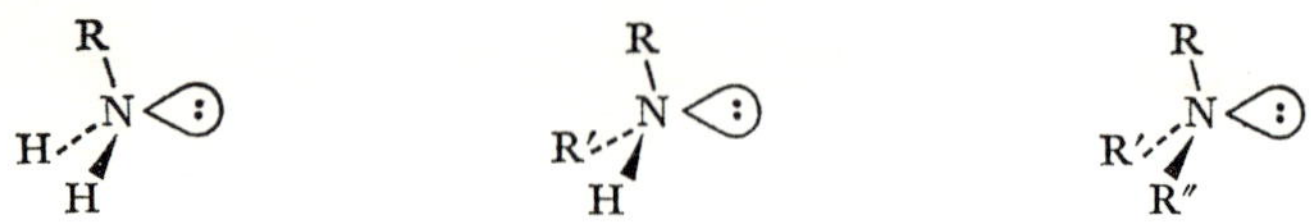

primary secondary tertiary

Of the five electrons in the valence shell of the nitrogen atom, three are involved in covalent bonds, and two are unshared. Ammonia has a pyramidal configuration with approximately tetrahedral angles between the three sp^3 bonds, and this arrangement also prevails in amines, as the above diagrams indicate (see also p. 156).

Amines are named either by specifying the alkyl groups attached to nitrogen, or in the case of primary amines by

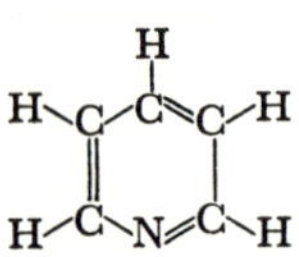

pyridine

* Heterocyclic amines, e.g. pyridine, contain one or more nitrogen atoms as part of a ring system. Pyridine is present (*ca.* 0·01%) in coal tar (p. 187), but most is obtained by synthesis. Vitamin B_6 and nicotine are two of the many biological compounds that contain pyridine rings.

regarding the amino-group as the substituent in a hydrocarbon chain.

CH_3NH_2	methylamine or aminomethane	b.p. −6°
$CH_3CH_2CH_2NH_2$	n-propylamine or 1-aminopropane	b.p. 49°
$(CH_3)_2CHNH_2$	isopropylamine or 2-aminopropane	b.p. 33°
$CH_3NHCH_2CH_3$	ethylmethylamine	b.p. 36°
$(CH_3CH_2)_2N{-}CH_2CH_3$	triethylamine	b.p. 89°

The simpler amines are gases or liquids and are soluble in water. They can be identified by conversion into crystalline derivatives and several types of compound suitable for this purpose are referred to later in this Chapter.

Amines can be prepared from alkyl halides (see below) and by reduction of amides (p. 97) and nitriles (p. 102). Industrially, primary amines are often obtained from alcohols: a mixture of the alcohol and a large excess of ammonia is passed over an alumina catalyst at 300–350°; the reaction is of the S_N2 type:

$$ROH + NH_3 \longrightarrow RNH_2 + H_2O$$

Basicity. The most striking property of amines is their basicity, which may be defined as the ability to acquire a proton from an acid. Amines form salts with dilute aqueous acid, and are therefore much more basic than ethers, which only dissolve in concentrated acid (p. 52):

$$CH_3NH_2 + HCl \longrightarrow CH_3\overset{\oplus}{N}H_3\ \overset{\ominus}{Cl}$$

methylamine hydrochloride, m.p. 225°

$$(CH_3CH_2)_2NH + HBr \longrightarrow (CH_3CH_2)_2\overset{\oplus}{N}H_2\ Br^{\ominus}$$

diethylamine hydrobromide, m.p. 213°

$$(CH_3)_2NCH_2CH_3 + HClO_4 \longrightarrow (CH_3)_2\overset{\oplus}{N}HCH_2CH_3\ ClO_4^{\ominus}$$

ethyldimethylamine perchlorate

One way of naming these salts is shown above. An alternative method of nomenclature indicates the relation to the parent inorganic ammonium salts, e.g. methylammonium chloride, for methylamine hydrochloride.

Mechanistically, the formation of a salt can be expressed as:

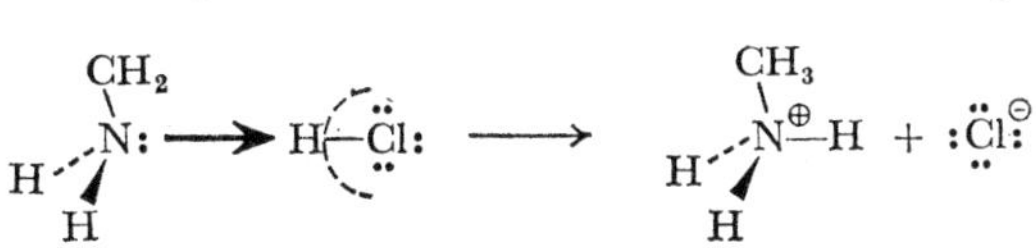

The unshared pair of electrons on the nitrogen atom is used to form a new covalent N—H bond; the three N—H bonds in the resultant cation are equivalent.

The reaction can be reversed by a base (e.g. sodium hydroxide) which has a greater affinity for a proton (i.e. is a stronger base) than the amine:

$$CH_3CH_2\overset{\oplus}{N}H_3\,Br^{\ominus} + Na^{\oplus}\,OH^{\ominus} \longrightarrow CH_3CH_2NH_2 + H_2O + Na^{\oplus}\,Br^{\ominus}$$

$$\text{cf.}\quad \overset{\oplus}{N}H_4\,Br^{\ominus} + Na^{\oplus}\,OH^{\ominus} \longrightarrow NH_3 + H_2O + Na^{\oplus}\,Br^{\ominus}$$

The amine salts of strong acids are usually crystalline solids and, as they are ionic compounds, are more soluble in water than in ether. The salts are sometimes useful for identifying amines, but they are often hygroscopic and have inconveniently high melting points. The ***picrates*** (salts of picric acid p. 209) do not have these disadvantages, and are employed frequently as crystalline derivatives of amines:

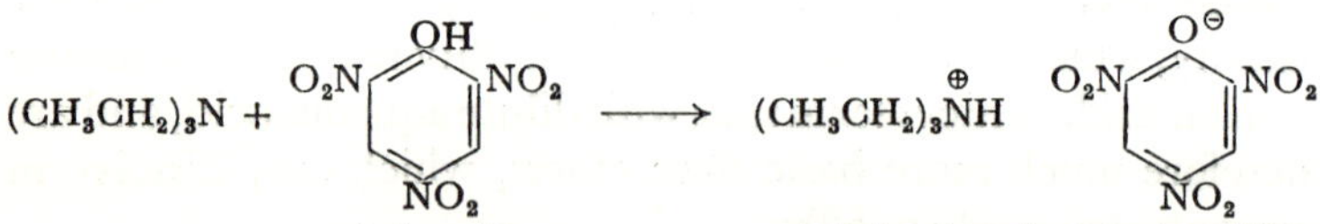

picric acid　　　　triethylamine picrate, m.p. 173°

Amine salts are invaluable for the purification of amines. When an ether or dichloromethane solution of a mixture of organic compounds containing an amine is shaken with aqueous sulphuric acid, the amine is extracted as its water-soluble sulphate. Addition of sufficient sodium hydroxide to the aqueous solution liberates the amine.

Reactions with Primary Alkyl Halides. Amines react as nucleophiles with primary alkyl halides, and displace the halogen as halide ion:

$$CH_3CH_2NH_2 + CH_3CH_2Br \longrightarrow (CH_3CH_2)_2\overset{\oplus}{N}H_2\,Br^{\ominus}$$

Such an equation does not represent the reaction adequately, as in practice a mixture of compounds results. This is best explained by considering the comparable reaction of a primary alkyl halide with ammonia (p. 36), in which a mixture of products is also formed.

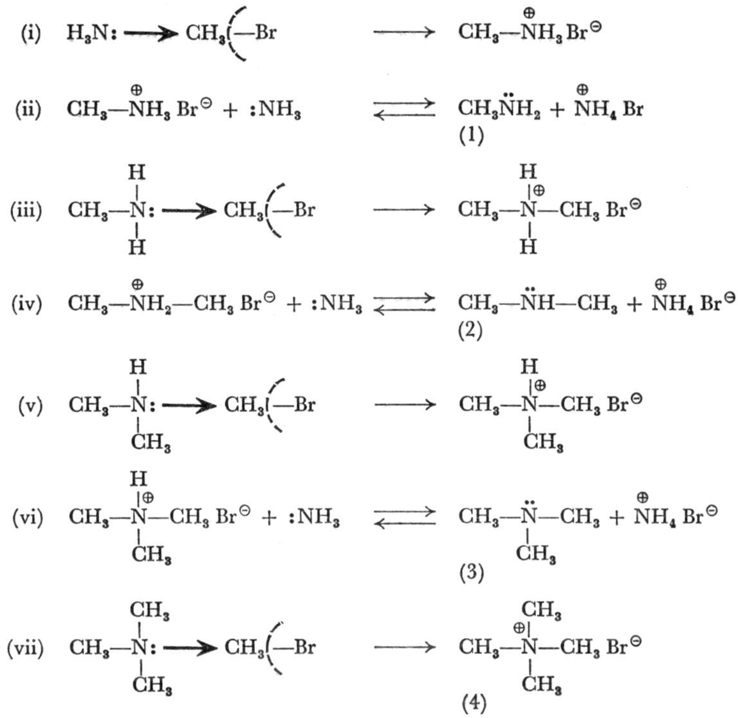

In reaction (i), ammonia displaces bromine (as bromide) from the alkyl bromide to give the hydrobromide salt of a *primary amine*. In the presence of unused ammonia, some of this hydrobromide is converted into the primary amine (1); this occurs because ammonia is a base of comparable strength to the primary amine, and the two bases compete, therefore, for the possession of a proton [reaction (ii)]. In reaction (iii) the primary amine (1) displaces bromide from unchanged alkyl bromide to give the *secondary amine* hydrobromide, which in the presence of ammonia comes into equilibrium with ammonium bromide and the secondary amine (2) [reaction (iv)]. The process continues; secondary amine reacts with alkyl bromide to give the *tertiary amine* hydrobromide [reaction (v)] which with ammonia gives some tertiary amine (3). The tertiary amine (3) is also nucleophilic and with alkyl bromide gives a new species (4), (*a*

quaternary ammonium salt) which is the final product of the reaction as it does not contain an N—H bond, and so cannot lose a proton to ammonia. Addition of excess of sodium hydroxide to the reaction mixture releases a mixture of the amines (1)–(3) from their salts. Thus, ammonia and primary alkyl halides give a mixture of primary, secondary and tertiary amines and quaternary ammonium compounds, and only by employing a large excess of ammonia can reasonable yields of primary amines be obtained.

In the reaction between a primary amine and an alkyl halide reactions (iii) to (vii) occur, because primary amines, as well as ammonia, can take part in the reversible reactions (iv) and (vi). Similar considerations apply to the reaction of secondary amines with alkyl halides. To summarise, the following products are obtained from amines and primary alkyl halides, after the reaction mixtures are made alkaline:

$$RNH_2 + R'X \longrightarrow RNHR' + RNR'_2 + R\overset{\oplus}{N}R'_3\,X^{\ominus}\ (\text{or }OH^{\ominus})$$

$$R_2NH + R'X \longrightarrow R_2NR' + R_2\overset{\oplus}{N}R'_2\,X^{\ominus}\ (\text{or }OH^{\ominus})$$

$$R_3N + R'X \longrightarrow R_3\overset{\oplus}{N}R'\,X^{\ominus}\ (\text{or }OH^{\ominus})$$

The group of equations (i) to (vii) represents a good instance of competitive and consecutive reactions in organic chemistry. The final stage gives the simplest kinetics; for example, the rate of quaternisation of triethylamine by ethyl iodide (in acetone) is dependent on the concentration of each of the two reactants, rate $= k[(C_2H_5)_3N]\,[C_2H_5I]$.

A more practical method of preparing a primary amine from an alkyl halide is exemplified later (in question 8, p. 249).

Reactions with Acid Derivatives. Amines containing $-NH_2$ or >NH groups react with many carboxylic acid and sulphonic acid derivatives; primary amines give secondary amides, secondary amines give tertiary amides. Acyl chlorides and acid anhydrides, which react with amines at room temperature, are particularly useful.

$$CH_3(CH_2)_2CH_2NH_2 + CH_3COOCH_2CH_3 \xrightarrow{100^\circ} CH_3(CH_2)_2CH_2NHCOCH_3 + CH_3CH_2OH$$

$$(CH_3CH_2)_2NH + (CH_3CO)_2O \xrightarrow{20°} (CH_3CH_2)_2NCOCH_3 + CH_3COOH$$

$$(CH_3)_2CHCH_2NH_2 + Cl{-}\overset{O}{\overset{\|}{C}}{-}C_6H_5 \xrightarrow{20°} (CH_3)_2CHCH_2NH\overset{O}{\overset{\|}{C}}{-}C_6H_5$$

benzoyl chloride (p. 223) m.p. 57°

$+$ HCl ($\rightarrow$amine salt)

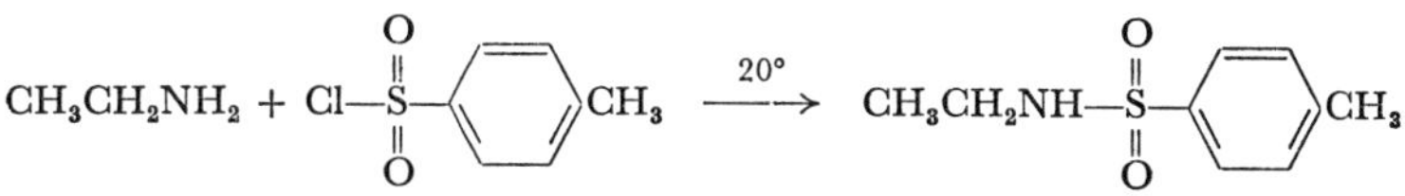

toluene-*p*-sulphonyl chloride (p. 232) m.p. 62°

$+$ HCl ($\rightarrow$amine salt)

On a larger scale it is economical to heat the salt of the acid and the amine under conditions where water is removed as it is formed:

$$(CH_3)_2NH + HCOOH \longrightarrow (CH_3)_2NH_2^{\oplus}HCOO^{\ominus} \longrightarrow$$

$$(CH_3)_2N{-}C{\begin{smallmatrix} {/\!\!/}O \\ \diagdown H \end{smallmatrix}} + H_2O \uparrow$$

dimethylformamide, b.p. 153°

These reactions of carboxylic acid derivatives are mechanistically analogous to the corresponding reactions with ammonia (p. 94) in which primary amides are formed. The nucleophilic amino-group attacks the carbonyl group of the acid derivative, leading eventually to the replacement of a hydrogen atom on nitrogen. Tertiary amines do not contain replaceable hydrogen, and do not react in this way.

Some of these derivatives are crystalline solids which are used to identify individual amines. Primary and secondary amines can be recognised, and distinguished from tertiary amines, by the infrared absorption of the N—H bond at 3600–3100 cm.$^{-1}$. Alcohols also absorb in this region, but confusion can be avoided by reaction with acetic anhydride, whereby alcohols yield esters (infrared absorption, 1750–1700 cm.$^{-1}$) and primary and secondary amines afford amides (infrared absorption, 1650 cm.$^{-1}$).

The reactions of primary and secondary amines with aldehydes and ketones have been discussed elsewhere (p. 77). When formaldehyde is employed in the presence of the reducing agent formic acid, tertiary amines are obtained by the introduction of methyl groups (*methylation*):

$$CH_3CH_2NH_2 + 2HCHO + 2HCOOH \longrightarrow CH_3CH_2N(CH_3)_2 + 2CO_2 + 2H_2O$$

This is an important practical procedure for the methylation of primary and secondary amines, and avoids the formation of quaternary ammonium salts.

Quaternary Ammonium Salts. The halides, e.g. (4), p. 117, are formed from tertiary amines and alkyl halides:

$$(CH_3)_3N + CH_3I \longrightarrow (CH_3)_4\overset{\oplus}{N}\,I^{\ominus}$$

tetramethylammonium iodide

$$CH_3CH_2CH_2N(CH_3)_2 + CH_3CH_2Br \longrightarrow CH_3CH_2CH_2\overset{\oplus}{N}(CH_2CH_3)(CH_3)_2\,Br^{\ominus}$$

ethyldimethylpropylammonium bromide

Quaternary ammonium halides are converted into the corresponding bases, quaternary ammonium hydroxides, by an aqueous suspension of silver oxide:

$$CH_3(CH_2)_2CH_2\overset{\oplus}{N}(CH_3)_3\,I^{\ominus} \xrightarrow[H_2O]{Ag_2O} CH_3(CH_2)_2CH_2\overset{\oplus}{N}(CH_3)_3\,OH^{\ominus} + AgI\downarrow$$

butyltrimethylammonium hydroxide

Removal of the silver halide by precipitation encourages the reaction to go in the direction indicated. The quaternary ammonium hydroxides, $R_4\overset{\oplus}{N}\,OH^{\ominus}$, are comparable to ionic metal hydroxides, e.g. $Na^{\oplus}\,OH^{\ominus}$, in base strength, and can be isolated in contrast with the structurally analogous ammonium hydroxide, $H_4\overset{\oplus}{N}\,OH^{\ominus}$, which only exists in solution. They are sometimes preferred to alkali hydroxides as basic catalysts in organic reactions because of their greater solubility in organic solvents.

Elimination Reactions. When a quaternary ammonium hydroxide is heated, an alkene is formed by elimination of a tertiary amine and water:

$$(CH_3CH_2)_4\overset{\oplus}{N}\,OH^{\ominus} \longrightarrow CH_2{=}CH_2 + (CH_3CH_2)_3N + H_2O$$

$$CH_3CH_2\underset{(CH_3)_3\overset{\oplus}{N}\,OH^{\ominus}}{\underset{|}{C}H}CH_2CH_3 \longrightarrow CH_3CH_2CH{=}CHCH_3 + (CH_3)_3N + H_2O$$

The decomposition of a quaternary hydroxide is an elimination reaction that can be formulated in the following way:

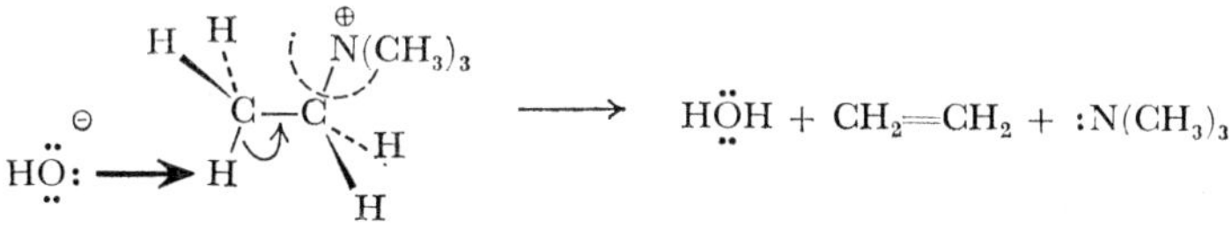

In this reaction three bond changes are occurring simultaneously within one elementary process: (*a*) a hydrogen–carbon bond is broken as the hydroxide ion removes the proton, (*b*) an electron pair moves to form a carbon–carbon double bond, and (*c*) the electron pair of the carbon–nitrogen bond becomes the lone pair of the eliminated amine. The hydrogen (as a proton) is removed from the β-carbon atom.* Consequently, compounds without β-hydrogen, e.g.

$$(CH_3)_4N^{\oplus}\,OH^{\ominus} \quad \text{and} \quad (CH_3)_3CCH_2N^{\oplus}(CH_3)_3\,OH^{\ominus},$$

cannot undergo elimination in this manner.

The elimination reaction of t-butyl bromide described earlier (p. 40) has a mechanism analogous to that of the quaternary ammonium hydroxide decompositions:

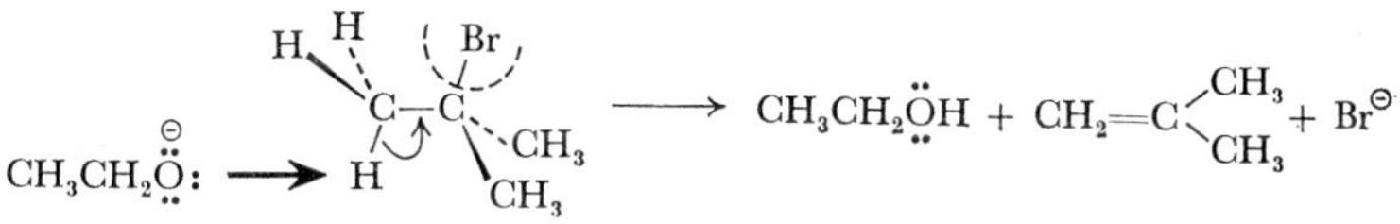

The observation that this is a second-order reaction with the rate proportional to the product of the concentrations of base and of bromide is consistent with the mechanism. The energy profiles for the eliminations of this type are similar to those for S_N2

* The carbon atom adjacent to a functional group (in this case a nitrogen cation) is designated α, the next carbon in the chain β, the next γ, etc.; substituents (e.g. hydrogen atoms) are given the same Greek letter as the carbon atoms to which they are attached. The quaternary ammonium cation (5) contains one γ-carbon, two β-carbons, and four α-carbons.

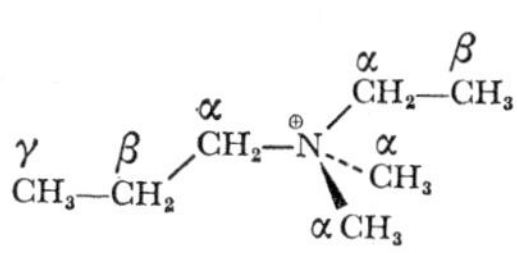

(5)

reactions (p. 34), and substitution and elimination (designated E2) often occur simultaneously (p. 40). A further example is the acid-catalysed elimination of water from an alcohol (p. 46), in which the positively charged group of intermediate (6) is transformed into a neutral group (water) during the reaction:

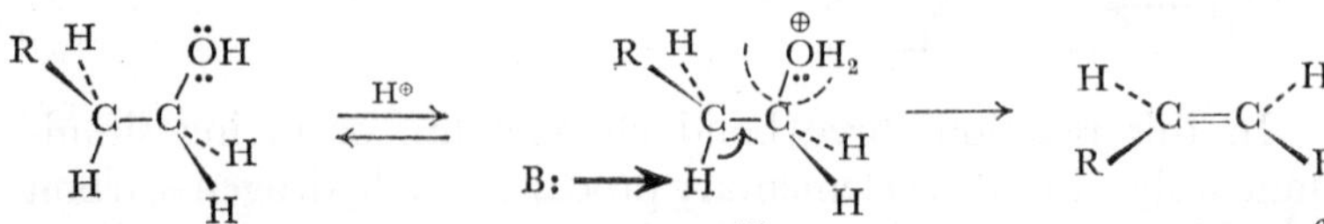

(6) $+ H_2O + HB^{\oplus}$

When phosphoric acid is the catalyst, the base, B:, may be either of the neutral nucleophiles, water or phosphoric acid, or the anion, $H_2PO_4^{\ominus}$.

The acid-catalysed dehydration of alcohols (elimination of water) and the hydration of alkenes each proceed especially readily when tertiary carbonium ions are formed as intermediates, e.g.,

$$(CH_3)_3COH \underset{H_2O,\ -H^{\oplus}}{\overset{H^{\oplus},\ -H_2O}{\rightleftharpoons}} \left[(CH_3)_3C^{\oplus}\right] \underset{H^{\oplus}}{\overset{-H^{\oplus}}{\rightleftharpoons}} (CH_3)_2C{=}CH_2$$

By appropriate choice of conditions either t-butyl alcohol or isobutene can become the major product. Hydration occurs at low temperatures (e.g. 20°), but at higher temperatures (e.g. 83° or above), the alkene is obtained because of the entropy effect (p. 37), and because it can be readily removed from the system as a gas (b.p. $-7°$). A related example is the addition of hydrogen bromide to isobutene at 20° to give t-butyl bromide, and the first order elimination of hydrogen bromide from this halide at temperatures above 300° (in the gas phase):

$$(CH_3)_3CBr \underset{20°}{\overset{300°}{\rightleftharpoons}} (CH_3)_2C{=}CH_2 + HBr$$

Pharmacology of Amines and Quaternary Ammonium Salts. The poisonous nature of some plants is often due to the presence of a group of amines known as *alkaloids*. Socrates died in 399 B.C. from a draught of hemlock juice (containing the poisonous alkaloid, coniine), and the alkaloid tubocurarine (curare) is one of the active principles of the arrow-poison used by South American Indians. In small doses many alkaloids have useful physiological effects, and for many years played a dominant role in medicine. Some well-known examples are morphine (an analgesic from opium poppy), quinine (an anti-malarial from cinchona bark) and cocaine (a local anaesthetic from coca leaves). Many alkaloids have now been replaced by synthetic drugs, but some are still used, e.g. morphine, atropine and ephedrine, and some have found new applications in recent years, e.g. lobeline in anti-smoking preparations, tubocurarine as a muscle-relaxant in anaesthesia, and reserpine as a tranquilliser.

The elucidation of alkaloid structure has often proved to be an extremely difficult task. Morphine, for example, has a comparatively simple molecular formula of $C_{17}H_{19}NO_3$ but has a very complicated ring system. Although it was the first pure alkaloid to be obtained (in 1805) its correct structure was not suggested until 1925 and not finally verified by synthesis until 1950.

Many successful synthetic drugs have been obtained by studying the chemical structure of an alkaloid and trying to discover the features of the structure that are essential for physiological activity. Local anaesthetics were evolved in this way.

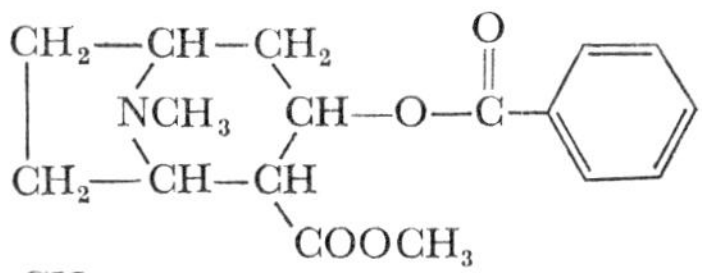

(7)

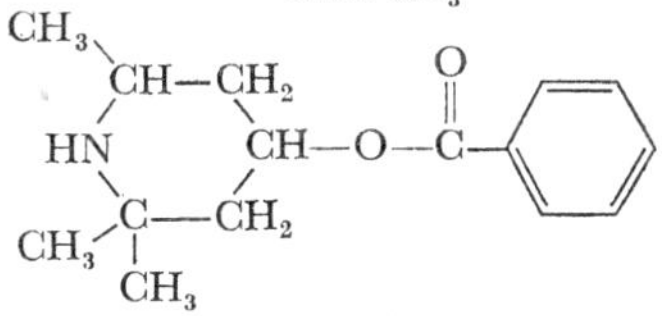

(8)

Cocaine, first introduced in 1884 as a local anaesthetic in ophthalmology and dentistry, was shown later to have the structure (7). Cocaine has certain disadvantages (it is rather toxic when injected, and can cause addiction), and synthetic analogues were investigated. *β*-Eucaine (8), obtained in 1897,

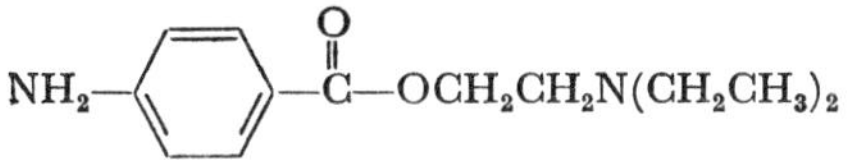

(9)

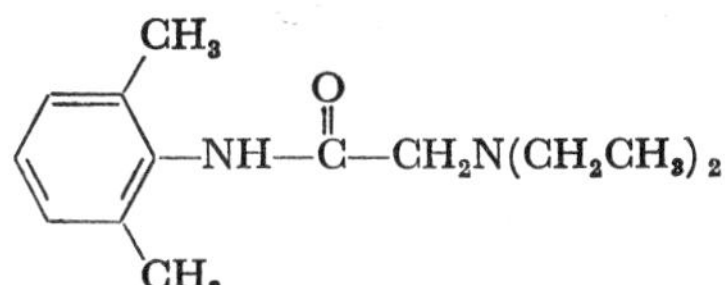

(10)

is clearly related structurally to cocaine, and the modern procaine (9) (p. 224) and lignocaine (10) contain both the amino-group and ester or amide group found to be important for local anaesthetic activity.

Many quaternary ammonium compounds show physiological activity (cf. acetylcholine, p. 135). Tetraethylammonium chloride, for example, lowers blood pressure when injected as an aqueous solution, but has some unpleasant side effects. Furthermore, its action is short lived because its great solubility in water results in rapid excretion. These disadvantages are overcome in quaternary salts with more complex alkyl groups.

QUESTIONS

1. Write structural formulae (*a*) for compounds of molecular formula $C_4H_{11}N$, and (*b*) for the following compounds:

trimethylamine, di-n-propylamine, tetramethylammonium hydroxide, 3-aminohexane, triethylamine hydrochloride.

2. An aqueous solution of a quaternary ammonium iodide, $C_8H_{20}IN$, was shaken with silver oxide, the mixture was filtered, and the aqueous solution was evaporated, The residue was heated at 150°, and the volatile distillate (b.p. 30–100°) was collected and shaken with a mixture of ether and dilute hydrochloric acid. Evaporation of the acid layer gave a solid, m.p. 278° (with decomposition) identified as trimethylamine hydrochloride. The ether solution was shown by gas chromatography to contain 2-methylbut-1-ene.

Write balanced equations for the reactions and suggest two possible structures for the quaternary ammonium iodide.

Why is silver oxide rather than sodium hydroxide used for the first stage? What chemical method might be used to establish the structure of the olefin? What derivative might be more suitable than the hydrochloride for characterising the amine, and for what reasons? How could a pure dry sample of trimethylamine be obtained from its hydrochloride?

3. (*a*) Which one of the following amines is most likely to give a good second order plot when the kinetics of its reaction with methyl iodide is studied: ethylamine, diethylamine, triethylamine. Outline a practical procedure for studying the kinetics of one of these reactions.

(*b*) Explain how the reaction of 1,2-dichloroethane with ammonia gives the hydrochlorides of four compounds: ethylene diamine (cf. p. 136); diethylene triamine, $C_4H_{13}N_3$; triethylene tetramine, $C_6H_{18}N_4$; tetraethylene pentamine, $C_8H_{23}N_5$.

(*c*) Suggest a practical method for determining the partition coefficient [$K_t = C_A/C_B$, where C_A and C_B are the concentrations of the substance in solvents A and B at temperature t] of $(CH_3)_2CHCH_2NH_2$ between water and *m*-xylene (cf. p. 185).

4. An excess of a compound, A, C_7H_5ClO, showing strong infrared absorption at 1760 cm.$^{-1}$, was shaken with an amine, B, C_2H_7N, in aqueous sodium hydroxide. Extraction with dichloromethane gave a liquid, C, $C_9H_{11}NO$, which absorbed strongly in the infrared at 1660 cm.$^{-1}$ and did not contain an N—H bond. Acidification of the aqueous alkaline solution yielded a solid, D, $C_7H_6O_2$.

Suggest structures for A, B, C and D, and write balanced equations for the reactions involved. Why is sodium hydroxide used in this reaction, and why is an excess of A employed?

5. (*a*) What is the gaseous hydrocarbon that is evolved when 1,2-dibromoethane (p. 132) is added slowly to a solution of an excess of potassium hydroxide in boiling ethanol?

(*b*) Suggest procedures, including the necessary apparatus, (i) for following the kinetics of the first-order thermal decomposition of t-butyl bromide in the gas phase, (ii) for preparing isobutene from t-butyl alcohol using a catalytic amount of a strong acid, and (iii) for obtaining t-butyl alcohol from isobutene and N aqueous acid.

Books must follow sciences, and not sciences books.

FRANCIS BACON (1561–1626)

CHAPTER 10

ORGANOMETALLIC COMPOUNDS

A mixture of organoarsenic compounds was obtained as long ago as 1760 by heating arsenious oxide with potassium acetate. Later investigations by Bunsen and others led to the preparation of tetramethyldiarsine, $(CH_3)_2As\text{–}As(CH_3)_2$, a vile-smelling flammable liquid. In attempts to prepare the free ethyl radical by dehalogenation of ethyl iodide with zinc, Frankland (1849) obtained diethylzinc, $(CH_3CH_2)_2Zn$, another highly flammable liquid, b.p. 118°. This pioneering work paved the way for later developments in which more easily handled organometallic compounds were found to be of great value in organic synthesis.

Carbon–metal Bonds. In organometallic compounds the metal atom is attached directly to carbon. Organic derivatives of less electropositive metals are volatile and soluble in hydrocarbons since the metal–carbon bonds are covalent; examples are dimethylmercury, $(CH_3)_2Hg$, b.p. 96°, and tetraethyl-lead, b.p. 200° (p. 130).

Electropositive metals, however, form bonds to carbon that have considerable ionic character, estimated to be 52% for C–K, 43% for C–Li and 35% for C–Mg. These organometallic compounds are often involatile solids and are less soluble in hydrocarbons; their hybrid structures (1), illustrated by methyl-lithium, indicate that they are effectively *carbanions* (p. 57).

$$H_3\overset{\ominus}{C}{:}\ Li^{\oplus} \longleftrightarrow H_3C{-}Li \quad (1) \qquad\qquad H_3\overset{\ominus}{C}{:}\!\longrightarrow Li^{\oplus} \quad (2)$$

Alternatively, the structure can be expressed as (2), in which a carbanion acts as a *ligand* (an atom, ion or molecule, e.g. bromine, alkoxide, or ether, that can form a donor bond to a metal atom).

ORGANOMETALLIC COMPOUNDS IN SYNTHESES

The first three topics will be exemplified by the properties of readily available organomagnesium compounds. These were studied by Grignard (1900 onwards), following observations by Barbier (1899), and are now often referred to as Grignard reagents.

A typical member of the series, ethylmagnesium iodide, is formed when magnesium is added to a mixture of ethyl iodide and diethyl ether; the reaction is performed under an atmosphere of nitrogen in order to prevent oxidation of the Grignard reagent:

$$CH_3CH_2Br + Mg + 2(CH_3CH_2)_2O \xrightarrow[\text{ether}]{\text{in}} CH_3CH_2MgBr[O(CH_2CH_3)_2]_2$$

The product, containing a carbon–magnesium bond, is stabilised by two molecules of ether joined as ligands to the metal. The organometallic compound is soluble in ether, and this solution is used directly for further reactions.

Reactions with Acids. Grignard reagents are converted into the corresponding alkane, when treated with acids. Strong (mineral) acids may be used but the following reactions exempify the use of weaker acids:

$$CH_3CH_2CH_2CH_2MgI + H_2O \longrightarrow CH_3CH_2CH_2CH_3 + IMgOH$$

$$(CH_3)_2CHMgBr + CH_3CH_2OH \longrightarrow (CH_3)_2CH_2 + BrMgOCH_2CH_3$$

$$CH_3CH_2MgBr + CH_3NH_2 \longrightarrow CH_3CH_3 + BrMgNHCH_3$$

$$CH_3MgI + HC{\equiv}CCH_3 \longrightarrow CH_4 + IMgC{\equiv}CCH_3$$

(cf. $NaC{\equiv}CCH_3$, p. 68)

The reaction sequence, $RX \rightarrow RMgX \rightarrow RH$, provides a general method for reducing alkyl halides to the corresponding hydrocarbons. Methane, which can be measured volumetrically, is produced in the reactions of methyl magnesium halides with alcohols, primary and secondary amines or acetylenes; by using a known weight of one of these compounds the number of acidic (or 'active') hydrogens in the compound may be determined.

A simplified picture of these reactions shows the electron rich carbon of the C—MgX bond attacking an acidic (electron deficient) hydrogen in the other molecule. Grignard reagents may be regarded as carbon nucleophiles.

Reactions with Carbonyl Compounds. The alkyl group (effectively R:$^{\ominus}$) of a Grignard reagent attacks the electron-deficient carbon atom of a carbonyl group. Aldehydes and ketones give alcohols; primary alcohols are obtained from formaldehyde, secondary alcohols from other aldehydes, and tertiary alcohols from ketones.

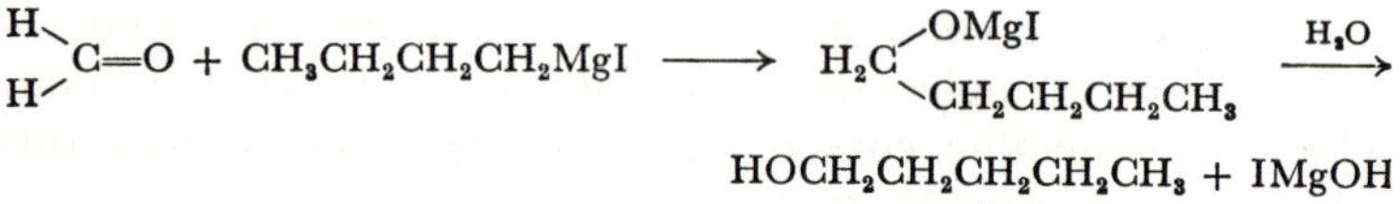

pentan-1-ol

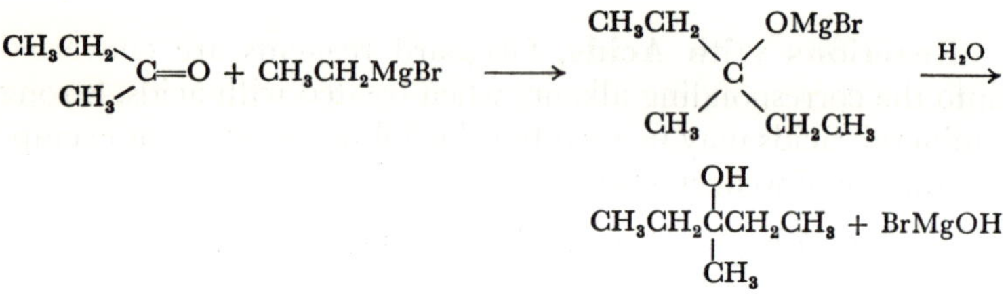

3-methylpentan-3-ol

The general formula for the inorganic products, XMgOH, represents a mixture of magnesium halide and magnesium hydroxide. Although water can be used for hydrolysis, dilute acid is employed in practice, in order to prevent the separation of magnesium hydroxide.

Esters react with *two* molecules of Grignard reagents to give tertiary alcohols containing at least two identical alkyl groups: In the example, the adduct (3) decomposes, in the reaction mixture, to methyl isopropyl ketone, which with further Grignard reagent affords a complex (4) hydrolysed to a tertiary alcohol. It is not usually possible to prepare ketones by using one molecule of Grignard reagent in this reaction.

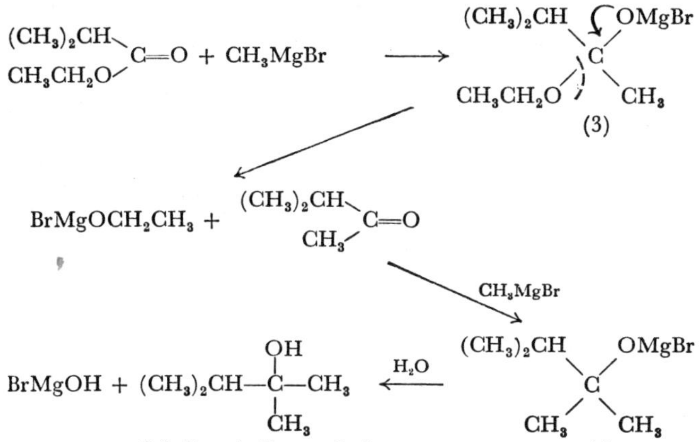

2,3-dimethylbutan-2-ol (3) (4)

Carboxylation. When a solution of a Grignard reagent is treated with carbon dioxide (the gaseous or the solid form may be used) and the resultant product hydrolysed, a carboxylic acid is obtained:

$$O{=}C{=}O + (CH_3)_3CMgCl \longrightarrow (CH_3)_3C{-}C({=}O){-}OMgCl \xrightarrow[HCl]{H_2O} (CH_3)_3C{-}C({=}O){-}OH + HOMgCl$$

trimethylacetic acid

Yields in these carbonyl addition reactions are in the region of 60–90%.

Reactions with Alkenes. Although Grignard reagents do not usually add to olefinic bonds at 20°, reactions between *ethylene* and some organometallic compounds take place surprisingly readily:

$$(CH_3)_3CLi + C_2H_4 \xrightarrow[-45^\circ]{\text{ether}} (CH_3)_3CCH_2CH_2Li \xrightarrow[\text{then acid}]{CO_2} (CH_3)_3CCH_2CH_2COOH$$

When ethylene is heated with a catalytic amount of triethylaluminium at 120°, a chain-lengthening process occurs by insertion of ethylene units:

$$(C_2H_5)_2AlCH_2CH_3 \xrightarrow{C_2H_4} (C_2H_5)_2AlCH_2CH_2CH_2CH_3 \xrightarrow{nC_2H_4} (C_2H_5)_2Al(CH_2CH_2)_{n+1}CH_2CH_3$$

(for simplicity, reaction in only one chain is indicated)

Aerial oxidation of the product yields the related alkoxide,

$(C_2H_5O)_2Al(OCH_2CH_2)_{n+1}CH_2CH_3$ which, on acidification, gives a mixture of long chain n-alkan-1-ols (C_8 to C_{16}) required for making esters and surfactants (p. 109).

One of the simplest looking reactions of ethylene, its conversion into butene, $2C_2H_4 \rightarrow C_4H_8$, has recently been accomplished. Ethylene under pressure is passed into ethanol at 40° containing a small amount (0·1 mole %) of rhodium trichloride in solution. Four different organorhodium species appear to be intermediates in this example of *homogeneous catalysis.*

Procedures for the more extensive self-addition of ethylene are discussed later (p. 142).

Other Organometallic Compounds. Tetraethyl-lead is added to petrol (gasoline) to improve its combustion performance. It provides ethyl radicals that react with the fuel, thereby initiating chain reactions. Tetraethyl-lead, b.p. 200°, is separated from the inorganic products of the following reaction by steam distillation.

$$4CH_3CH_2Cl + 4Na + Pb \longrightarrow (CH_3CH_2)_4Pb + 4NaCl$$

Some organomercury compounds are used as diuretics and spermicides, but many organometallic substances are too toxic to be employed in medicine. The early use of the arsenical, salvarsan, may be noted (p. 233).

QUESTIONS

1. Supply the missing reagent *x* and write balanced equations for the following reactions:

(*a*) $(CH_3)_2CHCH_2MgI + x(\text{then } H_2O) \longrightarrow (CH_3)_2CHCH_2CH_2OH$

(*b*) $CH_3COCH_3 + x(\text{then } H_2O) \longrightarrow (CH_3)_2C(OH)CH_2CH_2CH_2CH_3$

(*c*) $CH_3CH{=}CHCH_2CH_2MgBr + x(\text{then } H_2O) \longrightarrow CH_3CH{=}CHCH_2CH_2COOH$

(*d*) $CH_3CH(OH)CH_3 + x \longrightarrow (CH_3)_4C$

2. (*a*) Propose, including essential practical details, a method for making labelled acetic acid ($CH_3{}^{14}COOH$) from methyl iodide and barium carbonate ($Ba^{14}CO_3$). How is $Ba^{14}CO_3$ prepared?

(*b*) Starting from the hypothesis that treatment of an alkyl chloride in ether solution with sodium produces an alkane according to the equation:

$$2RCl + 2Na \xrightarrow[\text{ether}]{\text{in}} R\text{—}R + 2NaCl$$

suggest a suitable alkyl chloride that could be used in practice to find out if the alkane, R—R, is formed (remember that the product has to be separated from ether and also possibly from unchanged alkyl chloride, cf. b.ps. on pp. 3, 33).

3. (*a*) Calculate the molecular formula of a compound which contains by analysis: 64·6% carbon; 5·6% hydrogen; 30·1% iron, and whose molecular weight is 186·5.

(*b*) What methods are available for the determination of molecular weights? Give examples of compounds (one each) for which each method might not be suitable.

4. Heating silicon tetrachloride [in solution in pyridine (cf. p. 114)] with an excess of methylmagnesium iodide, gives a volatile compound, b.p. 26·5°. From these facts and from the composition by analysis (54·5% carbon; 13·7% hydrogen; 31·8% silicon) suggest a structure for the liquid product.

Give a likely equation for the complete combustion of this compound in an excess of oxygen.

How could the products from this combustion be quantitively separated and identified?

5. A mixture of 1-chlorobutane (2 mol.), propan-2-ol (1 mol.) and magnesium powder (2·5 mol.) in methylcyclohexane as solvent is heated under reflux (nitrogen atmosphere) until all the halide has reacted. An insoluble salt is removed by filtration (under nitrogen). Evaporation of the filtrate under reduced pressure gives an oily liquid, a sample of which (X) reacts with acid according to the equation

$$X + 2HCl\,(aq.) \longrightarrow C_4H_{10} + MgCl_2\,(aq.) + (CH_3)_2CHOH\,(aq.)$$

What is the molecular formula of X? What is the most likely structural formula for X?

The history of the last three hundred years is a story of the flowering of the creative powers of the human mind.

J. B. CONANT (1893–)

CHAPTER 11

POLYHALIDES, POLYALCOHOLS AND RELATED COMPOUNDS

In this and the two following chapters the chemistry of compounds containing more than one reactive group is considered. In a bifunctional compound, X—$(CH_2)_n$—Y, the two groups generally behave independently and like those in corresponding monofunctional compounds, but some special properties and reactions are encountered when X and Y are sufficiently close, especially when $n = 1$ or 2.

POLYHALIDES

Polyhalides are named as derivatives of alkanes, although specific names are also available, particularly for the derivatives of methane:

$BrCH_2CH_2Br$	1,2-dibromoethane
$CH_3CH_2CCl_2CH_3$	2,2-dichlorobutane
$(CH_3)_2CHCH_2CHF(CH_2)_5Cl$	1-chloro-6-fluoro-8-methylnonane
$CHCl_3$	chloroform, trichloromethane
CHI_3	iodoform

The melting points of the tetrahalomethanes cover a wide span:

	CF_4	CCl_4	CBr_4	CI_4
m.p.	−185°	−23°	93°	171°

Chloroform, b.p. 61°, was first obtained in 1831 by the reaction of ethanol or acetone with bleaching powder. It was introduced as a surgical anaesthetic by Simpson in 1847, but on account of its toxic properties its use has now declined. Halothane, $CF_3CHClBr$, introduced in 1956, is more suitable; anaesthesia is induced by inhalation of oxygen or nitrous oxide–oxygen containing 0·5–5 % by volume of the polyhalide.

Dichloromethane and the more expensive chloroform are good solvents for extracting covalent compounds from aqueous solutions. They are also non-flammable, a property shared by most compounds containing more than 60% halogen.

Tetrafluoroethylene (p. 144) is made by the following route:

$$CHCl_3 \xrightarrow{Hg_2F_2} \underset{\text{b.p. } -40^\circ}{CHClF_2} \xrightarrow[\text{1 sec.}]{700^\circ} \underset{\text{b.p. } -78^\circ}{C_2F_4} + 2HCl$$

A special feature of the chemistry of 1, 2-dihalides is that both halogens are removed by treatment with magnesium or zinc:

$$CH_3CH_2CHBrCH_2Br + Zn \longrightarrow CH_3CH_2CH{=}CH_2 + ZnBr_2$$

The exothermic nature of the metal + Br_2 → metal bromide reaction (for zinc, $\Delta H = -74$ kcal./mole) provides thermodynamic 'driving force'.

POLYALCOHOLS, ETHYLENE OXIDE AND RELATED COMPOUNDS

Diols. A compound containing two hydroxyl groups is a dialcohol (usually abbreviated to *diol*) or *glycol*. Systematic names for individual compounds follow the method employed for alcohols, but the glycol terminology is also used for some of the simpler compounds:

$CH_2(OH)_2$	methanediol (formaldehyde hydrate)
$HOCH_2CH_2OH$	ethane-1,2-diol (ethylene glycol)
$HO(CH_2)_4OH$	butane-1,4-diol (tetramethylene glycol)
$HOCH_2CH(CH_3)CH_2CH(OH)CH_3$	2-methylpentane-1,4-diol

Alcohols and glycols have many reactions in common. Thus hydroxyl groups of glycols can be replaced by halogen, a stepwise reaction being possible in some instances:

$$HOCH_2CH_2OH \xrightarrow[160^\circ]{HCl} \underset{\text{2-chloroethanol}}{HOCH_2CH_2Cl} \xrightarrow[200^\circ]{HCl} ClCH_2CH_2Cl$$

Oxidation of diols containing two secondary hydroxyl groups affords diketones:

$$CH_3CH(OH)CH_2CH_2CH(OH)CH_3 \xrightarrow{CrO_3} \underset{\text{hexane-2,5-dione, b.p. } 194^\circ}{CH_3COCH_2CH_2COCH_3}$$

Diols in which the hydroxyl groups are attached to adjacent carbon atoms (α-glycols) are oxidised by certain reagents, e.g.

lead tetra-acetate, that do not readily attack alcohols or other diols:

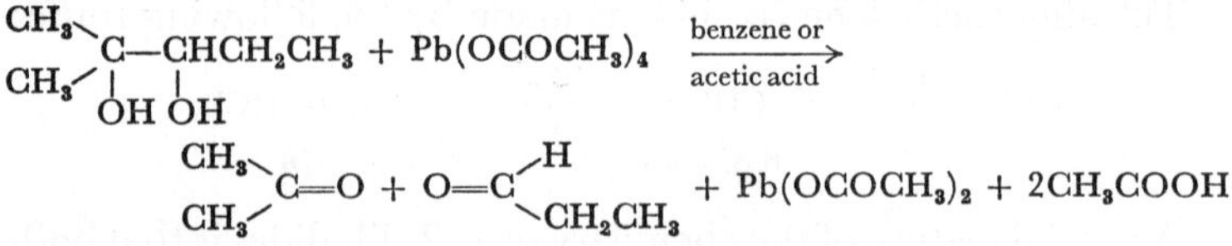

When the compound, e.g. a carbohydrate, is soluble in water, aqueous sodium periodate is a more suitable oxidant:

$$\text{-CHOHCHOH-} + NaIO_4 \longrightarrow \text{-CHO} + \text{OCH-} + NaIO_3 + H_2O$$

Epoxides. Ethers in which the oxygen atom is part of a three-membered ring are described as *epoxides*. The simplest member of the class, ethylene oxide (or more systematically epoxyethane) was a 'laboratory curiosity' for about 50 years, but is now made on a large scale by passing a mixture of air and ethylene (3 %; below the explosion limit) over a silver catalyst:

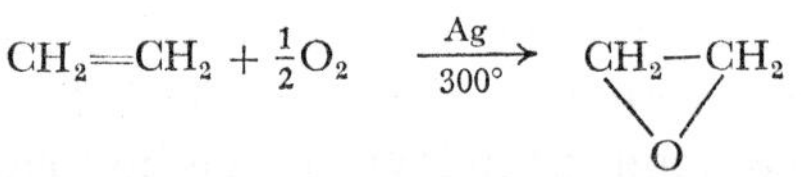

ethylene oxide, b.p. 11°

[Ethylene purity required: 98 %; yield of ethylene oxide: 60–70 %; heat of reaction, $\Delta H = -35$ kcal./mole]. The conditions necessary for the reaction are fairly critical for, although the oxidation was first attempted in 1904, it was not until 1931 that the epoxide was obtained in this way.

Ethylene oxide is an energy-rich and especially reactive compound, for in the three-membered ring the carbon–oxygen–carbon bond angle is much distorted from the normal value of *ca.* 110° for an ether. It readily reacts at 20° with water containing an acidic or basic catalyst to give ethylene glycol. The mechanism of the acid-catalysed hydration involves the intermediate formation of an oxonium ion:

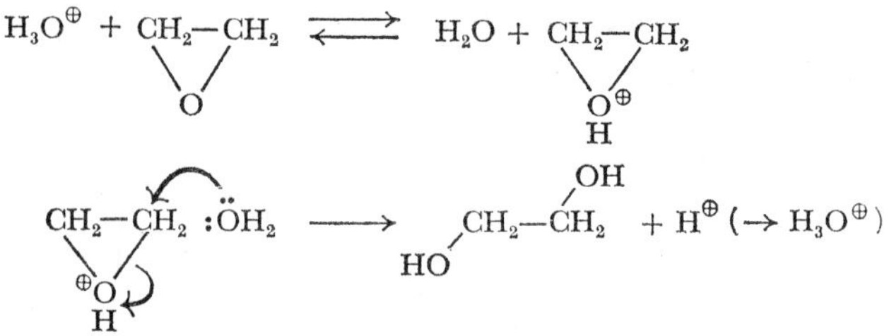

Ethylene glycol is miscible with water and ethanol (but not with hydrocarbons) and is a useful high boiling solvent (b.p. 197°). A large proportion of the industrial output of ethylene glycol is employed, added to water, as an anti-freeze (f.p. $-8°$ for a 20% solution and $-36{\cdot}5°$ for a 50% solution). An increasing quantity, however, is absorbed by the synthetic fibre industry, as the diol is the alcohol component of the polyester Terylene (p. 228).

Acid-catalysed additions of alcohols to ethylene oxide afford 2-alkoxyethanols, $ROCH_2CH_2OH$, which are used as solvents, for example, in varnishes. 2-Ethoxyethanol, b.p. 136°, obtained from the oxide and ethanol, is also known as ethyl cellosolve.

Reaction between ammonia and ethylene oxide gives a mixture of three products; the high yield of tertiary amine (65%) using equimolar proportions of reactants indicates that the rates of the later stages are higher than that of the initial reaction.

$$NH_3 \xrightarrow{C_2H_4O} \underset{\substack{\text{ethanolamine}\\ \text{(2-aminoethanol)}}}{NH_2CH_2CH_2OH} \xrightarrow{C_2H_4O} \underset{\text{diethanolamine}}{NH(CH_2CH_2OH)_2} \xrightarrow{C_2H_4O} \underset{\text{triethanolamine}}{N(CH_2CH_2OH)_3}$$

$NH_3:C_2H_4O$ ratio	yields, % based on C_2H_4O		
1:1	12	23	65
10:1	71	25	4

As expected, the yield of the initial product, ethanolamine, can be improved by limiting the amount of ethylene oxide. Ethanolamine is used to extract hydrogen sulphide from crude natural gas.

Ethanolamine and the quaternary ammonium derivative, *choline*, $[(CH_3)_3NCH_2CH_2OH]^{\oplus}X^{\ominus}$, occur as structural units in phospholipids (p. 109). Acetylcholine, $^{\ominus}OH\ (CH_3)_3\overset{\oplus}{N}CH_2CH_2OCOCH_3$, is the chemical messenger that transmits impulses between motor nerves and voluntary muscles. An impulse reaching a nerve ending liberates acetylcholine which then stimulates a receptor that causes for example a muscle fibre to contract.

Dioxan (1), b.p. $101{\cdot}5°$, is a cyclic ether obtained by treating ethylene oxide with a catalytic amount of acid. It is miscible with water, and is a powerful solvent for organic compounds. Compared with ethylene oxide, dioxan is chemically unreactive.

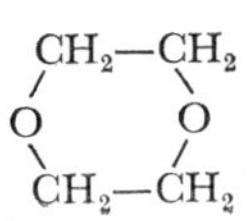

(1)

Glycerol. The triol, glycerol (2), is formed by hydrolysis of natural fats and oils and is therefore a product of soap manufacture (p. 108). About half the industrial supply of glycerol is now derived by synthesis, for example, from propylene *via* acrolein (p. 145).

(2) 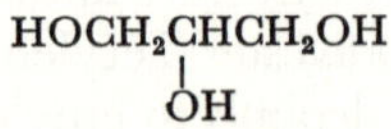 glycerol, b.p. 290° (propane-1,2,3-triol)

The viscosity of glycerol is 1490 centipoises at 20° compared with values of 19·9 and 1·0 for ethylene glycol and water respectively.

Glycerol is required for the production of polyesters (p. 226). Mining explosives, e.g. dynamite, contain ammonium nitrate and glyceryl trinitrate, often called 'nitroglycerine'. This ester is prepared by treatment of glycerol with a mixture of nitric and sulphuric acids.

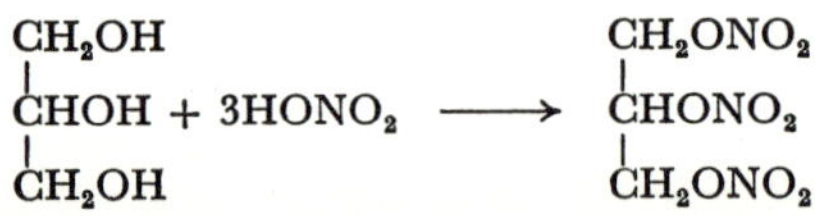

glyceryl trinitrate

BAL. Propan-1-ol-2,3-dithiol (3), better known as British anti-lewisite or simply BAL, was developed as an antidote to the arsenic-containing war gas, lewisite. BAL derivatives have a wider application, and are effective in treatment of poisoning by certain metals, especially arsenic, mercury or gold. These metals probably act as poisons by combining with thiol groups of cysteine (p. 50), an amino-acid component of proteins. Treatment with BAL yields a similar but more stable metal derivative, e.g. of partial structure (4), thus liberating the protein thiol groups.

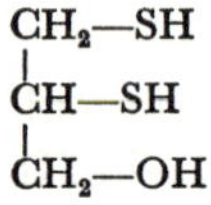 (3)

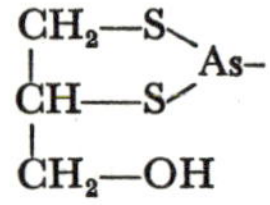 (4)

Diamines. Some examples of compounds containing two primary amino-groups are:

$NH_2CH_2CH_2NH_2$	1,2-diaminoethane or ethylene diamine, b.p. 118°
$NH_2(CH_2)_4NH_2$	1,4-diaminobutane or putrescine, m.p. 27°
$NH_2(CH_2)_6NH_2$	1,6-diaminohexane or hexamethylene diamine, m.p. 42°

These compounds are soluble in water and give disalts with strong acids. Putrescine received its name because it is formed in the decomposition of the proteins of flesh by bacterial action. Hexamethylene diamine is used in the production of the polyamide, nylon (p. 164).

QUESTIONS

1. (*a*) When sodium hypochlorite (NaOCl) is added to a solution of acetone and sodium iodide in water, a yellow solid separates (m.p. 119° after crystallisation from ethanol). What is the structure of this compound, whose composition is C, 3·05%; H, 0·25%; I, 96·7%; mol. wt. *ca.* 400?

Write a balanced equation for the reaction; sodium acetate is the other product from the acetone.

(*b*) Give names and structural formula to compounds of molecular formula, $C_4H_8Br_2$. Which of these compounds would you expect to react with zinc to give an olefin?

2. Write two possible structures for a compound, m.p. 70°, which has been isolated from shark liver oil, and which is believed to be a mono-octadecyl (n-$C_{18}H_{37}$) ether of glycerol. Suggest a chemical reaction that could be used to reach a decision between the two structures.

3. (*a*) Which compound is likely to be more hygroscopic: chloroform or glycerol?

(*b*) Why are the following desiccants unsuitable for contact drying of the given compounds? (i) sulphuric acid for oct-1-ene, (ii) sodium hydroxide pellets for 1,2-dibromobutane. Suggest drying agents that would be more suitable to use for these compounds.

4. A solvent recommended for cleaning certain components of a computer is stated to contain 'trichlorotrifluoroethane'. In how many structurally isomeric forms can this compound exist?

5. (*a*) How could the reaction between ethane-1,2-diol and concentrated hydrochloric acid (p. 133) be carried out in practice, and how could the ratio of 2-chloroethanol to 1,2-dichloroethane at various times be determined within an accuracy of 2%?

(*b*) Write an electronic mechanism for the reaction of ammonia with ethylene oxide in which 2-aminoethanol is formed.

6. Suggest a two-stage route from allyl alcohol (p. 146) to 1,2-dithioglycerol (BAL).

True art consists of habitually searching for the causes and meaning of everything that occurs.

CHARLES DARWIN (1809–1882)

CHAPTER 12

SUBSTITUTED ALKENES. POLYMERS

This chapter is concerned with compounds containing a carbon–carbon double bond and another reactive group; several biological molecules that include this feature are also considered. As in the chemistry of other bifunctional compounds, the double bond and the other group may behave independently but sometimes, especially when the two groups are close together, their properties are profoundly modified. We should also remember that some reagents can attack both groups, for example, allyl alcohol, $CH_2{=}CHCH_2OH$, contains two oxidisable positions, the double bond and the primary OH group. It is then desirable to find reagents that will react selectively with each of the groups (cf. p. 46).

The elaboration of certain types of large molecule *via* polymerisation will also be considered.

CONJUGATION. ULTRAVIOLET SPECTROSCOPY

Nomenclature. Dienes are compounds containing two ethylenic bonds:

$CH_2{=}C{=}CH_2$	propadiene or allene	allenes
$CH_3CH_2CH{=}C{=}CHCH_3$	hexa-2,3-diene	
$CH_2{=}CH{-}CH{=}CH_2$	buta-1,3-diene	conjugated dienes
$CH_2{=}\underset{CH_3}{\underset{\vert}{C}}{-}CH{=}CH_2$	2-methylbuta-1,3-diene or isoprene	
$CH_2{=}\underset{Cl}{\underset{\vert}{C}}{-}CH{=}CH_2$	2-chlorobuta-1,3-diene or chloroprene	
$(CH_3)_2CHCH{=}CH{-}CH{=}CHCH_2CH_3$	2-methylocta-3,5-diene	
$CH_3CH{=}CHCH_2CH_2CH{=}CH_2$	hepta-1,5-diene	non-conjugated diene

Dienes are classified (as shown above) according to the relative positions of the olefinic bonds. *Allenes* have the general structure >C=C=C<. In *conjugated* dienes single and double bonds alternate; *conjugation* is a general term implying this specific arrangement of multiple bonds and applies to >C=O, -C≡C-, -C≡N, etc., as well as >C=C< bonds. Dienes in which the double bonds are farther apart, e.g. hepta-1,5-diene, are described as non-conjugated dienes.

Unsaturated carbonyl compounds, including acids, are named systematically by an extension of familiar methods, although specific names are employed frequently:

$CH_2{=}CHCHO$	propenal or acrolein
$CH_3COCH{=}CH_2$	butenone or methyl vinyl ketone
$(CH_3)_2CHCH{=}CHCOCH_2CH_3$	6-methylhept-4-en-3-one
$CH_2{=}CHCOOH$	propenoic acid or acrylic acid
$CH_3CH{=}CHCOOH$	but-2-enoic acid or crotonic acid
$CH_2{=}CH{-}C{\equiv}N$	vinyl cyanide or acrylonitrile
$CH_3CH_2CH(CH_3)CH{=}CH(CH_2)_3COOH$	7-methylnon-5-enoic acid

All but the last of the above compounds have *conjugated* multiple bonds, and, like conjugated dienes, have special reactions. The alternative description, *αβ-unsaturated* compounds, refers to the position of olefinic bonds between carbon atoms α and β to carbonyl or cyano-groups. The geometrical isomerism of crotonic acid has been discussed (p. 58).

Ultraviolet Spectroscopy. Ultraviolet and visible spectroscopy are invaluable for investigating the structures of unsaturated molecules. All organic molecules absorb radiation, but only those containing double bonds and unbonded electron pairs absorb in the most accessible region of the ultraviolet (wavelength 1900–4000 Å). One effect of the absorption of this radiation is to produce an excited molecule in which one of the electrons of a C=C or C=O bond is promoted to a higher energy level. Ultraviolet data is usually presented in terms of the wavelength of maximum absorption (λ_{max}, measured in Å) and the intensity, ε (molecular extinction coefficient) at that wavelength. Values of λ_{max} are a measure of the energy required to effect transitions to excited states, and hence indicate the relative stability of normal and excited states; radiation at longer wavelengths provides less energy. A typical ultraviolet spectrum is presented in the diagram, and results for a number of unsaturated molecules are given in the following table:

		λ_{max}	ϵ
$CH_2{=}CHCH_3$	propene	1750	12,500
$CH_2{=}CHCH_2CH_2CH{=}CH_2$	hexa-1,5-diene	1850	20,000
$CH_2{=}CH{-}CH{=}CH_2$	buta-1,3-diene	2170	21,000
$CH_3CH{=}CHCHO$	but-2-enal	2170	15,650
$CH_2{=}CH{-}CH{=}CH{-}CH{=}CH_2$	hexa-1,3,5-triene	2570	34,700

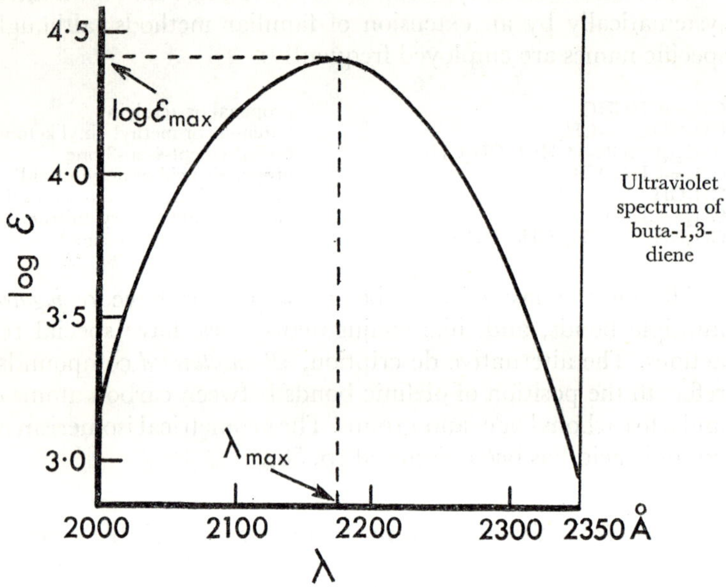

Fig. 12.1.

Compounds containing isolated double bonds, absorb at 1620–1900 Å, that is just outside the range of the normal spectrometer; reference to the data for propene and hexa-1,5-diene shows that the intensity of absorption is roughly additive. Conjugated molecules, e.g. buta-1,3-diene and but-2-enal, absorb above 2100 Å, and are therefore readily distinguished from compounds containing single or non-conjugated multiple bonds. Extension of the conjugated system, e.g. in hexa-1,3,5-triene, intensifies the absorption peak and shifts it to higher wavelength. The higher values of λ_{max} for these molecules indicates that conjugation diminishes the energy difference between normal and excited states.

One of the causes of colour in organic compounds is the presence of a long system of conjugated double bonds, and the consequent shift of the absorption maximum into the visible region of the spectrum (above 4000 Å). The orange pigment, β-carotene, present in carrots, absorbs at λ_{max} 4500 Å (ϵ 148,000) in hexane.

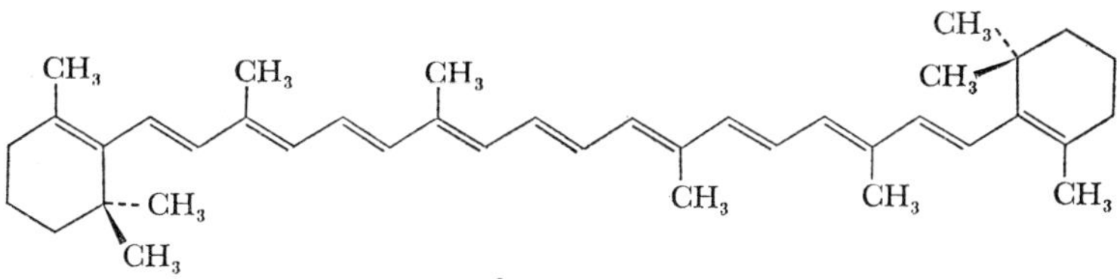

β-carotene

The early development and use of absorption spectroscopy in the 1930's was greatly stimulated by its value in the detection and identification of compounds present in low concentrations in biological materials, for example, the very pale yellow, air-sensitive vitamin A.*

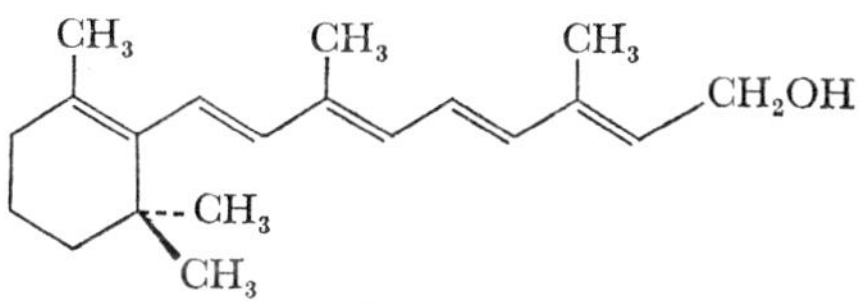

vitamin A
λ_{max} 3250 Å (ϵ 51,000) in ethanol

VINYLIC COMPOUNDS AND POLYMERS

The name vinyl bromide for the molecule, $CH_2{=}CHBr$, illustrates the use of the word vinyl to denote a $CH_2{=}CH-$ group. An alternative name for buta-1,3-diene could be divinyl. An atom or group is described as *vinylic* when it is joined to a carbon atom of an olefinic bond; for example, 3-chlorohept-3-ene, $CH_3(CH_2)_2CH{=}CClCH_2CH_3$, is a vinylic chloride.

In contrast with alkyl halides, vinylic halides do not readily

* Vitamins, supplied in the diet, are required in small amounts for the normal growth and health of organisms. Hormones have a similar general function but are produced in endocrine glands.

undergo hydrolysis or other nucleophilic substitution reactions. They will, however, react with lithium to give vinylic lithium compounds.

Polymers. The most important property of compounds containing $CH_2{=}CH{-}$ groups is the relative ease with which many molecules undergo self addition reactions (polymerisation) to give high molecular weight substances called *polymers*, e.g.

$$n\mathrm{CH_2{=}CH{-}R} \longrightarrow \left(-\mathrm{CH_2}-\underset{\mathrm{R}}{\underset{|}{\mathrm{CH}}}-\mathrm{CH_2}-\underset{\mathrm{R}}{\underset{|}{\mathrm{CH}}}- \right)_{n/2}$$

Polymers can be built up from one kind of unit, as indicated in the above equation; from two components (e.g. the general purpose rubber from butadiene and styrene); or from three or more different types of molecule (ABS polymer, p. 191). In the 'normal' reactions of vinyl compounds, the formation of higher molecular weight materials often has only the nuisance value of a side reaction. However, under appropriate conditions, polymerisation can become essentially the sole reaction, and products (plastics) useful for a variety of constructional purposes are obtained.

In the following sections the polymerisation of ethylene and mono-substituted ethylenes will be considered.

Polyethylene. From a consideration of bond energy values it can be predicted that one molecule of a butene will be thermodynamically more stable than two molecules of ethylene; therefore reaction, if it occurs, will proceed in the direction $2C_2H_4 \rightarrow C_4H_8$. Conditions for effecting this reaction have recently been found (p. 130), although a much more extensive version of the reaction was encountered earlier in the formation of polyethylene from ethylene:

$$n\mathrm{CH_2{=}CH_2} \longrightarrow (-\mathrm{CH_2{-}CH_2}-)_n$$

Each chain-like molecule of polyethylene is built up from 3000 to 15,000 ethylene units. The polymerisation reaction is, as expected, exothermic ($\Delta H = ca. -24\ n$ kcal./mole at 25°), and careful control of temperature is therefore required in large-scale preparations.

The discovery of polyethylene demonstrates the value of speculative research. In 1931, R. Robinson submitted a memorandum to I.C.I. Ltd. that contained the opening words 'It is desired to test the hypothesis that by the use of pressure of a very high order it should be possible to bring about organic chemical reactions, which at lower pressures require the aid of catalysts'. The initial phase of the ensuing work required the design and construction of equipment that could withstand an internal pressure of 3000 atmospheres. In 1933 the reaction between ethylene and benzaldehyde (p. 214) was attempted at 170° and 2000 atmospheres. In thc first experiment much of the starting material was lost through a leak that developed in the apparatus after a day or two. However some substance remained, the observation being recorded in a notebook as 'Waxy solid found in reaction tube'. The product had m.p. 113°, and analysis showed that it was derived entirely from ethylene; the benzaldehyde had not been utilised.

In the present day production of polyethylene by the high pressure method, ethylene of 99·9 % purity is mixed with a trace of oxygen (0·6 parts per thousand), and heated at 190° under pressure (2000 atmos.). The liquid polyethylene formed solidifies on cooling, m.p. 100–110°. A product with a higher melting point (130° or above) is obtained by treating ethylene with a 'co-ordination catalyst' (titanium chloride + a trialkyl-aluminium) at ordinary pressures below 100° (Ziegler, Natta, 1954). Either product is a tough, flexible material that has many uses (hygienic wrapping, water pipes, cable sheaths) since it is stable towards most chemical agents and is a poor conductor of electricity.

Polypropylene. Before 1954, propylene could only be converted into liquid, low molecular weight polymers. Introduction of co-ordination catalysts dramatically changed the situation, for the polymer produced was not only solid but was also crystalline, due to polymerisation taking place in a *stereo-regular* way:

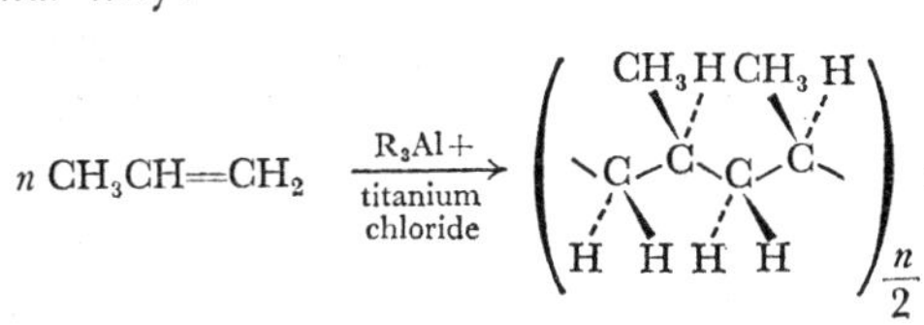

polypropylene, m.p. 165° or above

Because of its high m.p., low density (0·90–0·91), and durability, polypropylene is used for the manufacture of lightweight cases and crates, and of textiles, including carpets and sacks. Production has also started of the stereoregular polymer from 4-methylpent-1-ene, which has a m.p. above 200° and a density of 0·85.

The mechanism of co-ordination catalysis, illustrated below for ethylene, involves recurrent addition and rearrangement reactions:

$$\longrightarrow [Cl_4TiCH_2CH_3] \xrightarrow{C_2H_4} [Cl_4Ti(C_2H_4)CH_2CH_3] \xrightarrow{\text{rearrangement}}$$
$$[Cl_4TiCH_2CH_2CH_2CH_3] \longrightarrow$$

Fluon or Teflon, $(\text{-}CF_2\text{—}CF_2\text{-})_n$, obtained by polymerisation of tetrafluoroethylene, is exceptionally resistant to heat and chemical agents, including solvents, but it is also relatively expensive.

Vinyl Compounds. Vinyl chloride is prepared from acetylene or from ethylene, the lesser cost of the latter compound offsetting the more complex route required:

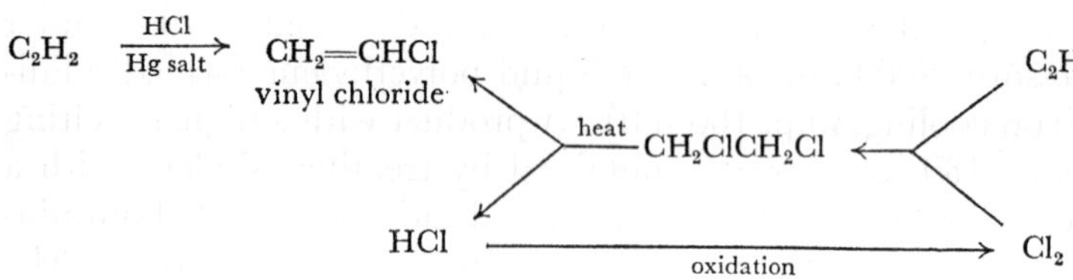

Vinyl chloride, b.p. −14°, is readily polymerised *via* a free radical mechanism.

$$\xrightarrow[\text{initiator}]{\text{radical}} (\text{-}\dot{C}H_2\text{—}CHCl) \xrightarrow{CH_2\text{=}CHCl}$$
$$(\text{-}CH_2CHClCH_2\text{—}\dot{C}HCl) \xrightarrow[CH_2\text{=}CHCl]{n} \text{polyvinyl chloride}$$

The hardness of the polymer is reduced by the addition of plasticisers, e.g., high-boiling dialkyl phthalates; the material obtained, PVC, is used for floor and seat coverings, raincoats and records.

Previous methods for making vinyl cyanide (acrylonitrile) are being superseded by a process in which a mixture of propylene, ammonia and air is passed over molybdenum oxide at 400°. Polyacrylonitrile is a component of some textiles, e.g. Orlon, and plastics (p. 191).

Molecules with Two Double Bonds. *Buta*-1,3-*diene*, b.p. −2·5°, is present (35–40%) in the C_4-fraction (10% of total product) from the steam cracking of naphtha (p. 27, cf. p. 60); it is isolated by distillation and solvent extraction. At temperatures above 1100° it is the most stable C_4-hydrocarbon.

Chloroprene (p. 71), b.p. 59·5°, polymerises very readily to neoprene, which is more resistant to oil and petrol than natural rubber.

Rubber is a high molecular weight hydrocarbon, $(C_5H_8)_n$, obtained from *Hevea brasiliensis*. Ozonolysis reveals that it is a polymer of *isoprene*, $CH_2{=}C(CH_3)CH{=}CH_2$. Infrared and n.m.r. (p. 186) studies show that at least 99% of the double bonds in rubber contain the methylene groups in a *cis*-arrangement:

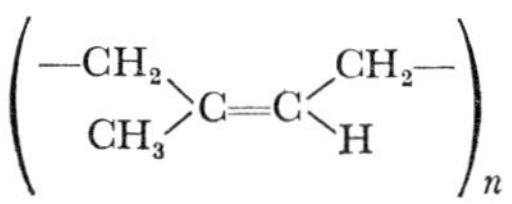

A material closely similar to natural rubber is obtained by polymerising isoprene with a lithium catalyst. A different mode of polymerisation occurs when one of the other alkali metals is used.

C_5-Isoprene-like units are present as building blocks in many biological compounds; for example, vitamin A (p. 141) can be dissected (in theory) into four of these units.

The hydrocarbon, *squalene*, $C_{30}H_{50}$, is composed of six isoprene units. It was originally isolated from shark liver oil, but was later found to be widely distributed in mammalian tissue. Squalene is an intermediate in the biosynthesis of cholesterol and sex hormones from acetate units.

$$H[CH_2C(CH_3){=}CHCH_2]_3[CH_2CH{=}C(CH_3)CH_2]_3H \quad \text{squalene}$$

Acrolein, a sharp-smelling liquid, b.p. 52°, made from propylene by catalytic air oxidation, is an intermediate in the synthesis of glycerol:

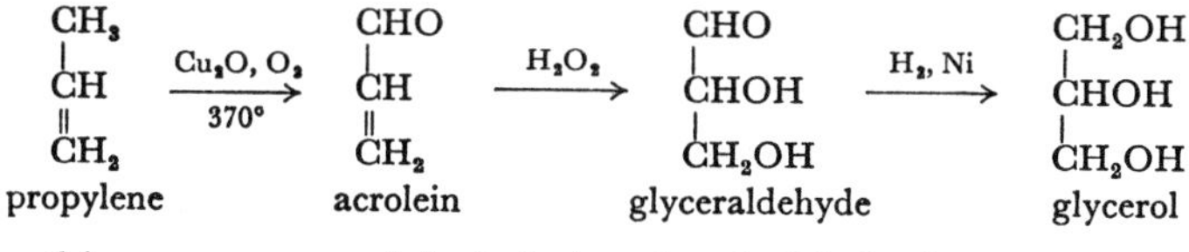

In this process, acrolein is *hydroxylated* with hydrogen peroxide, and the product, glyceraldehyde, is reduced to glycerol.

Crotonaldehyde, b.p. 104°, is prepared from acetaldehyde *via* aldol:

$$2CH_3CHO \longrightarrow CH_3CH(OH)CH_2CHO \xrightarrow[\text{trace } H^{\oplus}]{\text{heat}} CH_3CH{=}CHCHO$$

n-Butanol is obtained by catalytic hydrogenation of both double bonds of crotonaldehyde; this is an alternative route to that involving propylene and the oxo-reaction.

OLEFINIC ALCOHOLS

Vinylic alcohols are the enol forms of aldehydes and ketones (p. 81). Vinyl alcohol, for example, is only a trace constituent of acetaldehyde, and effectively the first compound of this series is allyl alcohol, $CH_2{=}CHCH_2OH$, for which the systematic name is prop-2-en-1-ol.

Allylic compounds in general contain the structural feature

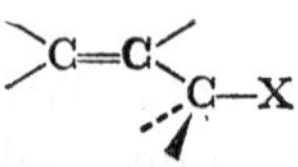

(where X = OH, Cl etc.). Allyl chloride can be obtained in 90% yield by mixing streams of propylene and chlorine at 550°; at this temperature chlorine atoms are formed which initiate a homolytic reaction mechanism.

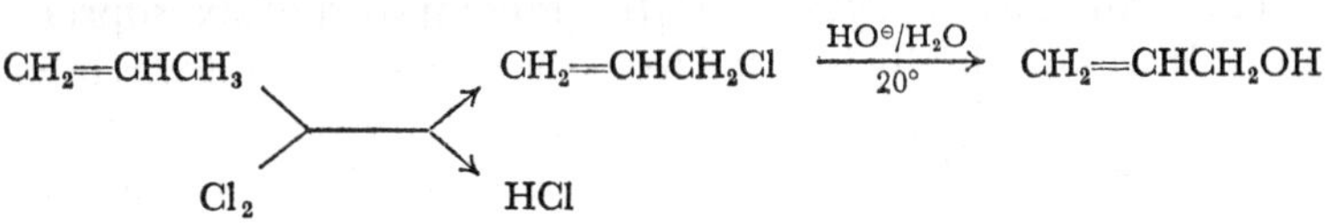

In contrast with vinyl chloride, allyl chloride is very readily hydrolysed by aqueous alkali.

By using appropriate reagents, allyl alcohol can be selectively oxidised at the double bond or at the alcohol group:

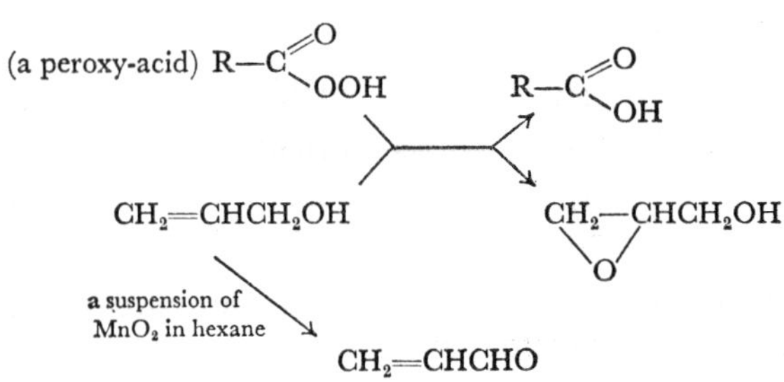

Many unsaturated alcohols have been isolated from biological sources, e.g. vitamin A, geraniol (both allylic alcohols), and *cis*-hex-3-enol.

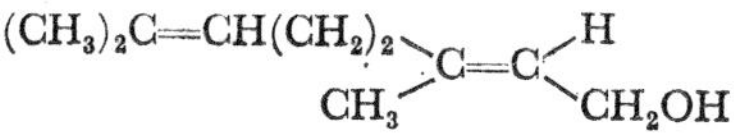

geraniol (in palmarosa oil)

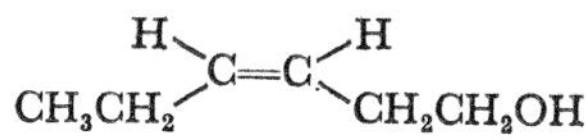

cis-hex-3-enol ('leaf alcohol')

These two compounds are representative of many that impart fragrant odours to plants.

Some organisms emit trace amounts of substances that help to regulate their environment by affecting the behaviour of other organisms. Such compounds, *pheromones*, constitute therefore a type of chemical language. For example, bombykol, produced by virgin silkworm moths, will attract the male of the species at a concentration of 10^{-19} g. per ml. of air. After isolation (12 mg. from 500,000 moths), bombykol was found to have the following structure:

$$HOCH_2(CH_2)_8\overset{10}{C}H{=}\overset{11}{C}H{-}\overset{12}{C}H{=}\overset{13}{C}HCH_2CH_2CH_3$$

The geometrical configurations of the double bonds (*trans* C_{10}/C_{11}; *cis* C_{12}/C_{13}) was established by rational syntheses of each of the four possible isomers. The isomer containing two *trans* double bonds is 10^{14} times less active than bombykol.

QUESTIONS

1. (*a*) Give systematic names for the following compounds:

(i) $CH_3CHBrCH_2CH{=}CHCH_3$ (ii) $(CH_3)_3CCH{=}CHCH_2OH$
(iii) $CH_3CH_2CH{=}CHCHO$ (iv) $CH_2{=}C(Cl)CH{=}CHCH_3$
(v) $HOOCCH_2CH_2CH{=}C(CH_3)_2$ (vi) $CH_3CH{=}CHCH_2CH{=}CHCH_2I$

Specify which of these compounds contains a conjugated system of bonds.

(*b*) Show how the formulae of geraniol and *β*-carotene can be divided into C_5 isoprenoid units.

2. (*a*) Comment on the difference between the carbon to chlorine bond lengths in chloroethane (1·77 Å) and in chloroethene (1·69 Å).

(*b*) What is unsatisfactory about the following statement? Vinyl alcohol can be prepared by hydrolysis of vinyl acetate ($CH_2{=}CHOCOCH_3$).

3. Give structures and names for three compounds of molecular formula, C_4H_6, which do not show an absorption maximum at 2200 Å or above.

4. (*a*) How may the isomeric compounds, methylacrylamide, $CH_2{=}C(CH_3)CONH_2$, and acetone cyanohydrin (p. 77), be differentiated by infrared absorption spectroscopy? Which compound is more likely to be a solid?

(*b*) Suggest a method for converting isobutene into methylacrylonitrile, $CH_2{=}C(CH_3)CN$.

5. (*a*) How could you find out if methyl methacrylate $CH_2{=}C(CH_3)COOCH_3$, is lighter or heavier than its polymer, Perspex?

(*b*) Combustion analysis of a fragment (3·821 mg.) of a contact lens gave 8·406 mg. of carbon dioxide and 2·751 mg. of water. The infrared absorption of a powdered sample of the lens contained strong bands at 1725 and 1250 cm.$^{-1}$. Suggest one material from which the lens could have been made.

(*c*) Interpret the following observations:

A suspension of powdered polyvinyl acetate

$$(-CH_2—CH(OCOCH_3)-)_n$$

in water containing 1% of sulphuric acid is heated at 95°. After some time a clear solution is obtained. When steam is passed through the solution, the condensed vapour gives an acidic reaction.

6. (*a*) Reservoirs may be lined with a plastic material to prevent loss of water. Propose a procedure by which the relative permeability of sheets of plastic might be examined.

(*b*) How might the following statement be tested experimentally? The chemical inertness of Teflon is greater than that of gold or platinum.

There are no sciences, only science and its applications.

LOUIS PASTEUR (1822–1895)

CHAPTER 13

SUBSTITUTED CARBOXYLIC ACIDS. ENANTIOMERS. PROTEINS

DICARBOXYLIC ACIDS AND HYDROXY-ACIDS

The simplest saturated dicarboxylic acid, oxalic acid, was isolated from wood-sorrel in the eighteenth century, and its name is derived from the botanical name of this plant, *Oxalis acetosella.* It also occurs in rhubarb leaves, and its poisonous properties are due to the removal of calcium ions from the body by the precipitation of calcium oxalate, $(COO^{\ominus})_2Ca^{\oplus\oplus}$. Succinic acid was obtained earlier by the distillation of the fossil pine resin, amber (Latin, *succinum*). In fact, saturated dicarboxylic acids are amongst the earliest known organic compounds, and the specific names given at the time of discovery are still in use.

$COOH$ / $\vert$ / $COOH$	or $(COOH)_2$	oxalic acid	m.p. 189°
$CH_2\langle{}^{COOH}_{COOH}$	or $CH_2(COOH)_2$	malonic acid	m.p. 135°
CH_2—$COOH$ / $\vert$ / CH_2—$COOH$	or $HOOC(CH_2)_2COOH$	succinic acid	m.p. 186°
$HOOC(CH_2)_3COOH$		glutaric acid	m.p. 97°
$HOOC(CH_2)_4COOH$		adipic acid	m.p. 153°

It is necessary to employ systematic or semisystematic names in some instances:

CH_3—CH—$COOH$ with $(CH_2)_2COOH$ on the CH — 1-methylpropane-1,3-dicarboxylic acid or α-methylglutaric acid.

Hydroxy-acids are named as hydroxyl derivatives of carboxylic acids; the position of the hydroxyl group is indicated by a number (the carbon atom of the carboxyl group is designated

1) or by a Greek letter if the specific name of the acid is employed:

$CH_3CH(OH)COOH$	2-hydroxypropanoic acid α-hydroxypropionic acid or lactic acid
$CH_3CH(OH)CH_2COOH$	3-hydroxybutanoic acid β-hydroxybutyric acid
$CH_3\overset{\delta}{C}H(CH_3)\overset{\gamma}{C}H(OH)\overset{\beta}{C}H_2\overset{\alpha}{C}H_2COOH$	4-hydroxy-5-methylhexanoic acid (a γ-hydroxy acid)

Dicarboxylic acids are often prepared by adaptions of methods used for monocarboxylic acids:

$$HOCH_2CH_2CH_2OH \xrightarrow{HBr,\ H_2SO_4} BrCH_2CH_2CH_2Br \xrightarrow[\text{aq. ethanol, reflux}]{\text{NaCN in}} NCCH_2CH_2CH_2CN \xrightarrow{\text{reflux with aq. HCl}} HOOCCH_2CH_2CH_2COOH$$

α-Hydroxy-acids can be synthesised from aldehydes or ketones:

$$RCHO \xrightarrow[CN^{\ominus}\text{ catalyst}]{HCN} RCH\begin{matrix}OH\\CN\end{matrix} \xrightarrow[\text{heat}]{\text{aq. HCl}} RCH\begin{matrix}OH\\COOH\end{matrix}$$

Dicarboxylic acids form the expected derivatives (diesters, diamides etc.) ,and hydroxy-acids have many chemical properties typical of alcohols and carboxylic acids, as the following reactions indicate:

$$\underset{\substack{\text{hydroxyacetic}\\\text{acid}\\\text{(glycollic acid)}}}{HOCH_2COOH} \xrightarrow{PCl_5} \underset{\substack{\text{chloroacetyl}\\\text{chloride}}}{ClCH_2COCl} \xrightarrow{H_2O} \underset{\substack{\text{chloroacetic}\\\text{acid}}}{ClCH_2COOH}$$

$$CH_3CH_2CH(OH)CH_2CH_2COOH \xrightarrow{CrO_3} CH_3CH_2COCH_2CH_2COOH$$

Subsequent discussion is mainly confined to special reactions that depend on the relative positions of the functional groups.

Formation of Cyclic Compounds. Dicarboxylic acids and their derivatives form ring compounds by internal reaction (an example of *cyclisation*, e.g. p. 172) when the two groups are separated by *two* or *three* carbon atoms. Heating the acids alone or with acetic anhydride gives cyclic anhydrides, which have reactions similar to those of open-chain anhydrides; succinic anhydride and ethanol for example give a half-ester, ethyl hydrogen succinate.

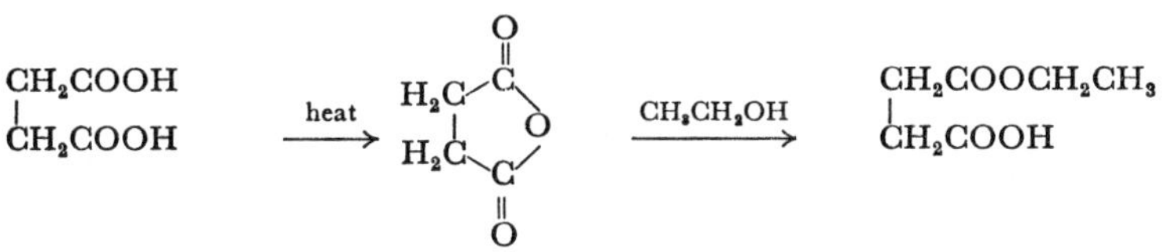

succinic anhydride, m.p. 119°

ethyl hydrogen succinate

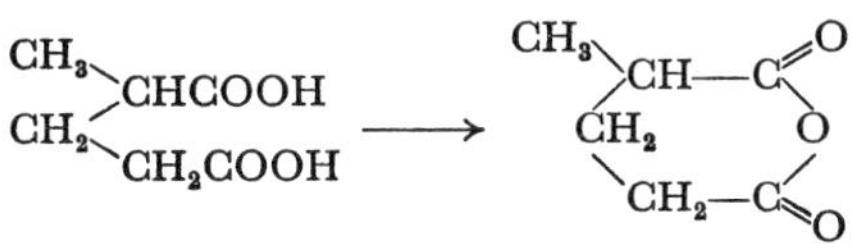

α-methylglutaric anhydride

Diamides (prepared as usual from diesters) also cyclise on heating, when the carbonyl groups are separated by two or three carbon atoms. This reaction involves the loss of ammonia and the formation of cyclic *imides* (imides contain the group -CO—NH—CO-):

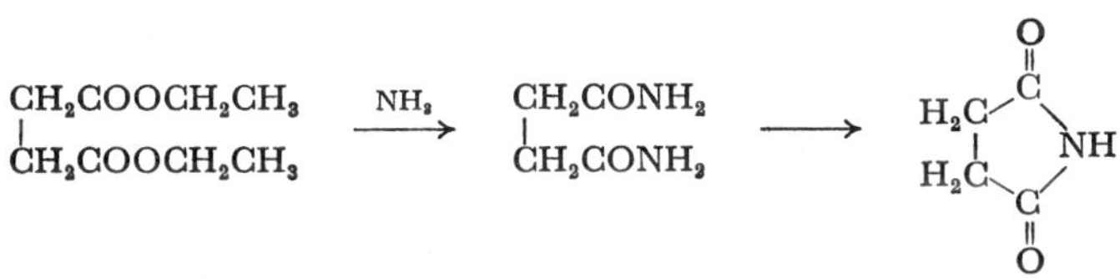

diethyl succinate, b.p. 216°

succinamide, m.p. 243°

succinimide, m.p. 125°

Succinimide is weakly acidic, like the analogous phthalimide (p. 226).

γ- and δ-hydroxy-acids also form ring compounds when heated alone or when treated with an acid catalyst; an internal esterification reaction occurs to give γ- and δ-*lactones* (cyclic esters), respectively:

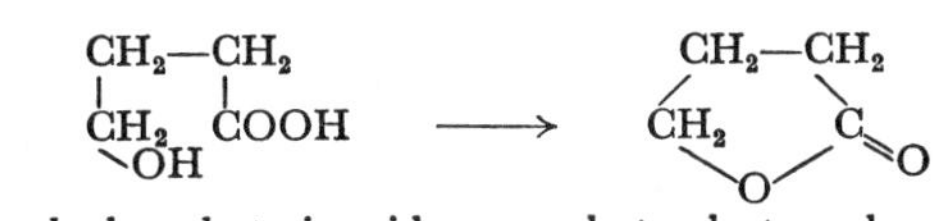

γ-hydroxybutyric acid

γ-butyrolactone, b.p. 204°

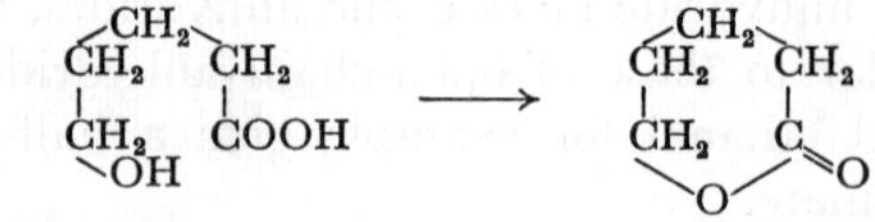

5-hydroxypentanoic acid or δ-hydroxyvaleric acid — δ-valerolactone (polymerises on heating)

These examples indicate that molecules containing *five* or *six* atoms in the ring are formed easily. Compounds containing smaller and larger rings are more difficult to prepare by the reactions given above, and other approaches are usually employed. Malonic acid and substituted malonic acids lose carbon dioxide (are *decarboxylated*) exceptionally readily (100–180°):

$$CH_2(COOH)_2 \longrightarrow CH_3COOH + CO_2$$

Maleic and Fumaric Acids. When bromosuccinic acid is heated with aqueous alkali, hydrolysis to malic acid is accompanied by elimination of hydrogen bromide giving a mixture of maleic acid (1), m.p. 130°, and fumaric acid (2), m.p. 287°. Maleic and fumaric acids are geometrical isomers (p. 58),

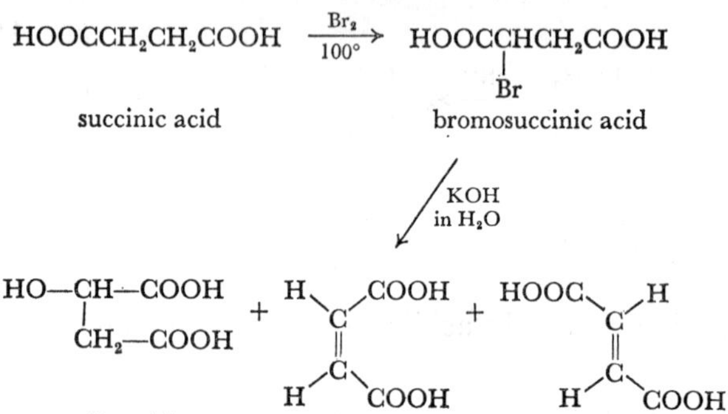

succinic acid; bromosuccinic acid; malic acid; *cis-* (1); *trans-* (2)

and each is reduced to succinic acid by hydrogen in the presence of a platinum catalyst. In this instance *cis-* and *trans-*configurations can be assigned from chemical evidence. Thus, maleic acid (1), when heated above its melting-point, readily forms an anhydride (cf. succinic acid, p. 151), which by alkaline hydrolysis and acidification of the alkaline solution is converted back into maleic acid:

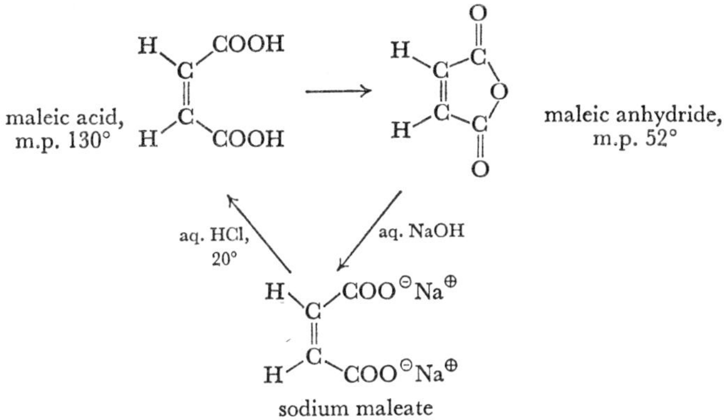

The ready formation of a cyclic anhydride suggests that maleic acid is the *cis*-isomer (1) in which two carboxyl groups are on the same side of the double bond and therefore sufficiently close for interaction. Fumaric acid is therefore the *trans*-isomer (2); it does not form an anhydride of its own, but on heating is converted partly into maleic anhydride.

Stereoisomers of Lactic Acid. Lactic acid, $CH_3CH(OH)COOH$, was isolated from sour milk in 1780 and from lean muscle in 1807. It is obtained industrially by microbial fermentation of molasses, which contain the carbohydrates lactose, sucrose and glucose. Lactic acids can be obtained in three forms which possess the following characteristics:

(*a*) lactic acid from muscle	m.p. 26°	$[\alpha]_D + 4°$
(*b*) lactic acid from glucose by fermentation with *B. acidi laevolacti*	m.p. 26°	$[\alpha]_D - 4°$
(*c*) synthetic lactic acid (from hydrolysis of acetaldehyde cyanohydrin)	m.p. 18°	$[\alpha]_D$ 0°

The three products exhibit the same chemical properties, and (*a*) and (*b*) have many identical physical properties (e.g. m.p., solubility). The principal difference between them is their behaviour towards polarised light.

The electric field component of ordinary light vibrates in all planes perpendicular to the direction of the light. This situation is illustrated in diagram (3), in which the central dot represents the axis of a light beam progressing in the direction of the observer.

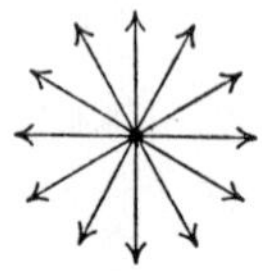

(3)

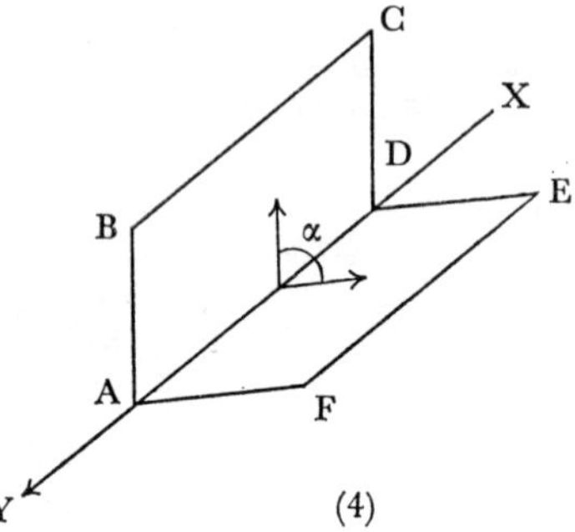

(4)

Passage through a special set of prisms (the *polariser*) divides the beam into components and the emergent ray now vibrates in a single plane, for example, the plane ABCD for a light beam of direction XY [diagram (4)]. This is described as plane polarised light, and certain organic compounds have the property of rotating the plane of polarised light. The effect is measured in a *polarimeter* in which plane polarised light passes through a tube containing a solution of the compound. As a result, the plane of vibration, ABCD [diagram 4] is rotated to a new plane, ADEF, through an angle, α. Compounds that rotate the plane to the right (clockwise) are described as *dextrorotatory* and the rotation is given a positive sign; rotation to the left (anti-clockwise) is a property of *laevorotatory* compounds, and the angle of rotation carries a negative sign. The extent of rotation is a function of the wavelength of the light, the temperature, the nature of the solvent, and the number of molecules of compound encountered by the polarised light. The rotation is often expressed as the specific rotation, $[\alpha]_D$ (D refers to the wavelength of sodium light) so that

$$[\alpha]_D = \frac{\alpha}{lC}$$

α is the measured rotation, l the length of the polarimeter tube in decimetres, and C the concentration of the compound in grams per 100 millilitres of solution.

Lactic acid (*a*) (*dextro*-lactic acid) and lactic acid (*b*) (*laevo*-

lactic acid) rotate the plane of polarised light by the same amount, but in different directions. Compounds (*a*) and (*b*) are examples of *optical isomers.*

Asymmetry. The tetrahedral structure of methane was discussed earlier (p. 15), and the molecule was represented by a three-dimensional diagram. Similar diagrams can be drawn for any molecule containing a

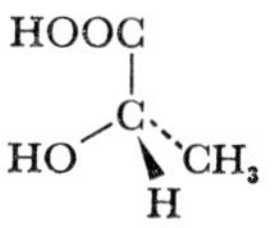

(5′)

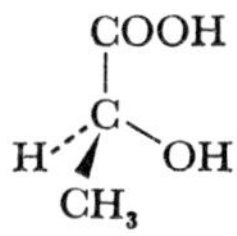

(5)

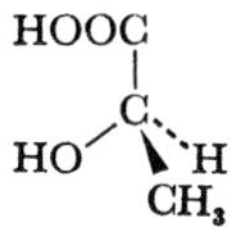

(6)

saturated (i.e. tetrahedral) carbon atom. Lactic acid, for example, contains four substituents (H, OH, CH_3, COOH) attached to a carbon atom, and can be represented as (5) or (6). The structures (5) and (6) are related as mirror images, which are *not superimposable* and are not interconvertible by rotation about carbon–carbon bonds; for example when (5) is rotated through 180° about the C–COOH axis, the resultant structure (5′) is clearly *not identical* with (6). Structures (5) and (6) represent different molecules, and are in fact the optical isomers of lactic acid. Such a *pair* of non-superimposable mirror images are called *enantiomers.*

Each of the enantiomers (5) or (6) is an example of an *asymmetric molecule* (one that does not contain a plane, centre or axis of symmetry). More generally, any molecule in which four *different* atoms or groups are linked tetrahedrally to a carbon atom is asymmetric in structure, and would be expected therefore to exhibit optical activity.

Propionic acid (7) is an example of a compound containing a tetrahedral carbon atom with only three different substituents; it cannot be obtained in enantiomeric (optically active) forms because the molecule is symmetrical (the plane passing through the methyl and carboxyl groups divides the molecule into two

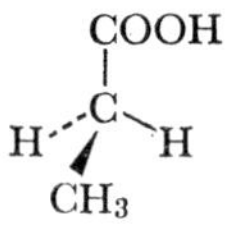

(7)

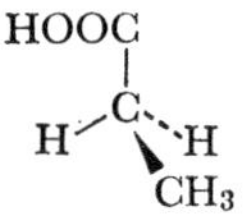

(8)

identical halves). The mirror images (7) and (8) are *identical*, and become superimposable by rotation about the C—COOH axis. These conclusions agree with the experimental observations that either (+)-lactic acid or (−)-lactic acid is reduced by hydriodic acid to optically inactive propionic acid.

$$CH_3CH(OH)COOH \xrightarrow{HI} CH_3CH_2COOH$$

As the two enantiomers of lactic acid differ only in the *sign* of their rotations, when equal quantities are mixed the product is optically inactive. This mixture is lactic acid (*c*). It has the same chemical properties as isomers (*a*) and (*b*), but differs in physical properties (e.g. m.p., solubility). Lactic acid (*c*) is called (±)-lactic acid (or *racemic*-lactic acid), and any equimolecular mixture of a pair of enantiomers is described as a *racemate*. In the synthesis of a compound Cabcd, e.g. acetaldehyde cyanohydrin from acetaldehyde, there is an equal chance of producing either enantiomer and thus the racemate is isolated. Racemates can be separated into enantiomers (by *resolution*), and enantiomers can be converted into racemates (by *racemisation*, see below), but a general discussion of the methods used is beyond the scope of this book.

Enantiomeric forms of amines of the RR′R″N type, (9) and (10), cannot be isolated because of their very rapid umbrella-like inversion (*ca.* 10^4 times/sec.).

(9)  (10)

History of Optical Isomerism. In 1815 Biot showed that certain organic compounds, e.g. sugar, camphor, were optically active in *solution*; previously this property had been observed only with crystals. Subsequently tartaric acid, HOOCCH(OH)CH(OH)COOH, was obtained in two forms, one optically active [(+)-tartaric acid] and the other inactive [(±)-tartaric acid]. The French chemist, Pasteur, discovered in 1848 that the ammonium sodium salt of (±)-tartaric acid existed in two kinds of crystals, differing slightly in the arrangement of crystal faces. These were separated by hand and converted into (+)- and (−)-tartaric acids; this was the first resolution of a racemate. Pasteur noticed that the crystals of the salts of (+)- and (−)-tartaric acids were related as mirror-images, and suggested that each constituent molecule of tartaric acid was also asymmetric and that the optical isomers differed in the arrangement of atoms in space. This remarkable conclusion was reached before the development of the structural theory of organic compounds.

Wislicenus provided support for Pasteur's theory by proving (1863–73) that (±)-lactic acid and (+)-lactic acid were *not structural* isomers. In 1874 van't Hoff and Le Bel realised independently that optical isomerism was inconsistent with a planar carbon atom; the tetrahedral arrangement that they suggested led to a complete understanding of the phenomenon.

Optical Inversion. A consequence of the tetrahedral structure of four-bonded carbon is that a stereochemical change occurs during the nucleophilic substitution reactions of suitably substituted alkyl halides. When, for example, bromide displaces bromide from a molecule of *one* enantiomer of general formula, RR′CHBr, a molecule of the other enantiomer is formed; stereochemical inversion at the carbon atom has occurred:

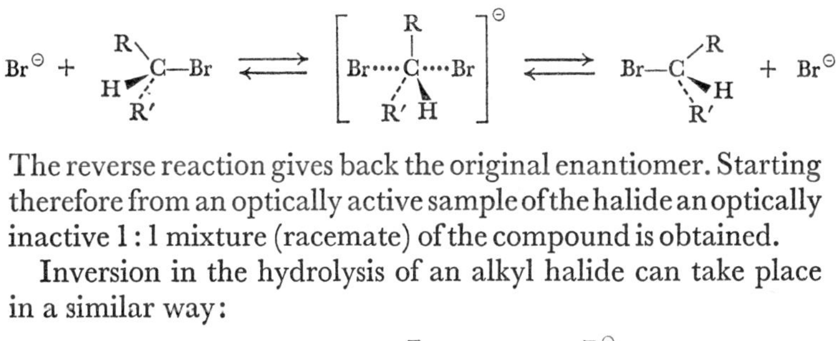

The reverse reaction gives back the original enantiomer. Starting therefore from an optically active sample of the halide an optically inactive 1:1 mixture (racemate) of the compound is obtained.

Inversion in the hydrolysis of an alkyl halide can take place in a similar way:

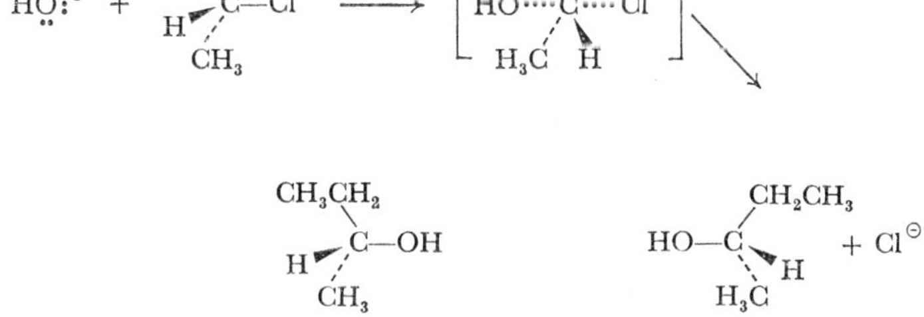

(11) (12)

In this case, as the product is structurally different from the starting material, it is not so easy to show that inversion has occurred, i.e. that the alcohol has the *absolute configuration* (12) rather than (11).

Historically, the demonstration that S_N2 reactions proceed with inversion provided powerful evidence for their general mechanism, showing that the attacking and leaving groups must be on opposite sides of the carbon at which reaction takes place.

AMINO-ACIDS AND POLYAMIDES

Amino-acids. The class name, amino-acid, embraces compounds containing both amino-groups (primary, secondary and tertiary) and carboxyl groups, but this account is confined to α-amino-acids; in these compounds, a carboxyl group and a primary amino-group are attached to the same carbon atom. Systematic nomenclature is analogous to that of hydroxy-acids, but many α-amino-acids have specific names:

NH_2CH_2COOH	glycine	(aminoacetic acid)
$CH_3CH(NH_2)COOH$	alanine	(2-aminopropanoic acid)
$NH_2CH_2CH_2COOH$		3-aminopropanoic acid or β-aminopropionic acid
$(CH_3)_2CHCH_2CH(NH_2)COOH$	leucine	(2-amino-4-methylpentanoic acid)
$C_6H_5CH_2CH(NH_2)COOH$	phenylalanine (p. 224)	
$HOOCCH_2CH_2CH(NH_2)COOH$	glutamic acid	
$NH_2CH_2CH_2CH_2CH_2CH(NH_2)COOH$	lysine	

Amino-acids exist in the solid state and in solution as internal salts, the acidic carboxyl group losing a proton to the basic amino-group:

$$NH_2CH_2COOH \rightleftarrows \overset{\oplus}{N}H_3CH_2COO^{\ominus}$$

glycine

Dipole moments support this ionic structure: glycine for example has a large dipole moment (13 D), whereas its ethyl ester, $NH_2CH_2COOCH_2CH_3$ which cannot be ionic, has a much lower value (2·1 D). Amino-acids, like other electrovalent compounds, are high-melting crystalline solids, soluble in water and insoluble in most organic solvents.

Amino-acids form salts with acids and bases:

$$\overset{\ominus}{Cl}\,\overset{\oplus}{N}H_3CH_2COOH \xleftarrow{HCl} \overset{\oplus}{N}H_3CH_2COO^{\ominus} \xrightarrow{NaOH} NH_2CH_2COO^{\ominus}\,Na^{\oplus}$$

glycine hydrochloride

The *isoelectric point*, at which the concentration of the dipolar form is at a maximum, falls within the range 4·8–6·3 for most α-amino-acids. Glutamic acid, however, is more acidic (isoelectric point 3·2) because it contains *two* carboxyl groups to one amino group, and lysine with *two* amino groups and one carboxyl group is basic (isoelectric point 9·7).

The first amino-acid to be discovered was glycine. The names glycine and glucose are derived from the same Greek word *glykys* (sweet) because, although unrelated structurally, both compounds have a sweet taste.

α-Amino-acids are of the greatest biological importance, as they are used by living organisms to elaborate proteins (p. 160), respiratory pigments and other essential organic compounds. Higher animals are unable to synthesise all the amino-acids they require (unlike plants), and certain of these compounds, called *essential amino-acids*, are acquired in food.

All the α-amino-acids listed above, except glycine, are asymmetric and are optically active when isolated from natural sources. Alanine, for example, contains a carbon atom with four different substituents; only one of its enantiomers [(13); $R=CH_3$], $[\alpha]_D + 14{\cdot}7°$, is obtained by hydrolysis of proteins. α-Amino-acids from proteins have the absolute stereochemistry represented in structure (13), and (+)-lactic acid present in muscle has the similar configuration (14).

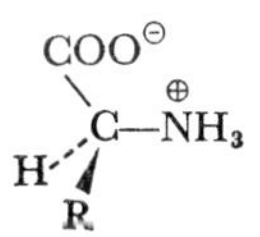

(13)

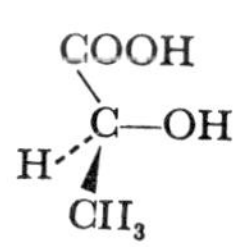

(14)

The amino-acid, ethylenediamine tetraacetic acid (EDTA, 15) forms exceptionally stable salts with heavy metals. It is consequently an important analytical reagent.

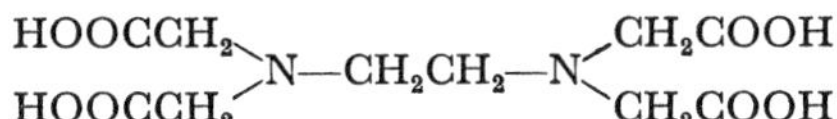

(15)

The calcium salt of EDTA is an antidote for lead poisoning, the lead salt being formed preferentially and then excreted.

Proteins. Proteins are derived from amino-acids, and have a great variety of essential biological functions. Hair is composed of the fibrous protein, keratin; more soluble proteins include insulin (a hormone secreted by the pancreas and required for carbohydrate metabolism) and the digestive enzyme, pepsin. Proteins are often associated with other types of compound; in haemoglobin, which transports oxygen in the blood stream, the protein globin, is joined to the iron-containing organic compound, haem (p. 228).

Proteins have molecular weights ranging from 12,000 to several millions, and are long-chain polyamides constructed from α-amino-acid units:

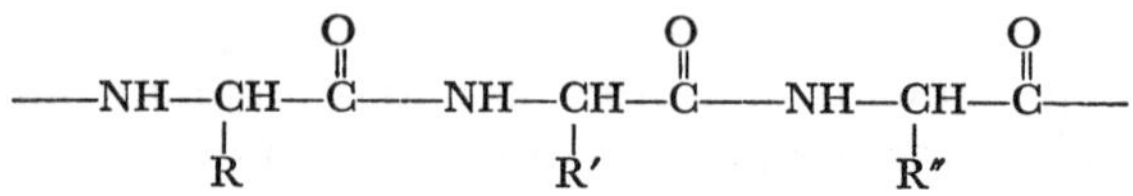

(Long CO—NH bonds are drawn in this structure and in other protein formulae to emphasise the amino-acid units.)

Each amide linkage is formed from the amino-group of one amino-acid molecule and the carboxyl group of another. Twenty α-amino-acids (some of which are listed on p. 158) commonly occur in proteins.

The diagram (16) represents a part of the insulin structure, and shows how one end of the molecule is derived from four different amino-acids.

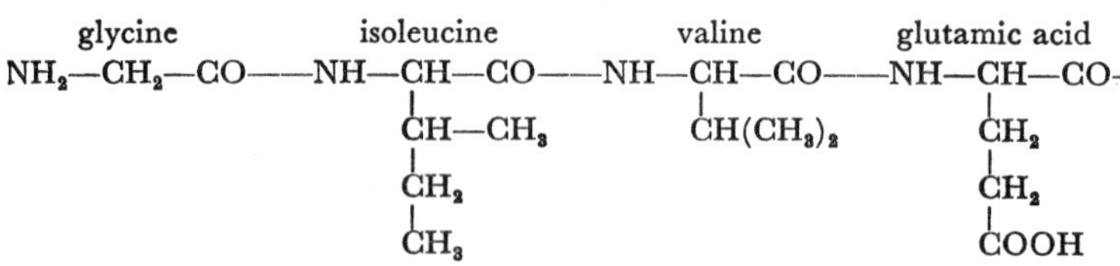

(16)

Proteins contain amino and carboxyl groups situated either at the end of the polyamide chain or in side-groups derived from basic and acidic amino-acids (e.g. lysine, glutamic acid). Because of the presence of these groups, proteins, like amino-acids, are ionic compounds with specific isoelectric points (p. 159). Proteins possessing a high proportion of amino-groups (derived, for example, from lysine) are basic and have high isoelectric points. Correspondingly, acidic proteins contain a preponderance of carboxyl, thiol or phenolic groups.

Peptides are similar in structure to proteins, but are composed of fewer amino-acid units. Oxytocin (17), a hormone that stimulates the muscles of the uterus, is a cyclic peptide derived from eight amino-acid molecules (i.e. it is a cyclic octapeptide). Oxytocin was synthesised in 1954 by du Vigneaud.

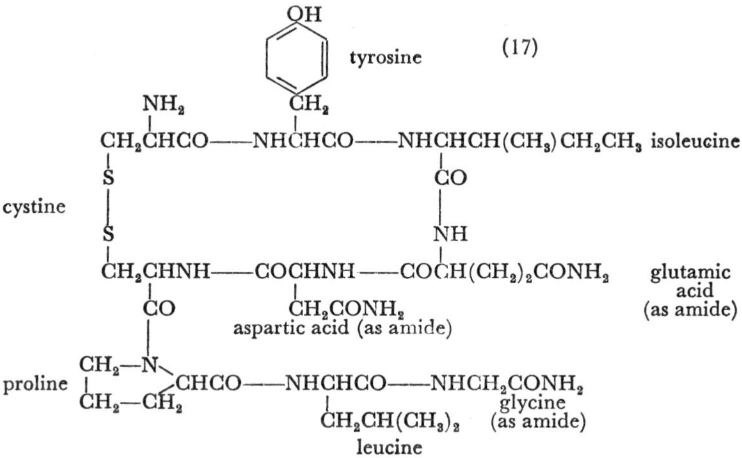

(17)

The elucidation of the structures of proteins is a formidable task, and has been accomplished only with a few of the simpler compounds. Bovine insulin, for example, has a comparatively low molecular weight (5734), and its detailed structure, involving 51 amino-acid units (cystine = 2 units, cf. p. 50) has been established (Sanger, 1955). Bovine and human insulin have slightly different structures, and both have been synthesised (1966).

The constitution of a protein can be investigated by acid hydrolysis. This process splits the protein chain at the amide group, affording α-amino-acids, which can be separated by ion exchange chromatography. The protein hydrolysate is added to a column of an acidic resin whereby more basic components are retained more firmly. The amino-acids are eluted by a buffer solution in an order that is characteristic of individual compounds; glutamic acid, for example, precedes the basic amino-acid, lysine. Each fraction is treated with ninhydrin reagent, which gives a colour test with amino-acids; the intensity of the colour is proportional to the concentration of the amino-acid. Equipment has been devised in which separation, collection of fractions and estimation of amino-acid concentrations is carried out automatically.

The second, and much more difficult stage of the investigation is the determination of the amino-acid *sequence*, in the protein. One useful procedure is partial hydrolysis, often achieved enzymically, to give peptides containing two or more amino-acid units. Many of these experiments may be necessary before a complete structure for a protein can be suggested. Thus, partial hydrolysis of insulin might give the tripeptide (18), but its identification would only show that the sequence isoleucine-valine-glutamic acid, cf. (16), occurs somewhere in the molecule, and not necessarily at one end of the chain.

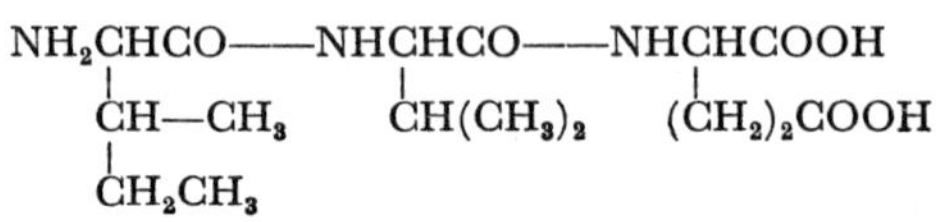

(18)

Other aspects of protein structure have to be considered, for example the conformation of the amino-acid units in the polypeptide chain (*secondary* structure) and the shape of the total molecule (*tertiary* structure). A secondary structure found in many fibrous proteins is the α-helix (19); as the diagram shows,

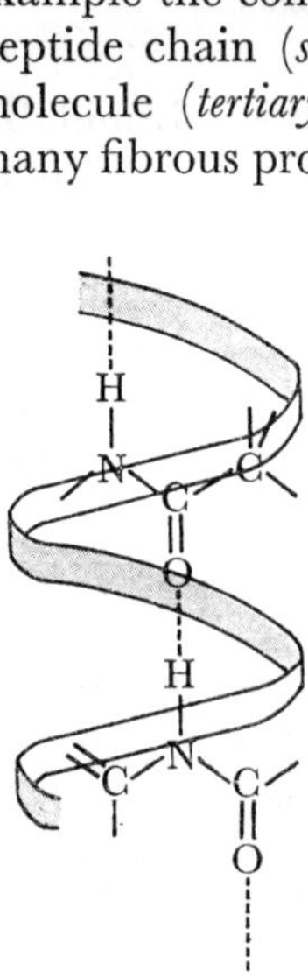

(19)

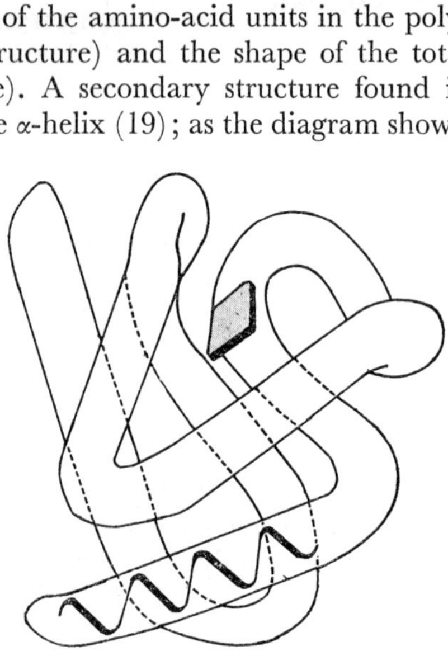
(20)

a hydrogen bond is formed between the NH group of one amino-acid residue, and the carbonyl group of a unit in an adjacent spiral of the helix. Advanced X-ray diffraction methods have led to the elucidation of the tertiary structures of some complex crystalline proteins. Myoglobin (20), a protein that acts as a temporary storehouse for oxygen in blood, consists of a single convoluted chain of 153 amino-acid units joined to the iron-derivative, haem (shown as a solid disc in the diagram); also illustrated is one of the helical segments of the chain. The poly-peptide compound of haemoglobin (*ca.* 600 amino-acid units) is similar, but is composed of two pairs of folded units.

Proteins are usually sensitive to heat. In the hardening or *denaturation* of egg white the polyamide molecules are converted from specific, hydrogen-bonded conformations into less soluble intermolecularly bonded forms.

Sulphur-containing Amino-acids. Cystine (p. 50) can be obtained by hydrolysis of the hair protein, keratin.

Human hair is heated with 25% hydrochloric acid at *ca.* 95° for 150 hours. The liquid is filtered from a small amount of undissolved material, and neutralised with a mixture of sodium acetate and hydroxide; crude cystine then separates slowly. This material is dissolved in 3% hydrochloric acid and the solution is heated several times with fresh portions of decolourising charcoal until most of the brown colour has gone. The purified amino-acid is reprecipitated by adding sodium acetate to the solution. The yield of cystine is 3–5% of the weight of the hair.

A cystine molecule can span two polyamide chains, which are thereby held together by disulphide bonds. Sections of the insulin molecule are connected in this way, and the two sulphur atoms of cystine form an integral part of the 20-membered ring of oxytocin (17).

Hair-curling lotion contains a reducing agent which severs some of the cystine disulphide bonds of the hair protein, keratin, and permits reorientation of the polyamide chains (curling). The new arrangement is made 'permanent' by the application of an oxidising agent (setting lotion), which converts thiol groups into disulphide cross-links.

Various proteins contain different proportions of cystine,

cysteine (p. 50), and methionine (p. 7). The biosynthetic relation of the latter two compounds is as follows:

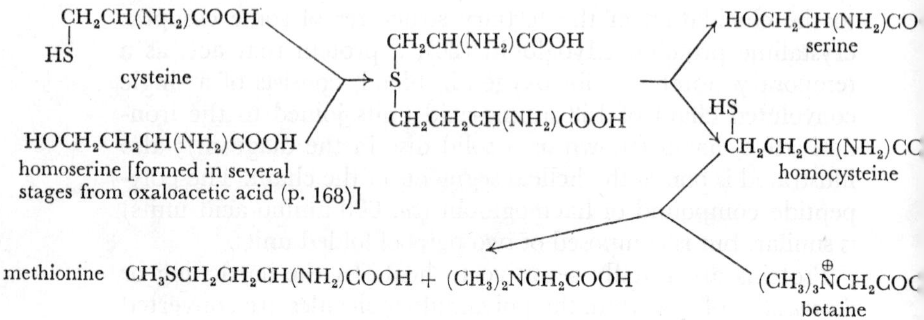

Apart from being a building unit in proteins, methionine functions as a reservoir of methyl groups for some biological methylation reactions. Methionine is synthesised on a moderate scale for use as a poultry feed supplement (cf. p. 249).

Synthetic Polyamides. α-Amino-acids often char on being heated, but under certain conditions a polyamide (a 'proteinoid') can be formed by elimination of water between molecules. This reaction may be a simplified version of one of those responsible for the origin of proteins, and hence life, on this planet.

In the important class of man-made fibre known as *nylon*, the amide groups are separated by several methylene groups. For example, nylon 6.6, first prepared by Carothers in 1935, has the structure

$$-\overset{O}{\overset{\|}{C}}-(CH_2)_4-\overset{O}{\overset{\|}{C}}-NH-(CH_2)_6-NH-\overset{O}{\overset{\|}{C}}-(CH_2)_4-\overset{O}{\overset{\|}{C}}-NH-(CH_2)_6NH-$$

One route for its preparation is:

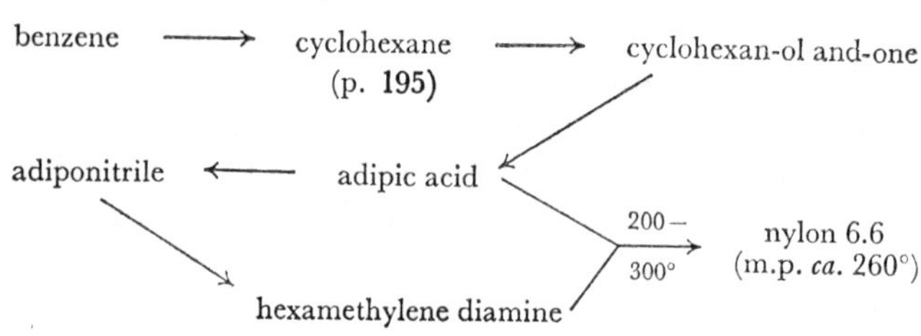

An alternative route to adiponitrile (thence the diacid and diamine) is based on the two recently discovered reactions:

$$\text{propylene} \longrightarrow \text{acrylonitrile (p. 144)} \longrightarrow \text{adiponitrile}$$

The second stage involves an electrochemical reduction:

$$2CH_2{=}CHCN + 2e + 2H^{\oplus} \longrightarrow NC(CH_2)_4CN$$

Nylon 6, m.p. *ca.* 215°, contains NH and CO groups in each of the repeating units:

$$[\text{-}NH(CH_2)_5CO\text{-}]_n$$

Fibres of nylon, obtained by drawing the melted polymer into filaments, are characterised by their strength and elasticity. The cost of nylon is being reduced by larger scale production.

A laboratory procedure for making nylon 6.10 involves a 'rope-trick':

A solution of 2·2 g. of hexamethylene diamine in 25 ml. water is added very carefully to a solution of 1·5 g. of sebacoyl chloride $ClCO(CH_2)_8COCl$, in 50 ml. carbon tetrachloride contained in a tall beaker. The film of polymer produced at the interface between the two liquids can be pulled out using tweezers. When it is removed continuously a rope of polymer is obtained, until the reactants are used up. The product is washed with ethanol–water (1:1 mixture) and dried at 50°; it has m.p. *ca.* 215°.

CARBONIC ACID AND ITS DERIVATIVES

Carbonic acid (21) is present in a solution of carbon dioxide in water, but cannot be isolated; it is related to dicarboxylic acids as it gives mono- and di-salts, e.g. sodium bicarbonate (22), sodium carbonate (23), and forms acid derivatives of general structure (24), in which X and Y (which may be the same) are Cl, NH_2, OR, etc.

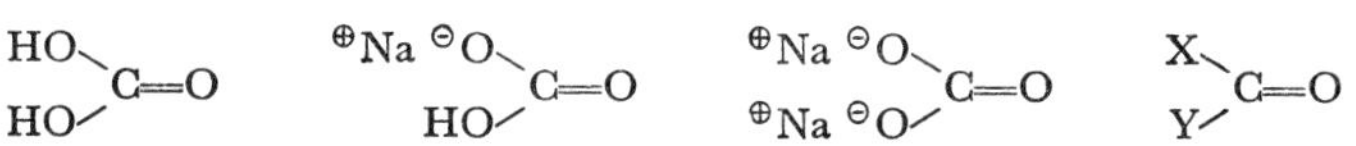

(21) (22) (23) (24)

Phosgene (carbonyl chloride), b.p. 8°, is the acyl chloride of carbonic acid. The reaction of phosgene with an alcohol yields first an *alkyl chloroformate* (named because of the structural relationship to alkyl formates, HCOOR) and then a *dialkyl carbonate*:

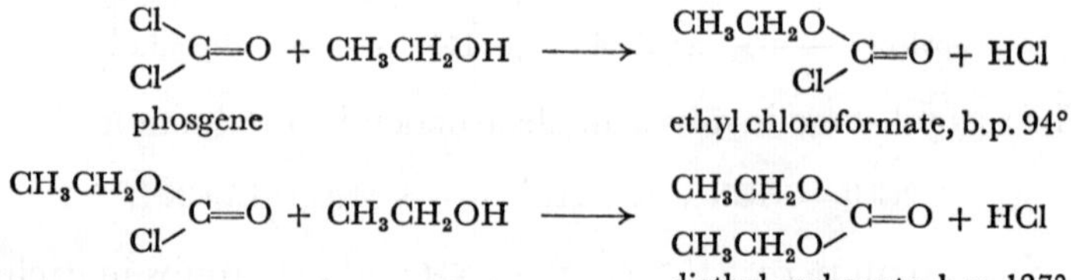

phosgene; ethyl chloroformate, b.p. 94°; diethyl carbonate, b.p. 127°

Unsymmetrical dialkyl carbonates can be obtained from alkyl chloroformates, for example, by employing an alcohol other than ethanol in the above reaction.

Urea, $CO(NH_2)_2$, is the diamide of carbonic acid, and is hydrolysed by heating with acid or alkali:

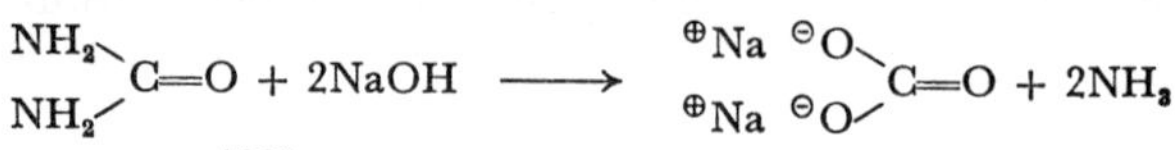

urea, m.p. 132°

Urea is more basic than other amides, and gives stable *mono*-salts with strong acids.

Urea is produced in the breakdown of proteins, and is the major organic component of urine (28–30 g. of urea may be excreted each day). Urea is used primarily as a fertiliser, but is also required for the preparation of barbiturates (sleeping tablets) and formaldehyde–urea plastics. It is obtained industrially from carbon dioxide and ammonia:

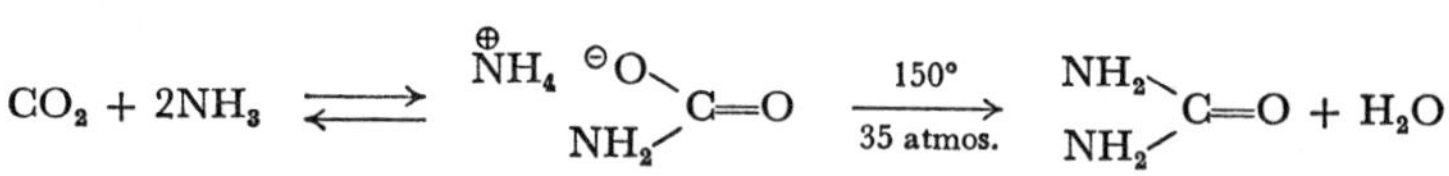

The initial addition product, ammonium carbamate, is the ammonium salt of the half-amide of carbonic acid and furnishes urea when heated under pressure.

Crystalline *inclusion* products may be formed from urea and many compounds containing an unbranched chain of carbon atoms by mixing concentrated solutions of the components in methanol. Chemical bonding is not involved; the compounds occupy channels in the crystal lattice of urea, and may be recovered by extraction with ether. By this means straight-chain hydrocarbons may be separated from hydrocarbons containing a branched chain of carbon atoms.

KETO-ACIDS, DIKETONES AND RELATED COMPOUNDS

Keto-acids and other compounds containing two carbonyl groups are often classified (using the symbols α, β, γ, etc.) according to the relative positions of the carbonyl groups:

$CH_3COCOOH$	pyruvic acid	(an α-keto-acid)
CH_3COCH_2COOH	acetoacetic acid	(a β-keto-acid)
$\mathrm{CHO{-}CHO}$	glyoxal	(the only α-dialdehyde)
$CH_3COCH_2COCH_3$	acetylacetone or pentane-2,4-dione	(a β-diketone; but see below)
$CH_3COCH_2CH_2COCH_3$	acetonylacetone or hexane-2,5-dione	(a γ-diketone)

In the respiration of normal cells, ATP is produced with the aid of energy provided by the oxidation of organic compounds, e.g. carbohydrates. The final stage of the whole process, the oxidation of acetic acid to carbon dioxide, involves the cycle of enzyme-catalysed reactions outlined on p. 168. The route from carbohydrates to acetic acid proceeds *via* pyruvic acid, and three other keto-acids are components in the cycle.

Some features of the chemistry of ethyl acetoacetate, $CH_3COCH_2COOCH_2CH_3$, b.p. 181°, illustrate how the properties of a keto-group may be modified by the presence of a β-carbonyl group (e.g. in an ester). As expected for a keto-ester, ethyl acetoacetate gives a semicarbazone,

$$CH_3C({=}NNHCONH_2)CH_2COOC_2H_5$$

m.p. 129°, in high yield, and its infrared spectrum shows strong absorption at 1724 ($>$C=O) and at 1748 cm.$^{-1}$ ($-COOC_2H_5$). On the other hand, there is a weak infrared band at 3300 cm.$^{-1}$ that could arise from a hydrogen bond between a hydroxyl and an ester group. Furthermore, when the ester is treated with bromine at 0° *ca.* 7 % of it reacts extremely rapidly to give the bromo-keto-ester, $CH_3COCHBrCOOC_2H_5 + HBr$. All this evidence suggests that ethyl acetoacetate exists as a mixture of two isomers:

$$\underset{\textit{keto}\text{-form (93\%)}}{CH_3{-}\overset{\displaystyle O}{\overset{\|}{C}}{-}CH_2{-}COOCH_2CH_3} \rightleftharpoons \underset{\textit{enol}\text{ form (7\%)}}{CH_3{-}\overset{\displaystyle OH}{\overset{|}{C}}{=}CH{-}COOCH_2CH_3}$$

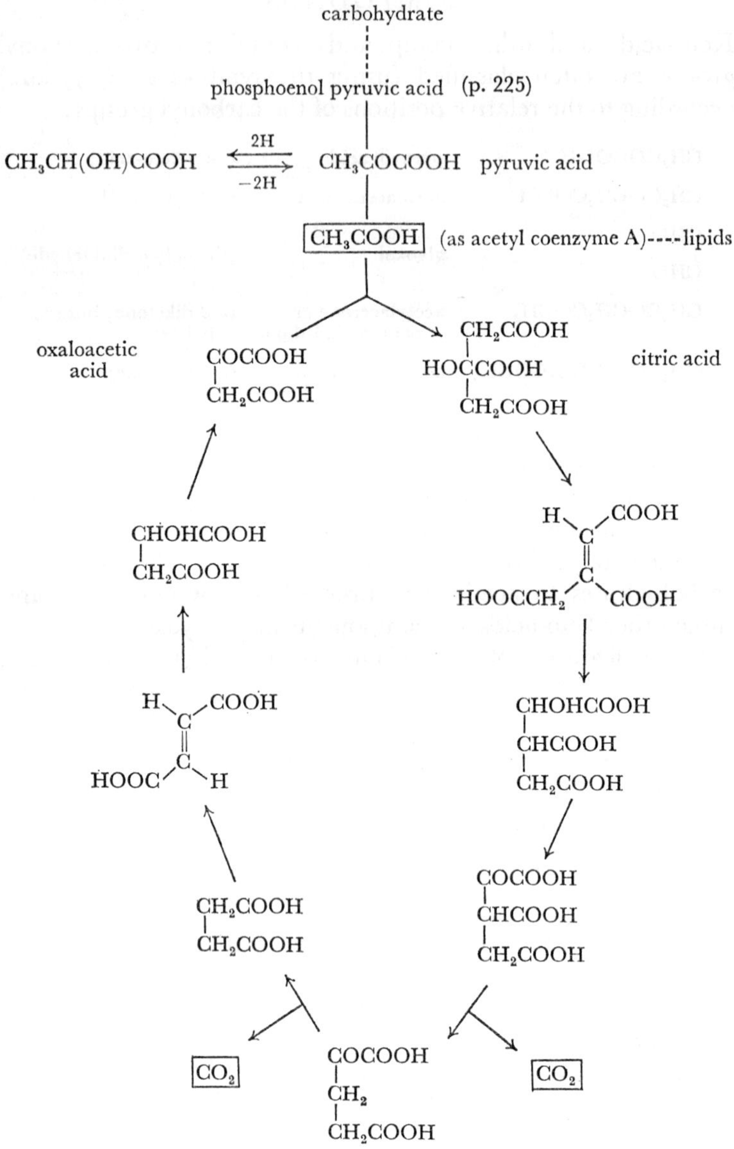
carbohydrate
phosphoenol pyruvic acid (p. 225)
CH₃CH(OH)COOH
2H
−2H
CH₃COCOOH pyruvic acid
CH₃COOH (as acetyl coenzyme A)----lipids
oxaloacetic acid
COCOOH
CH₂COOH
CH₂COOH
HOCCOOH
CH₂COOH
citric acid
H COOH
C
C
HOOCCH₂ COOH
CHOHCOOH
CH₂COOH
CHOHCOOH
CHCOOH
CH₂COOH
H COOH
C
C
HOOC H
COCOOH
CHCOOH
CH₂COOH
CH₂COOH
CH₂COOH
CO₂
COCOOH
CH₂
CH₂COOH
CO₂

Only the enol isomer reacts rapidly with bromine (p. 81), but when equimolar proportions of ethyl acetoacetate and the halogen are kept for a longer time a high yield of the bromo-derivative, $CH_3COCHBrCOOC_2H_5$ is obtained, indicating that the keto-form isomerises to the enol form, which is then brominated.

Acetylacetone contains even less of the keto-form (15% at 25°). Metal derivatives containing a cyclic structure related to that of the enolform are readily prepared from this compound: the blue, chloroform-soluble copper (II) derivative separates when aqueous solutions of acetylacetone and copper (II) acetate are mixed.

QUESTIONS

1. Which of the following molecules may exist in enantiomeric forms?

(*a*) $(CH_3CH_2)_2CHCOOH$ (*b*) $CH_3CH(CH_3)CH_2NH_2$

(*c*) $ClCH_2CH(CH_3)CH_2CH_2CH_3$ (*d*) $(CH_3)_2CHCH_2CH_2N(CH_3)_2$

(*e*) $ClCH_2CH(CH_3)CH_2F$ (*f*) $CH_3OOC\underset{\displaystyle |\atop \displaystyle CH_2CH_2OH}{C}(CH_3)COOCH_3$

2. Give structural formulae for:
dimethyl oxalate, adipic diamide, methylsuccinic anhydride, di-n-propyl carbamate, thiourea, and ethyl pyruvate.

3. What compounds would be produced on complete hydrolysis of the following peptide?

$$(CH_3)_2CHCH_2\underset{\displaystyle |\atop \displaystyle NH_2}{C}HCONHCH_2CONH\underset{\displaystyle |\atop \displaystyle (CH_2)_4NH_2}{C}HCOOH$$

Which of the products might exhibit optical activity?

4. (*a*) Ricinoleic acid (p. 107) occurs naturally in an optically active form. How many stereoisomeric forms are possible in a molecule with the structure, $CH_3(CH_2)_5CH(OH)CH_2CH{=}CH(CH_2)_7COOH$?

(*b*) One of the earliest emulsifying agents was made by treating castor oil (glyceryl triricinoleate) with sulphuric acid. Draw the structural formula of the product, $C_{57}H_{104}O_{18}S_3$.

5. (*a*) Would you expect the entropy of activation for the conversion of 5-hydroxypentanoic acid into δ-valerolactone to be positive or negative in sign?

(*b*) Write a formula for the polymer that is formed when δ-valerolactone is heated. The lactone and the polymer have the same elemental composition.

6. In what isolable stereochemical forms can the following compounds exist? Give diagrams that clearly represent individual structures: 1,2-dibromohexane; 2,5-dichlorohexane; 1,4-dichloropenta-1,4-diene; 1,5-dichloropenta-1,4-diene; 1,3-dichloropenta-1,4-diene.

7. Interpret the following observations:
Evaporation (on a steam bath) of a sample of urine to a small bulk gave a syrup which on treatment with cold nitric acid, afforded a colourless, crystalline solid containing 9·75% carbon, 4·1% hydrogen and 34·15% nitrogen by analysis. When a sample of this product was heated with an excess of aqueous sodium hydroxide, ammonia was slowly evolved.

8. (*a*) Which polymer would you expect to be less dense, nylon 6.6 or nylon 6.10?

(*b*) What products will be formed on complete hydrolysis of each of these polymers, and how could pure samples be isolated and identified?

9. Suggest reagents for the last three steps in the following synthesis:

buta-1,3-diene $\xrightarrow{Cl_2}$ 1,4-dichlorobut-2-ene $\longrightarrow$ 1,4-dicyanobut-2-ene $\longrightarrow$ adiponitrile $\longrightarrow$ adipic acid

10. (*a*) A molecule of 1,2-dichloroethane can be represented by the diagram

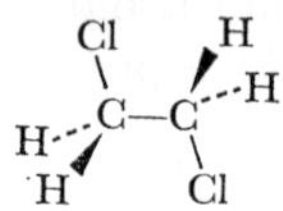

Explain why, although this structure does not contain a plane of symmetry, the compound has not been obtained in an optically active state.

(*b*) What is the likely structure of the yellow rubbery material (containing no chlorine) that separates when 1,2-dichloroethane is slowly added to a stirred solution of sodium tetrasulphide, Na_2S_4, in water at 70°?

11. What method could be used to find out which is the longer molecule, cyanogen $(CN)_2$ or diacetylene $(C{\equiv}CH)_2$? Both compounds are gases at ordinary temperatures.

12. Identify the compounds obtained in the following ways.

(*a*) The protein gelatine is heated with concentrated hydrochloric acid for 4 hours, and the solution is then evaporated to remove water. The residue is dissolved in ethanol, saturated with hydrogen chloride, and then refluxed for 30 min. On cooling, colourless needles separate: 34·4% carbon; 7·2% hydrogen; 10·0% nitrogen; 25·4% chlorine; remainder oxygen. The compound gives an immediate precipitate (AgCl) with silver nitrate in dilute nitric acid.

(*b*) Ammonium thiocyanate, NH_4CNS, is heated at 140° for 6 hours. Water dissolves the unchanged salt, leaving a sparingly soluble isomeric compound (m.p. 170°).

All opposition is good for you.

BARNES WALLIS (1887–   )

CHAPTER 14

CARBOHYDRATES

Cellulose and starch, the sugars sucrose, glucose and fructose, and many other compounds of natural origin are polyhydroxy substances containing potential aldehyde and ketone functions. They are called *carbohydrates* (i.e. 'hydrates of carbon') because the general formula $(CH_2O)_n$ was thought at one time to apply to all members of the group.

The simplest carbohydrates (*monosaccharides*, e.g. glucose) usually contain 3–7 carbon atoms and are not broken down by hydrolysis. *Disaccharides*, e.g. sucrose, are composed of two monosaccharide units, and *polysaccharides*, e.g. cellulose, are polymers derived from many molecules of monosaccharides.

Structure and Reactions of Glucose. Glucose (1) is a cyclic compound, the six-membered ring being constructed from five carbon atoms and an oxygen atom. Each ring carbon is substituted, four with hydroxyl groups and one with a $-CH_2OH$ group. Although glucose does not contain a formyl group (–CHO)

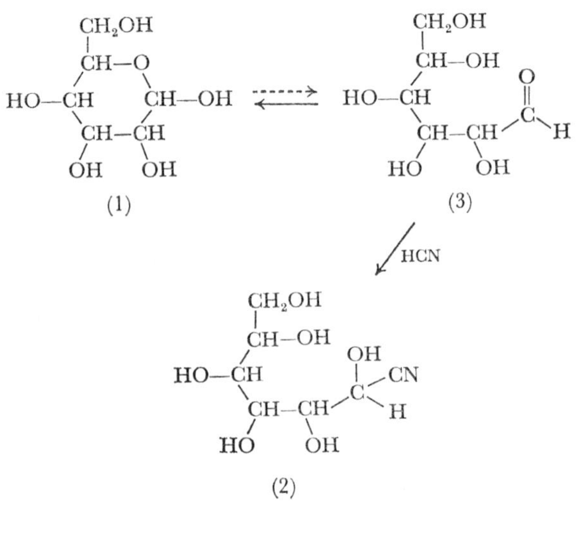

(1) (3) (2)

it has the characteristic reactions of an aldehyde and with hydrogen cyanide, for example, affords the cyanohydrin (2). Apparently glucose is in equilibrium with a trace of the aldehyde (3) which reacts irreversibly with hydrogen cyanide. In fact for many years glucose was believed to have structure (3), but it is now known that in water at least 99·9% glucose is present in the cyclic form (1).

When aldehydes are treated with alcohols, *hemiacetals* are formed in solution, but cannot usually be isolated (cf. hydration, p. 76). In the presence of an acid catalyst hemiacetals also react with a molecule of an alcohol to yield *acetals*. Acetals are stable under neutral or basic conditions, but are hydrolysed with acid to the corresponding aldehydes:

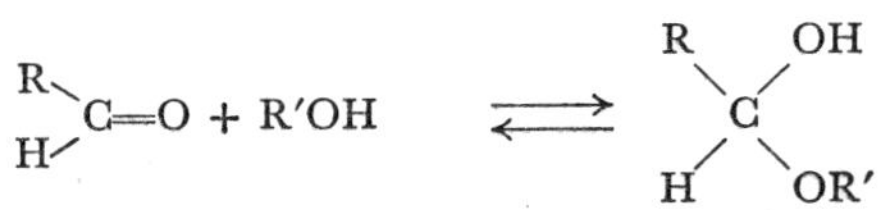

a hemiacetal

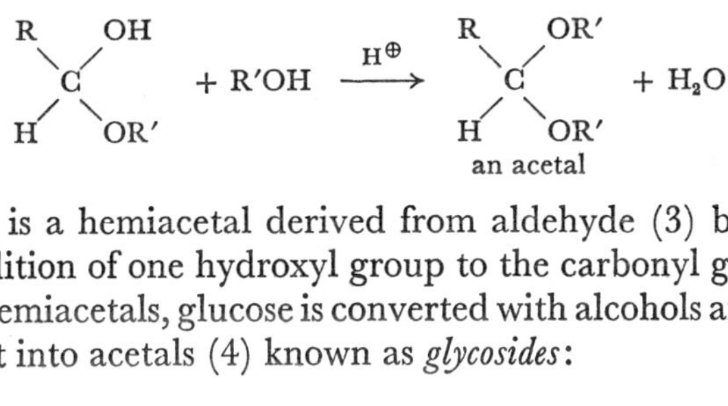

an acetal

Glucose (1) is a hemiacetal derived from aldehyde (3) by the internal addition of one hydroxyl group to the carbonyl group. Like other hemiacetals, glucose is converted with alcohols and an acid catalyst into acetals (4) known as *glycosides*:

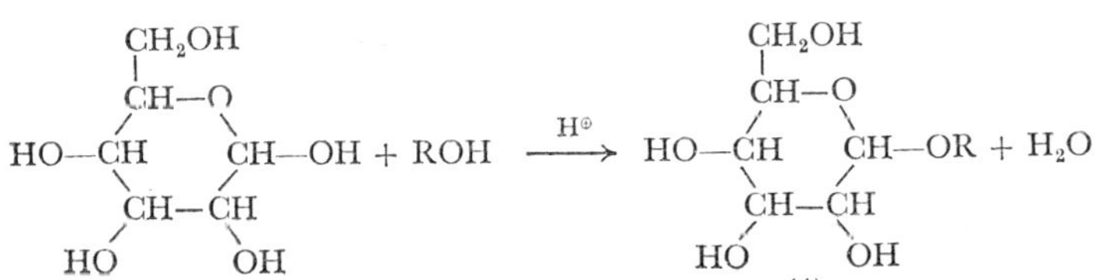

(4)

Natural glycosides are carbohydrate derivatives of complex alcohols, phenols or other hydroxy-compounds (cf. amygdalin, p. 214). Glycosides, in accord with their acetal structure, are unaffected by hydrogen cyanide, Fehling's solution or other reagents which attack aldehydes, but are hydrolysed to monosaccharides when heated with aqueous acid.

Stereochemistry of Glucose. Formula (1) does not represent adequately the structure of glucose, because there are five centres of asymmetry in the molecule (at the five-ring carbon atoms) and hence many possible stereoisomers. The stereochemistry of glucose is best understood by first considering a simple cyclic compound such as cyclohexane-1,4-diol (5).

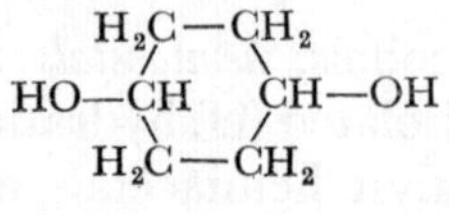

(5)

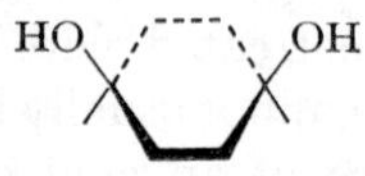

(6) *cis-*

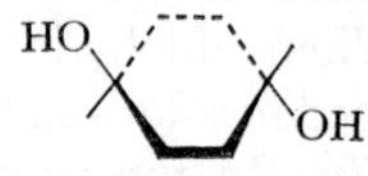

(7) *trans-*

Although the cyclohexane ring is in fact non-planar, it will be regarded in this discussion as a flat rigid structure, as depicted in diagrams (6) and (7); the thick lines represent the part of the ring that projects from the plane of the paper. There are two stereoisomers of cyclohexane-1,4-diol, the *cis*-isomer (6) in which the two hydroxyl groups are on one side of the plane of the ring, and the *trans*-isomer (7) in which the hydroxyl groups are on opposite sides (cf. p. 58). Each stereoisomer is symmetrical (e.g. about the plane passing through the hydroxyl groups) and therefore cannot be optically active (p. 155).

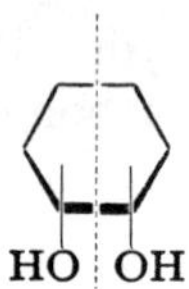

(8) *cis-*

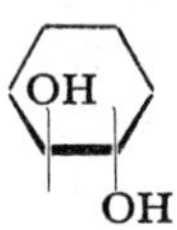

(9)

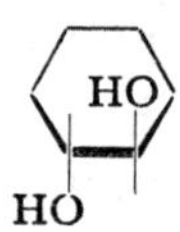

(10)

trans-

There are three stereoisomers (8, 9, 10) of cyclohexane-1,2-diol. The *cis*-isomer (8) is optically inactive because it is symmetrical about the plane indicated by the dotted line. When the hydroxyl groups are placed on opposite sides of the ring no plane of symmetry exists, and two arrangements, (9) and (10), are possible. These structures are related as mirror images and are not superimposable; they represent the two enantiomers of *trans*-cyclohexane-1,2-diol, which, like other enantiomers, have specific rotations of equal magnitude but of opposite sign.

The same stereochemical principles apply to monosaccharides of structure (1), but here the number of stereoisomers is enormously increased because there are *five* substituents in the ring. Glucose has structure (11), the arrangement of the hydrogen atom and hydroxyl group at the starred carbon atom remaining for the moment unspecified. Clearly structure (11) is asymmetric

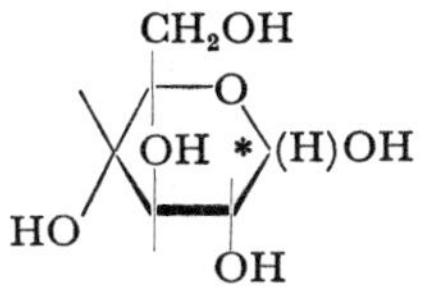

(11)

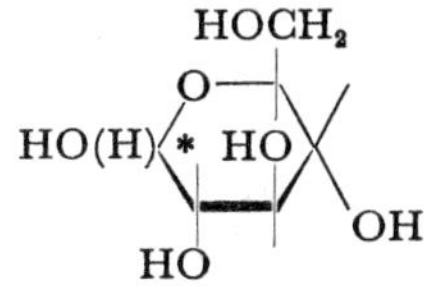

(12)

and is not superimposable on its mirror image (12); compounds (11) and (12) are enantiomers, and natural glucose is the *dextro* isomer, with the absolute stereochemistry as represented in structure (11). Certain other monosaccharides of natural origin, e.g. mannose and galactose, are stereoisomers of glucose:

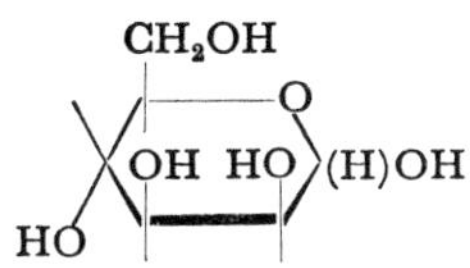

mannose

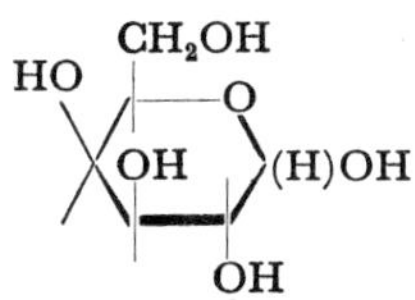

galactose

Crystallisation of natural glucose from methanol affords α-glucose, $[\alpha]_D$ +112°, but when acetic acid is used as a solvent β-glucose, $[\alpha]_D$ +19°, is obtained. In aqueous solution α- and β-glucose slowly come to equilibrium, the specific rotation of each isomer reaching the equilibrium value of +52°. α-Glucose (13) and β-glucose (14) are stereoisomers differing in the arrangement of the hydroxyl group at *C; this is the carbon atom that is part of the potential aldehyde group, and the equilibration of the isomers occurs *via* the aldehyde (15) (cf. structure 9). Rotation about the C-carbonyl bond in aldehyde

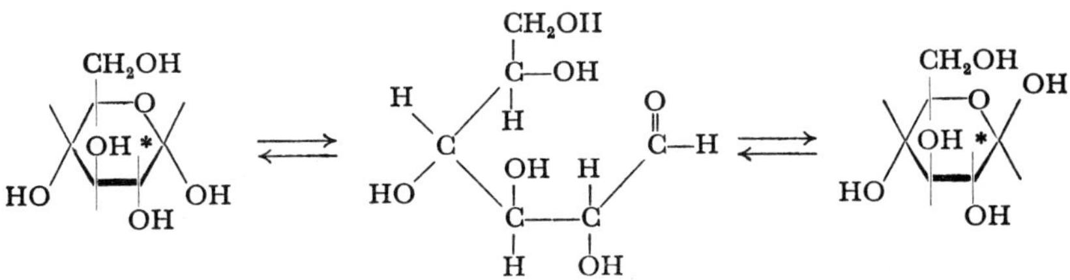

(13) α-glucose $[\alpha]_D$ + 112° (15) (14) β-glucose $[\alpha]_D$ + 19°

(15) does not affect the arrangement at the asymmetric centres, and union of hydroxyl and carbonyl groups can occur in two ways, (15) → (13) or (15) → (14).

The ready interconversion of α- and β-monosaccharides is characteristic of the hemiacetal structure, and can be followed by the change in optical rotation ('mutarotation'). Glycosides also exist as α- and β-isomers but, because of the presence of the stable acetal group, isomerisation does not occur easily; most natural glycosides, e.g. amygdalin (p. 214), have the β-arrangement.

Carbohydrates are more accurately represented by conformational structures. The diagram below, for example, depicts the most stable conformation of α-glucose.

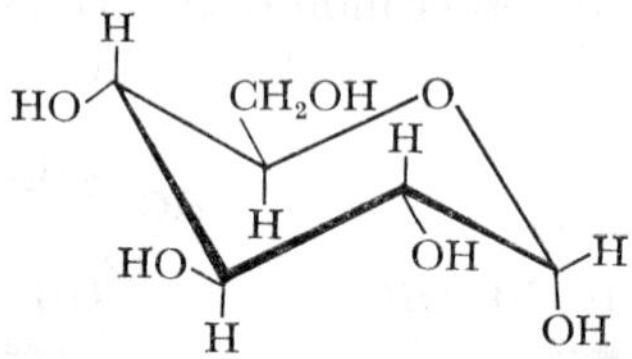

Fructose. Glucose and its stereoisomers are described as *aldohexoses* because they are potential aldehydes containing six carbon atoms. Fructose (16), like glucose, is a sugar component of many fruits and is a hydrolysis product of cane sugar. Fructose is an example of a *ketohexose* because, although it is isomeric with glucose, it is a potential ketone (cf. 17).

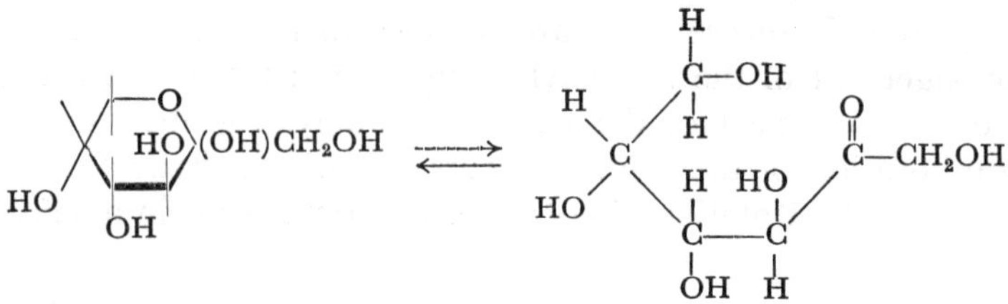

(16) (17)

Oxidation with Fehling's solution is a characteristic reaction of α-hydroxy-ketones as well as aldehydes. As fructose is a potential α-hydroxy-ketone, it reduces Fehling's solution, and both glucose and fructose are therefore classified as '*reducing sugars*'.

Fructose and glucose are interconverted *in vivo* in the form of mono-phosphates:

$$(HO)_2P(O)OCH_2(CHOH)_3COCH_2OH \rightleftharpoons (HO)_2P(O)OCH_2(CHOH)_3CHOHCHO$$

Pentoses. Ribose (18) is a C_5 monosaccharide (or pentose) that contains a six-membered ring, and in its reactions behaves as an aldehyde (cf. 19). Certain compounds of immense biological importance contain a ribose unit as part of their structure. One example is the cell constituent ribonucleic acid (RNA) which is intimately concerned with the synthesis of proteins. Desoxyribose (a pentose containing one less hydroxyl group than ribose) is similarly involved in the structure of desoxyribonucleic acid

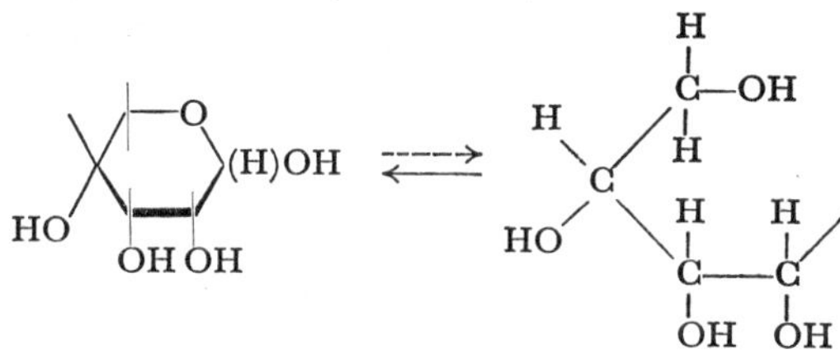

(18) (19)

(DNA), a cell component responsible for the transmission of hereditary features.

Disaccharides. Disaccharides consist of two monosaccharide units linked through an oxygen atom; they are related structurally to glycosides, the alcohol portion being derived from a second molecule of a monosaccharide.

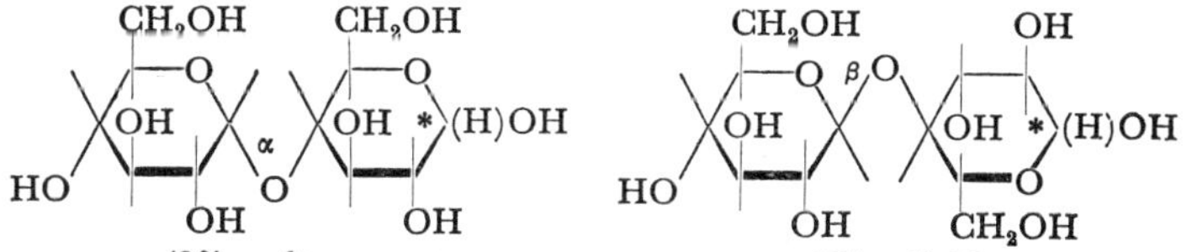

(20) maltose (21) cellobiose

The disaccharide (20) is formed from two molecules of glucose, and contains an α-glycosidic link, the C—O bond marked α being arranged in the same way as in α-glucose (13): in the isomer (21) the same hydroxyl groups of two glucose molecules are utilised to form the oxygen bridge, but the glycosidic bond has the β-arrangement. The glucose unit on the *right* of the structure (21) is derived from the conventional diagram for glucose (11) by rotation of 180° about an axis passing through the starred carbon atom and the carbon attached to the oxygen bridge. Structures (20) and (21) represent respectively the natural disaccharides maltose and cellobiose. Each exhibits mutarotation because of the hemiacetal hydroxyl group at *C, but they are not interconverted because they differ at the more stable acetal (glycoside) link. Acid hydrolysis of either maltose or cellobiose splits the compounds at the glycosidic bonds and affords two molecules of glucose.

Disaccharides do not necessarily contain *identical* monosaccharide units. Lactose (22), for example, is derived from glucose and galactose, and gives these monosaccharides when hydrolysed:

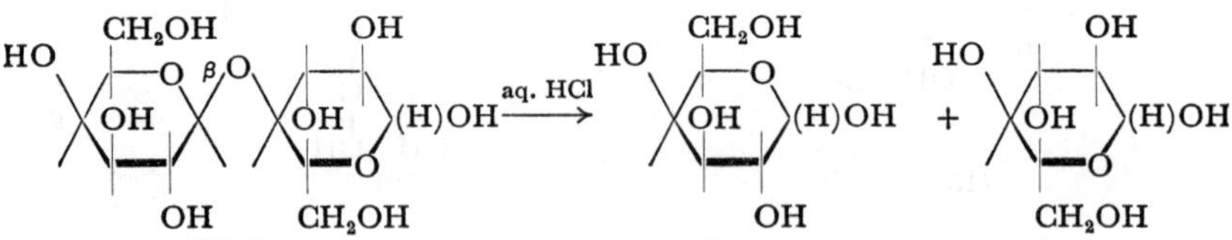

(22) lactose galactose glucose

Lactose is present in milk, and is prepared industrially from whey obtained during the manufacture of cheese.

Sucrose is a disaccharide composed of glucose and fructose units. Sucrose is not a 'reducing sugar' and does not therefore contain potential aldehyde or ketone functions; the glycosidic link in sucrose must be derived from the hemiacetal hydroxyl groups of *both* glucose and fructose. Sucrose is obtained either from sugar-cane or sugar-beet by pressing out the juice, removing protein impurities, and concentrating the clear aqueous solution until sucrose crystallises. Centrifugation separates sucrose from the mother-liquors (molasses), which contain lactose and glucose as well as sucrose. Sucrose is readily hydrolysed:

$$\underset{[\alpha]_D\ +67^\circ}{\text{sucrose} + H_2O} \xrightarrow{H^\oplus} \underset{[\alpha]_D\ +52^\circ}{\text{glucose}} + \underset{[\alpha]_D\ -92^\circ}{\text{fructose}}$$

In this way dextrorotatory sucrose is converted into glucose and fructose; the mixture is laevorotatory (as fructose has a higher specific rotation than glucose) and is therefore called 'invert sugar'. Honey is mainly 'invert sugar' since bees possess an enzyme which hydrolyses sucrose present in nectar.

Polysaccharides. The polysaccharide, *cellulose*, is the main structural material of plants. Cotton is almost pure cellulose, and together with wood is the most important industrial source of the polysaccharide; cellulose prepared from wood must first be separated from lignin.

Complete hydrolysis of cellulose affords only glucose, but the disaccharide cellobiose (21) is obtained by partial hydrolysis. Cellulose (23) in fact is a long-chain polymer (molecular weight 300,000–500,000) derived from glucose units, the isolation of cellobiose showing that these units are joined by β-glycosidic bonds.

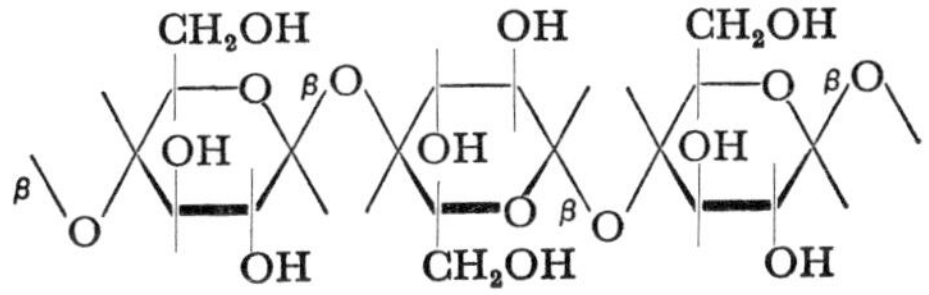

(23) cellulose

Synthetic fibres, e.g. celanese, viscose rayon, are prepared from cellulose. Each glucose unit of cellulose contains three hydroxyl groups, which are acetylated with acetic anhydride ($-OH \rightarrow -OCOCH_3$). Partial hydrolysis gives cellulose acetate containing approximately two ester groups per glucose unit, and this product in acetone solution is extruded through a capillary. Evaporation of the solvent affords threads of cellulose acetate (celanese). A method for spinning cellulose triacetate into fibres suitable for weaving has been devised (1955, Tricel).

Starch, the reserve carbohydrate of plants, is composed of amylose (*ca.* 20%) and amylopectin (*ca.* 80%). Amylose, extracted from the mixture by hot water, resembles cellulose in being a linear polymer of glucose, but differs from it in containing α-glycosidic links; maltose is obtained on partial hydrolysis. Adjoining glucose units are also held together by α-links in amylopectin and in the very similar glycogen (reserve polysaccharide of mammals). These products contain some α-links from terminal hexose units across to primary hydroxyl groups in non-terminal units in other chains.

Neuraminic acid (24) exemplifies how carbohydrate structures are sometimes further elaborated:

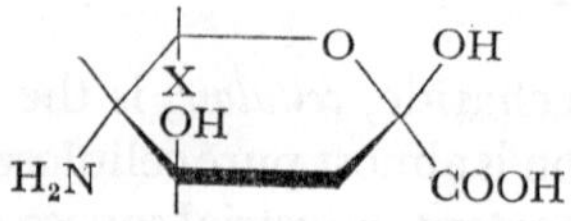

(24) $X = HOCH_2(CHOH)_2—$

Mucins, present for example in blood, brain, and semen, are built up from proteins, hexoses, hexosamines and sialic acids (*N*-acyl derivatives of neuraminic acid).

Metabolism. As essential components of living cells, carbohydrates are intimately concerned with the mechanisms of metabolic processes. Glucose, for example, is being continually produced in plants by photosynthesis and continually degraded in plants and animals into carbon dioxide and water by an energy-releasing oxidation process that is linked with the conversion of ADT into ATP (p. 103). ATP assists individual biochemical reactions to take place. For example, the unfavourable conversion of glucose into glucose 6-phosphate ($\Delta G = +3$ kcal./mole) proceeds when it is coupled with the energy-releasing reaction ATP + H_2O $\rightarrow$ ADP + phosphate ($\Delta G = -7$ kcal./mole):

$$\text{glucose} + \text{ATP} \longrightarrow \text{glucose 6-phosphate} + \text{ADP} \ (\Delta G = -4 \text{ kcal./mole})$$

Glucose 6-phosphate is the biosynthetic precursor of starch.

QUESTIONS

1. (*a*) Which compound, sucrose or sucrose octa-acetate, would you expect to be more soluble (i) in water, and (ii) in dichloromethane?

(*b*) By what procedures could the following properties of sucrose be studied? (i) Melting point, (ii) temperature of thermal decomposition, (iii) optical rotation, (iv) solubility in methanol, (v) density, (vi) reactivity towards acetic anhydride (used in excess).

(*c*) How could the relative sweetness of sucrose, glucose and fructose be satisfactorily investigated?

2. Carbohydrates are oxidised, but not very selectively, when they are heated with nitric acid. When 20 g. of sucrose was oxidised in this way, 7 g. of oxalic acid dihydrate, $(COOH)_2 . 2H_2O$, was obtained (by crystallisation from the partly evaporated reaction solution). What percentage of the total possible yield does this represent? How might the amount formed be estimated more precisely? Suggest a procedure for finding out the amount of carbon dioxide (if any) produced in the oxidation.

3. (*a*) In what isolable stereochemical forms may the following compounds exist? Iodocyclohexane; 1,4-dimethoxycyclohexane; 1,1-diethylcyclohexane.

(*b*) For galactose (i) write an open chain formula, and (ii) draw a possible conformation of a molecule.

(*c*) Explain how it is possible, under certain conditions, for glucose and fructose to be interconverted *via* an enolic form.

4. Propose a possible connection between the frost resistance of the prepupae of sawflies, and the presence in their body fluid of over 6% of the disaccharide, trehalose $(C_{12}H_{22}O_{11})$.

5. (*a*) Two methods for dyeing textiles are (i) pigmentation (dispersal of the colour in the material before it is spun), and (ii) reaction of the (spun) fibre with a coloured reagent to give a chemical bond between the two. Which fibre would be easier to dye by the second method, cellulose or polypropylene?

(*b*) Suggest a procedure for examining the colour stability of different textiles towards light.

(*c*) Comment on the suitability of the name, cellulose triacetate, for the material obtained by complete acetylation of cellulose.

One cannot explore new countries in express trains.

AUGUST KEKULÉ (1829–1896)

CHAPTER 15

BENZENE AND RELATED HYDROCARBONS

Benzene was discovered in 1825 by Faraday who isolated it from compressed illuminating gas. Benzene was obtained later from the distillation of benzoic acid with lime, and the molecular formula, C_6H_6, was established. The ratio of carbon: hydrogen in this formula suggested that benzene should have 'unsaturated' properties similar to those of alkenes (C_nH_{2n}) and alkynes (C_nH_{2n-2}). In most of its reactions, however, benzene does not behave as an unsaturated compound; for example, bromine does *not add* to benzene.

Benzene must therefore contain some special structural feature. From various strands of evidence Kekulé suggested, in 1865, that benzene had a ring or *cyclic* structure (1) with alternate single and double carbon–carbon bonds.

(1) 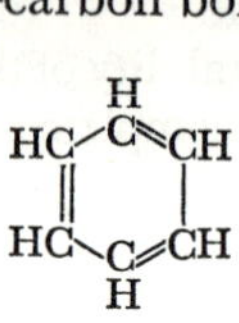

That benzene contains a ring of six carbon atoms received ample verification in the decades following Kekulé's suggestion. However, the relative lack of olefinic properties still remained puzzling. The idea developed gradually that the carbon–carbon bonds in benzene may be equivalent (identical) and not alternate in their bond order. In other words, each carbon–carbon bond may have a character somewhere between those of single and double bonds. The equivalence of the bonds and the symmetry of the benzene molecule was established beyond doubt by determination of carbon–carbon bond lengths by electron diffraction and by spectroscopic methods. All the bonds were found to be 1·39 Å in length, a value intermediate between those of single (1·54 Å) and double (1·35 Å) carbon–carbon bonds (cf.

the comparable situation with the carboxylate ion, p. 90). Furthermore, X-ray diffraction measurements showed that benzene was a planar molecule, with the carbon–hydrogen bonds in the plane of the ring and the C—C—C and H—C—C angles all 120°. In resonance terminology benzene has a hybrid bond structure, the two alternative Kekulé structures (1) and (2) being the main contributing forms.

(1) 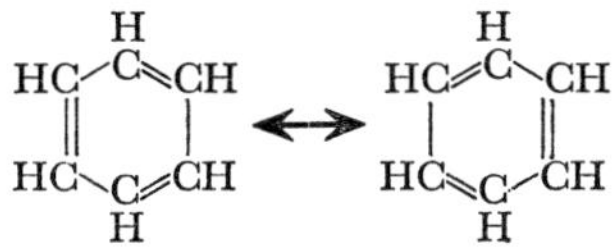 (2)

Substituted benzenes (one or more hydrogens replaced by other groups) also have the hybrid bond structure of the parent compound. The formulae are simplified by omission of the letters, and this convention will be used subsequently:

hybrid structure of benzene

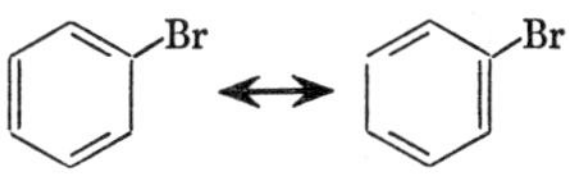

hybrid structure of bromobenzene

Benzene and substituted benzenes are more conveniently represented by *one* of their Kekulé structures, and this practice will be used except where the hybrid character requires special emphasis.

The structure of benzene can also be described in molecular orbital terms. The bond angles in benzene are 120°, close to the values in ethylene, and benzene resembles the alkene in being made up from sp^2 units, cf. p. 56. In a molecule of benzene, six CH units combine so that carbon and hydrogen atoms are in one plane, and the electrons in unbonded *p*-orbitals overlap to give a circular zone of electron density (π-electrons) on both sides of the plane; this is represented in diagram (3). The

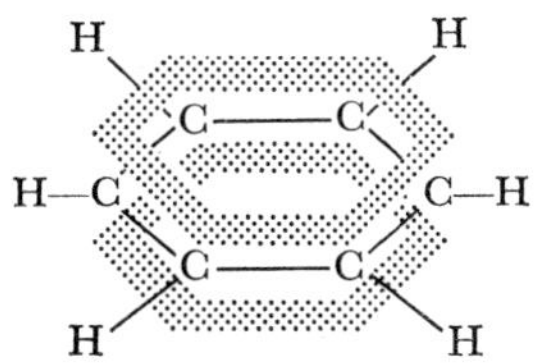

(3)

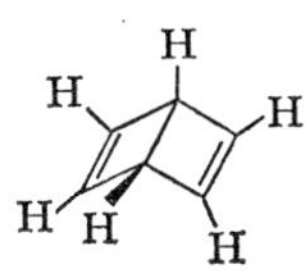

(4)

resonance approach also explains the equivalence of the carbon atoms and of the hydrogen atoms in benzene, but the molecular orbital picture has the advantage of emphasising that benzene rings have finite thickness; in graphite (p. 200) the distance between planes is 3·55 Å.

The special stability of benzene is not shared by the isomeric compound (4), which does not contain a cyclic conjugated structure. Substance (4), first synthesised in 1963, isomerises at 20° into benzene, the half life of the transformation being approximately two days.

Nomenclature. A number of special terms are commonly used in the description of the chemistry of benzene and related compounds. One of these is *aromatic*,* a term that referred originally to the pleasant smell of naturally occurring organic substances. Later, many of these compounds were found to be derivatives of benzene, and 'aromatic' was then used in a structural sense to denote compounds containing one or more benzene rings. The special feature of the chemistry of aromatic compounds, the preservation of the parent ring structures during reactions (p. 188–195), became known as 'aromatic stability'.

Other terms will be discussed by reference to the following compound, which has the specific name, cumene:

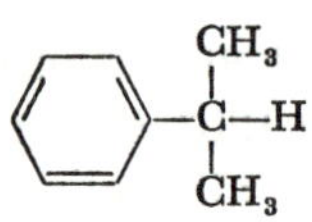

cumene, isopropylbenzene or 2-phenylpropane

The name, isopropylbenzene, shows that one hydrogen of benzene has been replaced by an isopropyl group. In the alternative name, 2-phenylpropane, *phenyl* refers to the C_6H_5 group which, in this compound, is linked to the second carbon of the propane chain. (A general name for aromatic groups is *aryl*, abbreviated to Ar: phenyl and tolyl ($CH_3C_6H_4$) are examples of aryl groups.)

Nucleus and *side-chain* are useful terms in aromatic chemistry

* A complementary term is *aliphatic* which refers to molecules or parts of molecules derived structurally from alkanes. Benzoic acid, C_6H_5COOH, is an aromatic compound, whereas butyric acid, lactic acid and succinic acid are all aliphatic compounds.

for distinguishing between the aryl ring and its substituents (other than hydrogen):

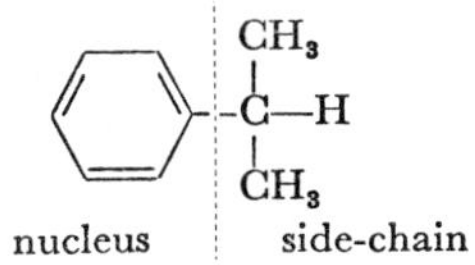

Structural isomerism is encountered when there are two or more nuclear substituents. Thus there are three dimethylbenzenes (xylenes), and three methylphenols (cresols):

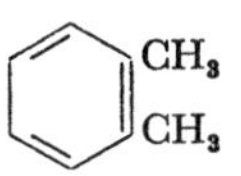

ortho-xylene or *o*-xylene (1,2-dimethylbenzene)

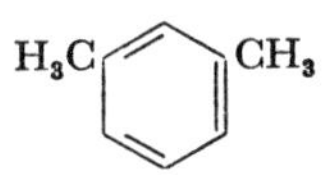

meta-xylene or *m*-xylene (1,3-dimethylbenzene)

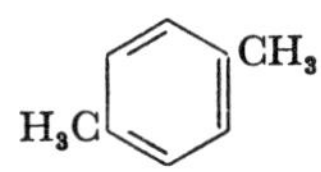

para-xylene or *p*-xylene (1,4-dimethylbenzene)

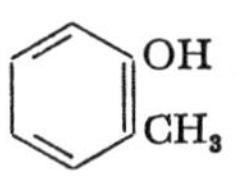

o-cresol

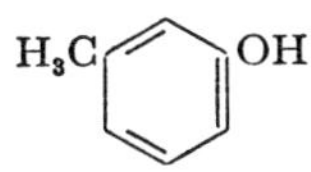

m-cresol

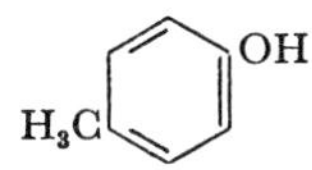

p-cresol

The words *ortho, meta* and *para*, abbreviated to *o*, *m* and *p*, indicate the three possible positions for a *pair* of substituents—which may be the same or different—on a benzene ring.

The alternative numbering procedure, given for the xylenes in parentheses, is used infrequently for disubstituted compounds, but becomes essential when there are three or more substituents on the ring:

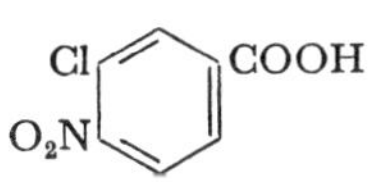

3-chloro-4-nitro-benzoic acid

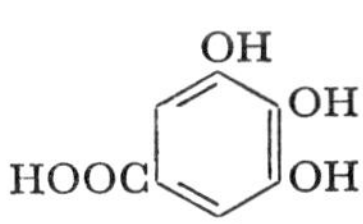

3,4,5-trihydroxy-benzoic acid (gallic acid)

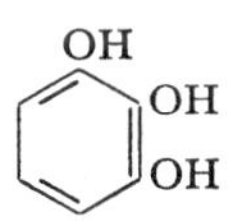

1,2,3-trihydroxy-benzene (pyrogallol)

Some of the polysubstituted derivatives of benzene, especially those from natural sources, have specific names, e.g. gallic acid (extracted from galls) and pyrogallol (obtained by heating gallic acid).

Spectroscopy of Aromatic Compounds. Infrared spectroscopy is used to detect the presence of an aromatic ring in an unknown compound. Absorption occurs at *ca.* 3000 (aromatic

C—H) and near 1600 and 1500 cm.$^{-1}$ (C=C). A series of intense peaks at 950–700 cm.$^{-1}$ is often characteristic of the substitution pattern in a benzene ring, e.g. *o*-, *m*- and *p*-.

π-Electrons of an aromatic nucleus, like electrons of conjugated polyenes (p. 139) are excited by ultraviolet radiation, and intense absorption occurs as shown in the Table (first band). The wave-length of maximum absorption of benzene is raised and intensified when the conjugated system is extended (cf. styrene) and the presence of certain other substituents, e.g. NH_2 as in

	First band		*Second band*	
	λ_{max} (Å)	ϵ_{max}	λ_{max} (Å)	ϵ_{max}
benzene, C_6H_6	1980	8000	2550	230
aniline, $C_6H_5NH_2$	2300	13,000	2800	1430
styrene, $C_6H_5CH{=}CH_2$	2440	12,000	2820	450

aniline, also increases the value of λ_{max}. Benzene produces a second weaker band not shown by conjugated polyenes, and this absorption is also shifted to longer wavelengths in substituted benzenes.

In *nuclear magnetic resonance* (n.m.r.) spectroscopy, the energy required to change the orientation of a magnetic nucleus (usually hydrogen) is measured by placing a compound in a powerful magnetic field. The positions of absorption bands in the resultant spectra are conveniently measured by reference to a standard in τ units. Aliphatic hydrogens, for example, absorb in the region of 9 τ, but the proximity of π-electrons reduces this value. Thus, olefinic hydrogen atoms absorb at 3–5 τ and aromatic hydrogen at 1·5–3·5 τ. The spectrum of toluene, for instance, contains two peaks of relative integrated intensity 5:3 at 2·8 τ (five aromatic protons) and 7·7 τ (three protons of the methyl group). Interaction between neighbouring hydrogen nuclei (*spin–spin coupling*) often produces splitting of the primary absorption bands. The nature and extent of the splitting, together with the position of the multiplet band, are of great value for the elucidation of the molecular environment of a hydrogen. Benzene gives a single band at 2·73 τ, which is not split by spin–spin coupling because the six hydrogens are structurally equivalent. The benzene isomer (4), however, gives multiplets of relative intensity 2:1 at 3·45 (four olefinic hydrogens) and 6·16 τ (two $>$CH— hydrogens, lowered from the aliphatic value of 9 τ by proximity to double bonds, cf. toluene). This spectrum is good evidence for the existence of a compound of structure (4).

Sources of Aromatic Compounds. The main elements in coal are carbon (*ca.* 85% by weight), hydrogen (5%) and oxygen (8%). The molecular composition is very complex and varied, but most of the carbon is contained in aromatic groups, and much of the remainder is in the form of aliphatic linkages and substituents. Hydroxyl and ether groups account for most of the oxygen present. The distillate from the decomposition of coal at a high temperature is partly aqueous and partly organic. The organic part, 'coal tar', consists mainly of a complex mixture of aromatic compounds. Many of the simpler compounds, e.g. benzene, toluene, the three xylenes, phenol and naphthalene, are present in appreciable amounts, and these substances are isolated in pure forms by a combination of fractional distillation and chemical methods.

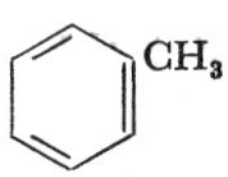

toluene (p. 196)
(methylbenzene)

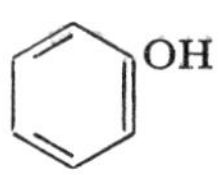

phenol (p. 206)
(hydroxybenzene)

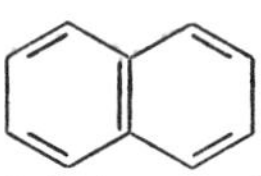

naphthalene (p. 198)

Demand for aromatic compounds, especially those of the benzene series, has outstripped their rate of supply from coal tar. Methods for their preparation from petroleum have therefore been developed, and more than 75% of aromatic compounds are now obtained from this source. Formation of benzene and alkylbenzenes from the alkanes of petroleum involves dehydrogenation and cyclisation reactions which, in practice, may be carried out in one operation. The conversion of n-heptane into toluene may be taken as an example:

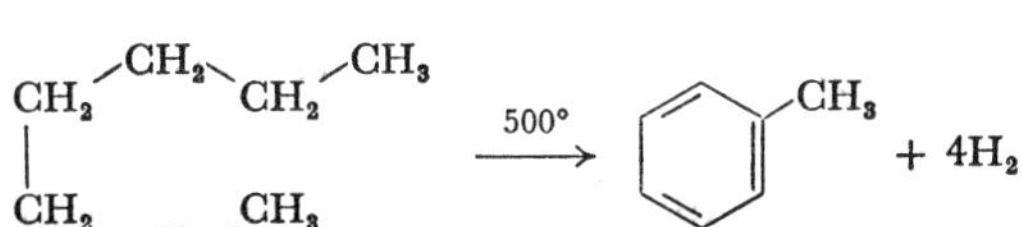

toluene, b.p. 110·5°

In practice, petroleum fractions rich in C_7 compounds are used, and the toluene is purified finally by distillation. Similar methods are used to convert C_6 alkanes (including cycloalkanes in the petroleum) into benzene, and C_8 alkanes into a mixture of xylenes. The boiling points of *m*- and *p*-xylene are very close (139° and 138° respectively), but the *o*-isomer (b.p. 144·5°) can be separated from the mixture of the three compounds by fractional distillation. The *m*- and *p*-xylenes are then separated by freezing out the *p*-isomer (m.p. 13°).

REACTIONS OF BENZENE

The most important and characteristic reactions of benzene are those in which a hydrogen is substituted (replaced) by another atom or group. These reactions are called *electrophilic substitutions* in view of the electron-seeking character of the attacking species. If the entering electrophile is designated as $Z^{\oplus}$, the general equation for the substitution of benzene is:

$$C_6H_6 + Z^{\oplus} \longrightarrow (C_6H_6Z)^{\oplus} \longrightarrow C_6H_5Z + H^{\oplus}$$

This two-stage mechanism and the nature of the intermediate, $(C_6H_6Z)^{\oplus}$, will be discussed in detail by reference to the bromination of benzene.

Bromination and Chlorination. An orange-red solution of bromine in benzene at 20° soon loses its colour when a small amount of aluminium bromide is added and the following reaction occurs:

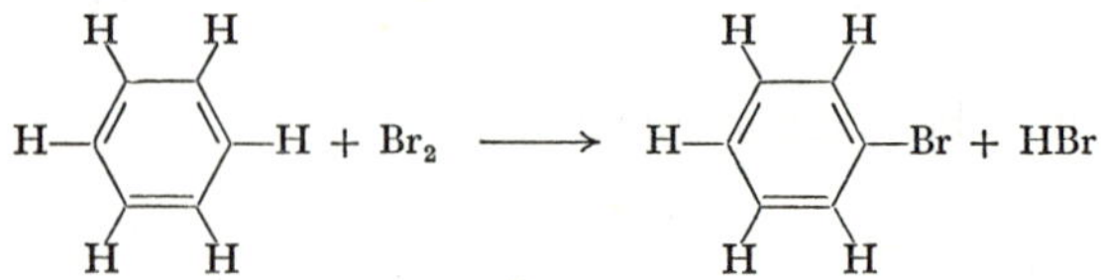

In the reactions of bromine with alkenes and alkynes the halogen *adds* to the unsaturated bonds (Chapter 5). In its reaction with benzene one of the hydrogens of the aromatic ring is *substituted* by bromine and hydrogen bromide is also formed.

Bromination of benzene proceeds by a two-step mechanism.

In the initial elementary step (*a*), bromine reacts with a molecule of bromine and a molecule of aluminium bromide:*

(*a*) 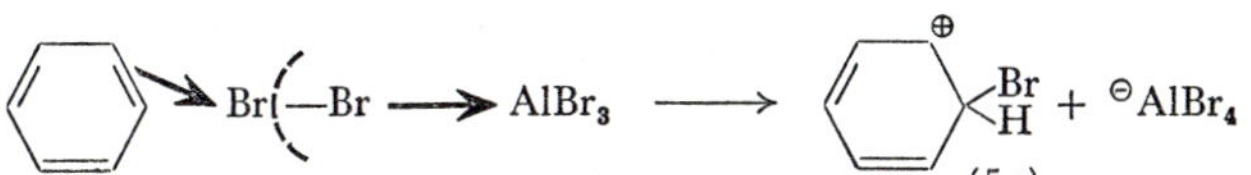

(5A)

The net result of this process is that $Br^{\ominus}$ adds to aluminium bromide and $Br^{\oplus}$ adds to benzene to give the cation $(C_6H_6Br)^{\oplus}$. The carbonium ion (5A) (p. 57) is a resonance hybrid to which two other similar carbonium ions (5B) and (5C) also contribute; alternatively, the hybrid structure can be represented as (6), a bromobenzenonium ion:

hybrid structure of $(C_6H_6Br)^{\oplus}$

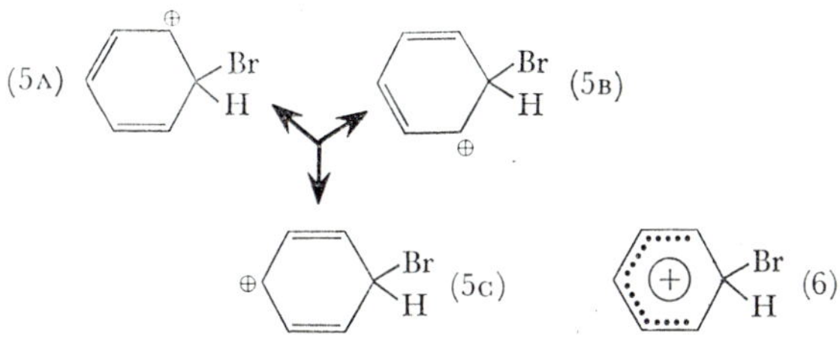

In the second step (*b*) of the bromination reaction, a proton is transferred from the intermediate cation (to $^{\ominus}AlBr_4$ to give HBr and $AlBr_3$), and bromobenzene is formed:

(*b*) 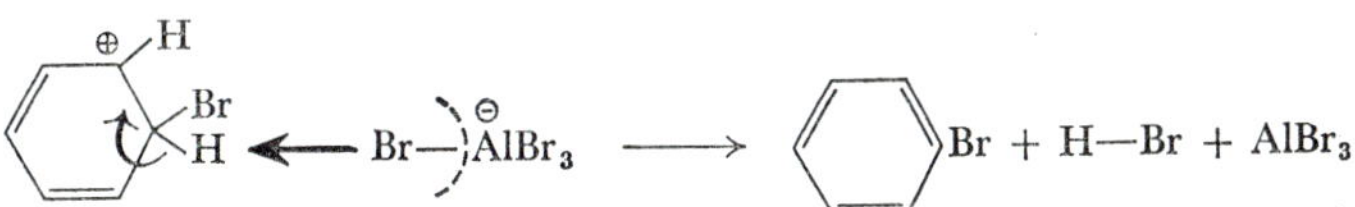

(5A)

The energy profile (7) represents the two elementary steps of the reaction mechanism. Formation of the bromobenzenonium ion intermediate is the slower stage, and the activation energy, E_A, is needed to attain the requisite transition state (A). Much less energy is required to reach the second transition state (B) in which the proton is being lost to give the final product, bromobenzene.

* Aluminium bromide, $AlBr_3$, has only a sextet of electrons in its outer shell. By acquisition of a bromide anion (from Br_2) aluminium bromide is converted into the complex anion, $^{\ominus}AlBr_4$, which contains an electronic octet around the aluminium.

As a possible alternative to step (*b*), the intermediate cation, $(C_6H_6Br)^{\oplus}$ could add $Br^{\ominus}$ (obtained from $^{\ominus}AlBr_4$) to yield a dibromide, $C_6H_6Br_2$, e.g. (8) (cf. the addition of bromine to an alkene). This compound, although it has two double bonds in

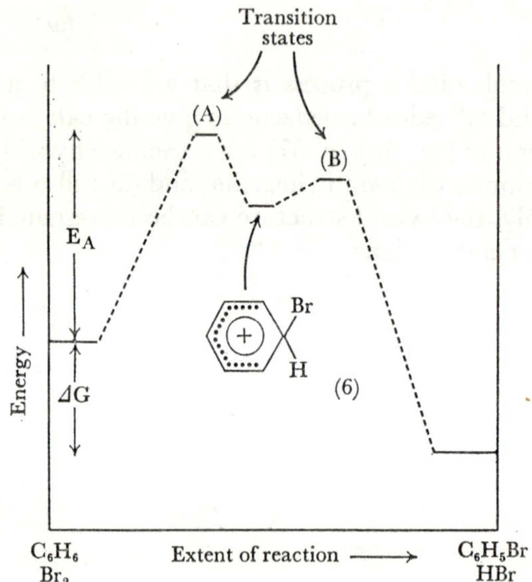

(7)

conjugation, does not have the cyclic conjugation of a benzene derivative that gives an especially large amount of resonance

(8) 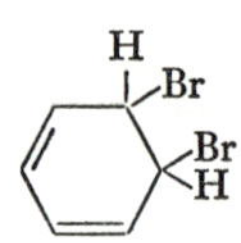

stabilisation. Loss of a proton from the intermediate cation (to give bromobenzene) occurs therefore in preference to the addition of a bromide ion.

Chlorination of benzene in the presence of aluminium chloride proceeds by an analogous mechanism to yield chlorobenzene:

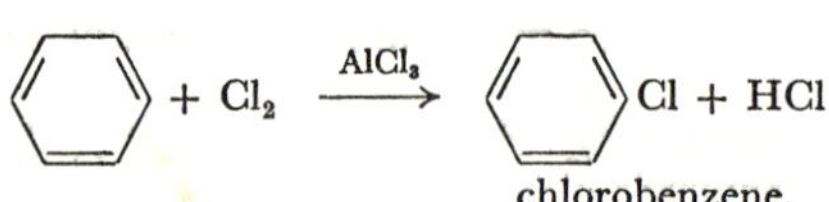

chlorobenzene,
b.p. 132°

In both of these electrophilic substitution reactions the conjugated structure of the benzene ring is preserved in the products (but not in the reaction intermediates). This phenomenon is one aspect of aromatic behaviour (aromaticity).

Alkylation. No reaction occurs between benzene and methyl chloride unless a catalyst, e.g. aluminium chloride, is added. The products are then toluene (methylbenzene) and hydrogen chloride:

$$C_6H_6 + CH_3Cl \xrightarrow{AlCl_3} C_6H_5CH_3 + HCl$$

The role of the aluminium chloride is similar to that of the aluminium bromide in the bromination of benzene, and a similar two-step mechanism is involved:

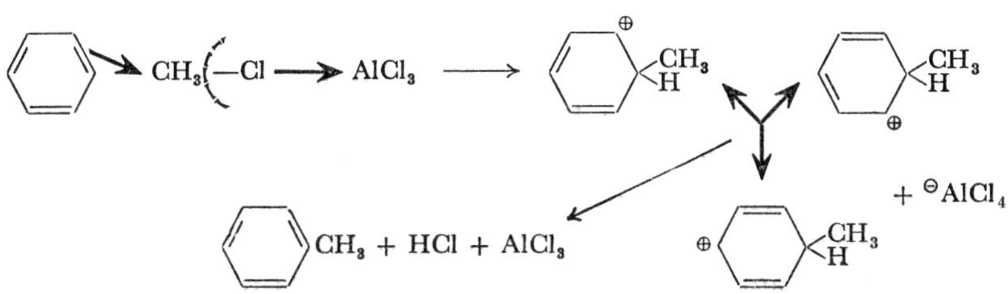

Alkylation of benzene with ethyl chloride yields ethylbenzene. On a large scale this is prepared more economically by the acid-catalysed addition of ethylene to benzene. Catalytic dehydrogenation of ethylbenzene gives styrene, $C_6H_5CH{=}CH_2$. Even on distillation at 145° styrene is partly converted into the non-volatile, polystyrene $[-CH(C_6H_5)CH_2-]_n$. Expanded polystyrene, a solid foam with a density of about one-fortieth that of the polymer, is used in life jackets and for light weight packaging. Polymers containing styrene as one of the building units are also manufactured; styrene–butadiene (SB) copolymer has rubber-like properties, and a crylonitrile–butadiene–styrene (ABS) copolymer is a tough general purpose plastic which is easy to prepare and to process.

Nitration. Nitrobenzene, $C_6H_5NO_2$, was first obtained by Mitscherlich in 1834 by the action of fuming nitric acid on benzene. The reaction proceeds more quickly and in higher yield when sulphuric acid is also present. The overall equation can then be written:

$$C_6H_6 + HONO_2 + H_2SO_4 \longrightarrow C_6H_5NO_2 + H_3O^{\oplus} + HSO_4^{\ominus}$$

The active species in this *nitration* of benzene is the nitronium cation, $^{\oplus}NO_2$, a highly reactive electrophile; the function of the sulphuric acid in the reaction is to produce an appreciable concentration of nitronium ions* from nitric acid:

$$HONO_2 + 2H_2SO_4 \rightleftharpoons {}^{\oplus}NO_2 + H_3O^{\oplus} + 2HSO_4^{\ominus}$$

These nitronium ions then add to benzene to give resonance-stabilised hybrid cations, which lose protons to give the product, nitrobenzene:

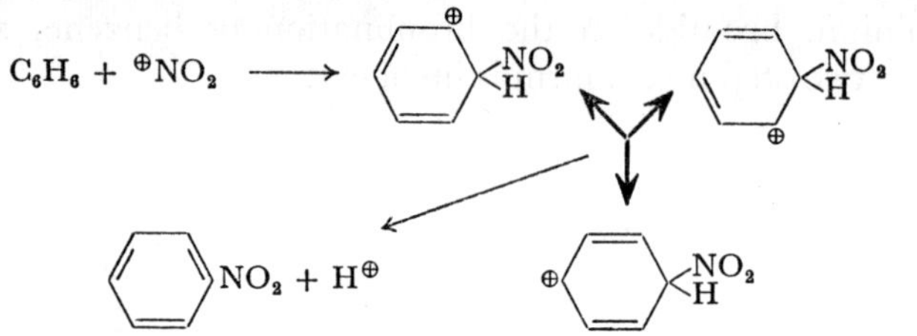

The bond formation and bond breaking processes taking place in the two stages of this electrophilic substitution reaction may be expressed alternatively as follows (only one of the contributing forms of the intermediate cation is written):

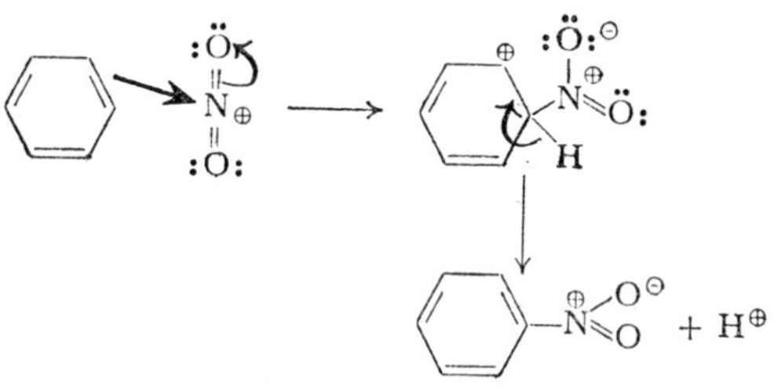

The nitration of benzene and substituted benzenes is a useful general reaction in the aromatic series because the nitro-group can be transformed into many other types of substituent (Chapter 18).

* The nitronium cation is known in the form of its stable crystalline salt, nitronium perchlorate, $^{\oplus}NO_2\ ClO_4^{\ominus}$. X-ray methods have shown that the cation has a linear ONO structure, similar to that of carbon dioxide, and absorption at 1400 cm^{-1} in the Raman spectrum indicates that a species of this structure is also present in solution.

Acylation. Reaction between benzene and acetyl chloride proceeds readily when aluminium chloride is added to the mixture. In this case the electrophile is the acetylium cation, $CH_3{-}\overset{\oplus}{C}{=}\ddot{O}: \leftrightarrow CH_3{-}C{\equiv}\overset{\oplus}{O}:$, produced in the following way:

The mechanism of the subsequent reaction with benzene can be written:

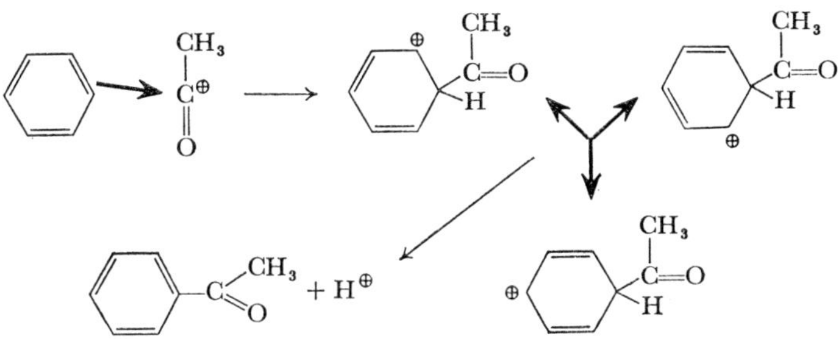

acetophenone
b.p. 202°

Diphenyl ketone (benzophenone) is formed in the similar reaction of benzoyl chloride with benzene:

$$C_6H_5COCl + C_6H_6 \xrightarrow{AlCl_3} C_6H_5COC_6H_5 + HCl$$

benzophenone, m.p. 48°

Reactions between aromatic compounds and acyl halides provide therefore a general route to aryl alkyl ketones, ArCOR and to diaryl ketones, ArCOAr.

Sulphonation. Benzene, when heated with sulphuric acid, is partially converted into benzenesulphonic acid:

$$C_6H_6 + HOSO_2OH \longrightarrow C_6H_5{-}\underset{\underset{O}{\|}}{\overset{\overset{O}{\|}}{S}}{-}OH + H_2O$$

The reaction is more rapid and more complete when fuming sulphuric acid (H_2SO_4 containing SO_3) is used. The electrophilic species in these reactions is believed to be SO_3, which, although it is a neutral molecule, contains a strongly electron-deficient sulphur atom.

In another method for the preparation of benzenesulphonic acid, benzene is treated with chlorosulphonic acid (the monoacid chloride of sulphuric acid) at room temperature, to give benzenesulphonyl chloride in good yield. This compound is readily hydrolysed to the sulphonic acid:

$$C_6H_6 + 2ClSO_2OH \longrightarrow \underset{\text{benzenesulphonyl chloride}}{C_6H_5{-}\overset{\overset{\displaystyle O}{\|}}{\underset{\underset{\displaystyle O}{\|}}{S}}{-}Cl} + H_2SO_4 + HCl$$

$$C_6H_5{-}\overset{\overset{\displaystyle O}{\|}}{\underset{\underset{\displaystyle O}{\|}}{S}}{-}Cl + H_2O \longrightarrow C_6H_5{-}\overset{\overset{\displaystyle O}{\|}}{\underset{\underset{\displaystyle O}{\|}}{S}}{-}OH + HCl$$

Chlorosulphonic acid can be used generally for the chlorosulphonylation of aromatic compounds:

$$ArH + 2ClSO_2OH \longrightarrow ArSO_2Cl + H_2SO_4 + HCl$$

Further aspects of the chemistry of aryl sulphonyl chlorides, $ArSO_2Cl$, are discussed later (p. 231).

Hydrogen Exchange. Treatment of benzene with 50% aqueous sulphuric acid results in exchange of hydrogen by protons acting as electrophilic species. This reaction can be detected by employing deuterosulphuric acid, $D_2SO_4 : D_2O$

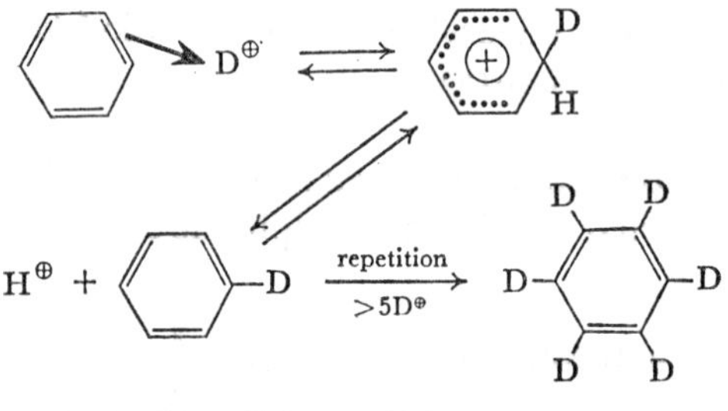

under these relatively mild conditions, sulphonation does not compete. Hydrogen–deuterium exchange also occurs when benzene and deuterium are brought into contact on the surface of a metal catalyst.

Sulphur trioxide is treated with enough deuterium oxide to give a 1 : 1-molar mixture of D_2SO_4 and D_2O. Benzene is shaken with 9 mols. of this acid for three days at 20°; it is then separated and shaken again with fresh acid. After four repetitions of the operation the benzene is recovered and distilled.

C_6H_6 m.p. 5·5°; b.p. 80·1°; density 0·879
C_6D_6 m.p. 6·8°; b.p. 79·3°; density 0·946

Addition Reactions. Chlorobenzene is formed when benzene is treated with chlorine in the presence of a polar catalyst, but the reaction takes a completely different course when chlorine is passed into benzene illuminated with ultraviolet light. Chlorine atoms are produced (p. 20) and these add to the benzene ring producing ultimately the saturated compound 'benzene hexachloride':

$$C_6H_6 + 3Cl_2 \longrightarrow$$

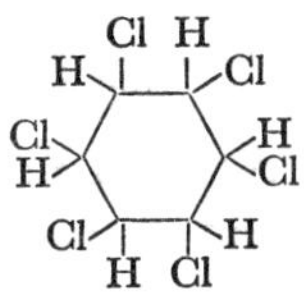

'Benzene hexachloride', better described as hexachlorocyclohexane, exists in a number of stereoisomeric forms, one of which is a powerful insecticide.

Benzene shows olefin-like properties in its reactions with ozone and with hydrogen/metal catalysts, because addition of these reagents to the unsaturated ring occurs in both instances. The triozonide from benzene can be decomposed to glyoxal and related C_2 compounds:

$$C_6H_6 + 3O_3 \longrightarrow \text{triozonide} \longrightarrow$$

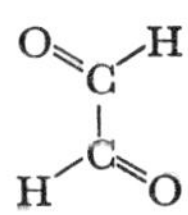

glyoxal

Hydrogenation of benzene over a platinum or rhodium catalyst gives cyclohexane, C_6H_{12}. The reaction proceeds less readily than the hydrogenation of alkenes, and the heat hydrogenation of benzene ($\Delta H = -50$ kcal./mole) is less than that for three isolated double bonds (-80 to -90), another manifestation of aromaticity.

TOLUENE AND ITS REACTIONS

The reactions of toluene can be divided into two classes depending on whether they occur in the nucleus (phenyl group) or in the side-chain (methyl group).

Nuclear Substitution. Like benzene, toluene is brominated by an electrophilic mechanism when it is treated with bromine and aluminium bromide. A bromine substituent is introduced into the nucleus, but although three structural isomers may, in principle, be formed in this process, in fact only a trace of *m*-bromotoluene is obtained (yield on p. 221):

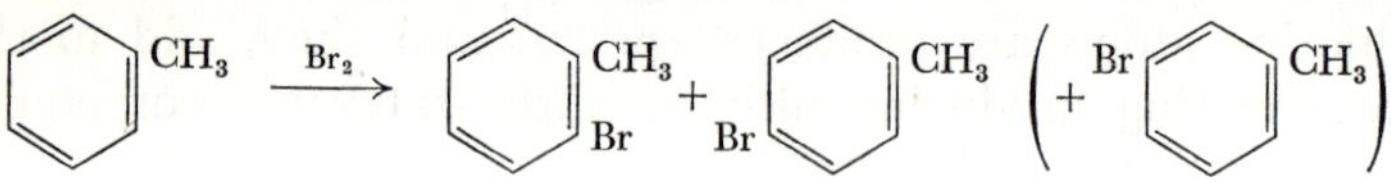

o-bromotoluene, b.p. 182° — *p*-bromotoluene b.p. 184·5° — *m*-bromotoluene (p. 244) b.p. 184°

The methyl group of toluene exerts an o,p-*directive effect,* not only in this bromination reaction, but also in all the electrophilic substitution reactions of toluene. Nitration of toluene at 20°, using a mixture of nitric and sulphuric acids, yields a mixture of *o*- and *p*-nitrotoluenes, separable by careful fractional distillation:

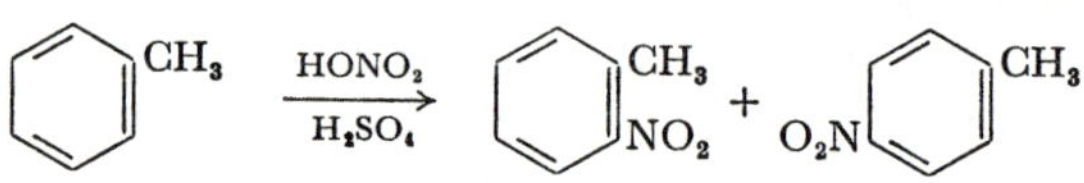

o-nitrotoluene, b.p. 222° — *p*-nitrotoluene, b.p. 238° (m.p. 51·5°)

The relative yields of *o*-, *m*-, and *p*-isomers from these reactions can be assessed (*ca.* 1% accuracy) by gas chromatography (p. 221).

Nitration of either isomer at higher temperatures gives 2,4-dinitrotoluene, then 2,4,6-trinitrotoluene, the well-known high explosive, TNT.

o-nitrotoluene or *p*-nitrotoluene

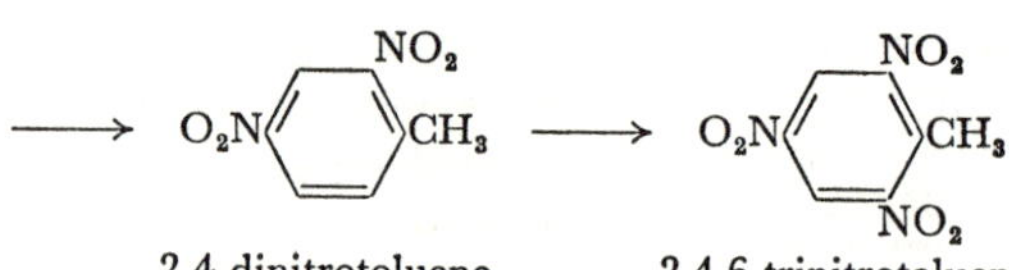

2,4-dinitrotoluene, m.p. 70° — 2,4,6-trinitrotoluene, m.p. 80°

The second and third nitro-substituents are introduced at positions that are also *o*- or *p*- to the methyl group. Actually a stronger influence in these last two reactions is the *m*-directive effect of the nitro-group or groups (p. 239) already present in the molecule. A combination of these effects leads to the eventual formation of 2,4,6-trinitrotoluene in high yield.

Directive effects in electrophilic substitution reactions of other monosubstituted benzenes are considered later (pp. 203, 218).

Substitution in the Side Chain. Chlorination and bromination of toluene takes place in the side-chain methyl group when the reaction is performed at the boiling point of the hydrocarbon in the presence of light. The reaction is similar to the chlorination of alkanes (p. 19) and also proceeds by a radical mechanism:

$$Cl_2 \xrightarrow{\text{light}} 2Cl\cdot \quad \text{initiation step}$$

$$\left.\begin{array}{l} C_6H_5CH_3 + Cl\cdot \longrightarrow C_6H_5CH_2\cdot + HCl \\ C_6H_5CH_2\cdot + Cl_2 \longrightarrow C_6H_5CH_2Cl + Cl\cdot \end{array}\right\} \text{propagation steps}$$

benzylchloride
b.p. 179°

Benzyl is a general term for the $C_6H_5CH_2$ group. The intermediate species $C_6H_5CH_2\cdot$ in the chlorination reaction is therefore a benzyl radical. This species is more stable than the methyl radical (p. 20) because it exists as a hybrid in which the unpaired electron is distributed (delocalised) through the benzene ring:

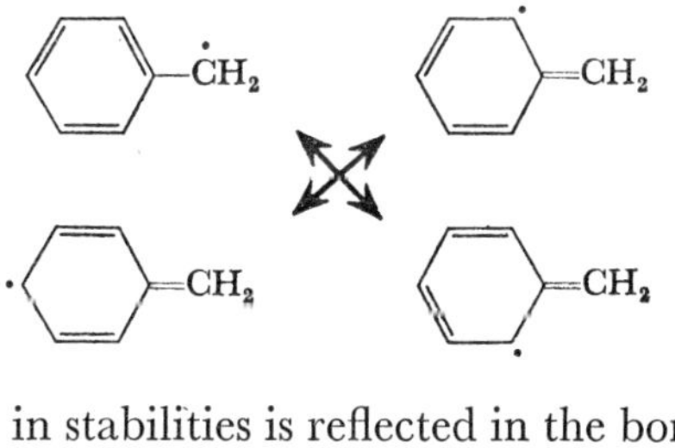

This difference in stabilities is reflected in the bond dissociation energies (78 and 101 kcal./mole for $C_6H_5CH_2$—H and CH_3—H respectively). The heat of reaction for the first propagation step in the photochlorination of toluene is calculated to be −25 (78–103) kcal./mole showing that this elementary process proceeds more readily than the corresponding stage in the chlorination of methane ($\Delta H = -2$ kcal./mole).

Triphenylmethyl. Methyl and benzyl radicals are transient reaction intermediates, but in 1900 Gomberg discovered that a yellow solution is formed when triphenylmethyl chloride, $(C_6H_5)_3CCl$, dissolved in benzene is shaken with zinc or silver powder at room temperature with the exclusion of oxygen. The species responsible for the yellow colour is the radical, triphenylmethyl, $\cdot C(C_6H_5)_3$. One reason for the stability of this radical and for its colour is that the odd electron is extensively delocalised (cf. the benzyl radical) by conjugation with the three phenyl groups p. 197).

Radicals in concentrations down to about 10^{-10} M can be detected by *electron spin resonance* (e.s.r.) spectroscopy. In this technique a sample placed in a magnetic field is subjected to microwave radiation. The unpaired electron acting as a magnet can take up two orientations with respect to the external field, corresponding to two energy levels. Transition between these levels induced by the microwave radiation produces an absorption peak. Further interaction of the odd electron with neighbouring hydrogen nuclei (which are themselves magnetic, cf. n.m.r., p. 186) causes splitting of the main absorption peak. The complex splitting pattern observed in the e.s.r. spectrum of triphenylmethyl provides evidence for the delocalisation of the unpaired electron that was suggested above.

NAPHTHALENE

Petroleum from some sources contains a small amount of naphthalene but the bulk of the world's supply comes from coal tar. Naphthalene was first isolated from coal tar in 1820 and the bicyclic formula, cf. (9A), was put forward in 1866, a year after Kekulé's proposal for the structure of benzene.

Isolation of the hydrocarbon from coal tar is performed by cooling the fraction of b.p. 170–270°, when crude naphthalene separates as a crystalline mass. This material is washed in succession with 5% sulphuric acid and aqueous sodium hydroxide, and finally purified by sublimation. Pure naphthalene is a colourless, flaky solid, m.p. 82°, with a characteristic smell.

In most of its reactions, naphthalene behaves as a typical

aromatic compound, and this, together with physical evidence, suggests that the compound has a resonance hybrid structure. Three structures (9A, 9B, 9C) with alternate single and double

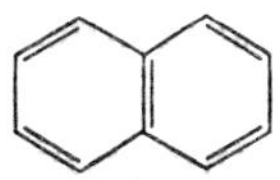
(9A)

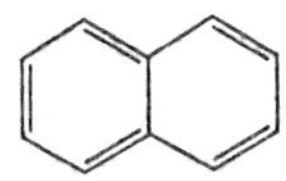
(9B)

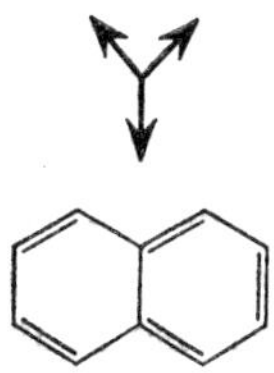
(9C)

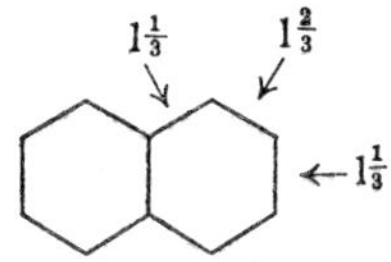

(10)

bond orders

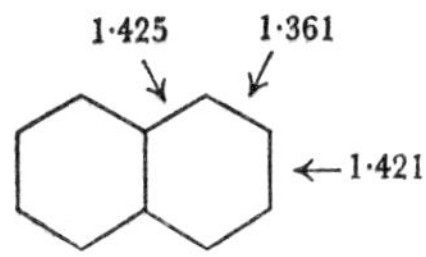

(11)

bond lengths, Å

bonds can be drawn; on the assumption that these three structures contribute equally to the hybrid, the bond orders shown in (10) are obtained. Qualitatively, these bond orders are in accord with the actual lengths of the carbon–carbon bonds (determined by X-ray methods) which are given against formula (11). The shortest bond is the one with the highest estimated bond order (cf. p. 66).

In the reactions of naphthalene with electrophiles, one or more hydrogens are replaced with preservation of the hybrid structure of the nucleus. These reactions, which proceed more readily than those of benzene, can give isomeric products depending on whether the initial point of attack is at the 1- or the 2-position, cf. (12). Reaction at the 1-position usually predominates as shown by the formation of 1-bromonaphthalene, 1-nitronaphthalene and 1-naphthalene sulphonic acid:

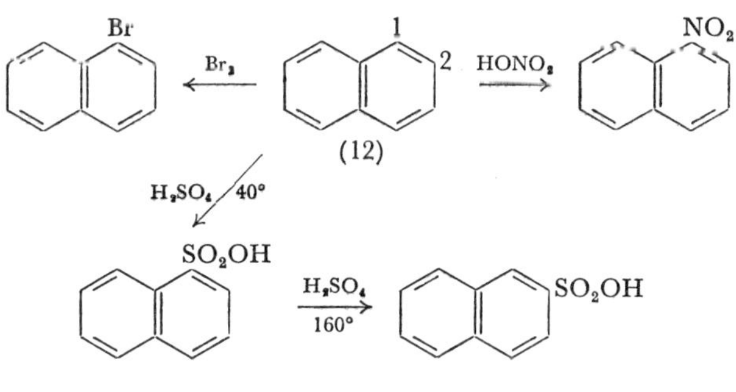

Although sulphonation of naphthalene occurs at the 1-position at 40°, the product is isomerised at higher temperatures to the 2-sulphonic acid. This valuable reaction provides a path to other 2-substituted naphthalenes e.g. 2-naphthol (p. 232).

The flow chart gives only a small selection of the many reactions possible in the naphthalene series. Additional substituents can be introduced into nuclear positions. Some of these compounds are intermediates for the production of dyestuffs, which often contain many benzene rings fused together (cf. benzene → naphthalene). The formation of polycyclic structures in this way results in a decrease in the proportion of hydrogen in the molecule; graphite, consisting of sheets of carbon atoms only, represents the final stage of this process. The layer structure of graphite (3·55 Å between planes) gives rise to its lubricant properties, although slip between layers may be assisted by the presence of water molecules. The use of graphite in pencils is well known (Gk. *graphein*, to write).

QUESTIONS

1. Give names for the following compounds:

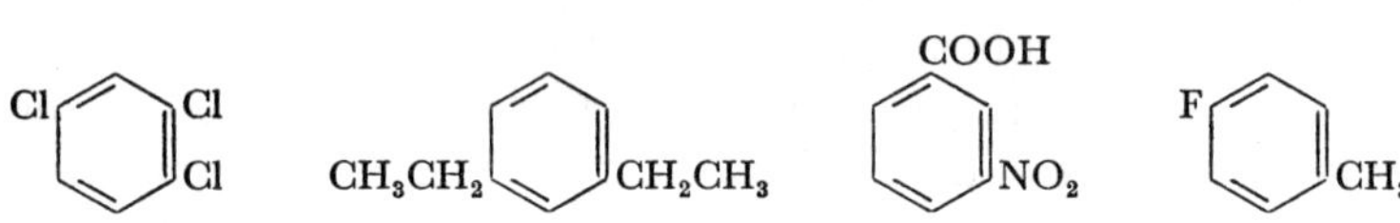

2. (*a*) What difference is there between the proportion of carbon in benzene and in hexadeuterobenzene?

(*b*) Air containing more than 1 mg. of benzene (as vapour) in 10 litres is poisonous to breathe. Devise a method for detecting the presence of benzene in air at such concentrations.

(*c*) Select, if necessary by experiment, pairs of solvents that are (i) miscible, and (ii) non-miscible from water, benzene, glycerol, carbon tetrachloride. In the non-miscible pairs indicate which component would form the upper layer.

3. What is unsatisfactory about the following statements?

(*a*) Benzene and nitrobenzene are both sparingly soluble in concentrated sulphuric acid at 20°;

(*b*) Nitration of benzene (26 g.) gives nitrobenzene (35 g.; 85·37% yield)?

4. (*a*) Will a methyl group attached to a benzene ring protrude above and below the effective level of the π-orbital on each side of the ring?

(*b*) What would you expect to be formed (i) by hydrogenation of toluene over a rhodium/alumina catalyst, and (ii) by ozonolysis of *o*-xylene?

5. (*a*) When an excess of methyl chloride (gas) is passed into a mixture of benzene and aluminium chloride, a colourless crystalline hydrocarbon, m.p. 165°, mol. wt. 162, is obtained. What is the structure of this compound? Outline a practical procedure for separating this product from the inorganic part of the reaction mixture.

(*b*) Comment on the following prices (per 100 g. for each compound).

benzyl bromide	18/–	*m*-bromotoluene	70/–
o-bromotoluene	45/–	*p*-bromotoluene	35/–

6. Explain the following observations:

(*a*) A compound, C_9H_{12}, gives the following spectral data: infrared absorption at 3030, 1610, 1500 and 690 cm.$^{-1}$; ultraviolet maxima at λ 200 (ϵ 8000) and 265 mμ (ϵ 260); n.m.r. absorption comprising a singlet at 2·8, a multiplet at 6·9 and a doublet at 8·75 τ of relative intensities 5:1:6.

(*b*) Reaction of cumene with bromine in the presence of light at 0° gives 2-bromo-2-phenylpropane in 94% yield.

7. (*a*) How many isomeric compounds have the molecular formula, $C_{18}H_{12}$, and contain four benzenoid rings?

(*b*) When naphthalene is reduced in boiling ethanol solution by the addition of sodium, a hydrocarbon of composition, $C_{10}H_{12}$, is formed. What is the most likely structure for this compound?

(*c*) Is there anything wrong with the following equation

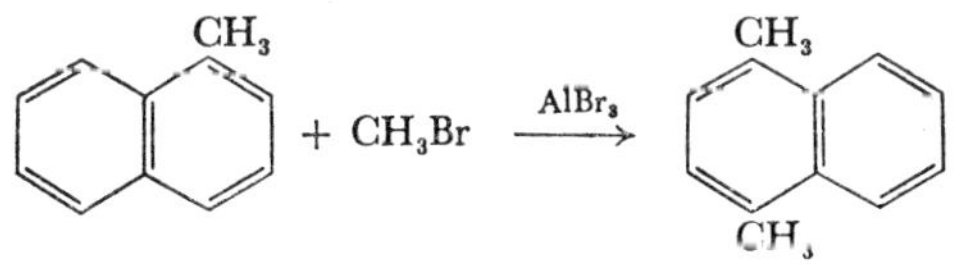

Every discovery opens a new field for investigation and shows us the imperfections of our theories.
HUMPHREY DAVY (1779–1829)

CHAPTER 16

ARYL HALIDES, PHENOLS AND RELATED COMPOUNDS

ARYL HALIDES

The four mono-halogen derivatives of benzene are liquids:

	b.p.	*Density*	*Refractive index*, n_D
fluorobenzene, C_6H_5F	87·5°	1·024	1·466
chlorobenzene, C_6H_5Cl	132°	1·107	1·525
bromobenzene, C_6H_5Br	156°	1·494	1·560
iodobenzene, C_6H_5I	188·5°	1·831	1·620

Chloro- and bromo-benzenes are prepared by electrophilic halogenation of benzene (pp. 188–191), and the fluoro- and iodo-compounds can be made from aniline, $C_6H_5NH_2$, *via* diazonium salts (p. 244). Electrophilic chlorination of toluene affords a mixture of *o*- and *p*-chlorotoluenes; the *m*-isomer is prepared by another route (p. 244). The formation of the isomeric dichlorobenzenes is discussed below, and the preparation of 1,3,5-tribromobenzene from aniline is given later (p. 244).

Reactions at the Carbon–Halogen Bond. *Metals.* Magnesium dissolves in a dry solution of bromobenzene in ether to give phenylmagnesium bromide. This Grignard reagent adds to carbonyl bonds, including that of carbon dioxide:

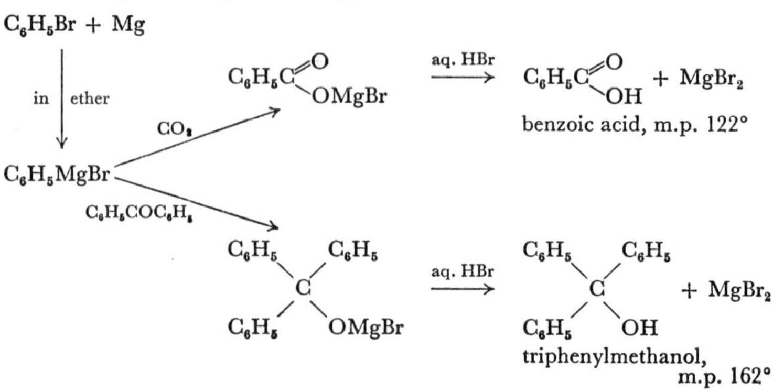

Nucleophilic reagents. Like vinyl halides (p. 141) monohalogenated benzenes do not react with nucleophilic reagents at ordinary temperatures. However, hydrolysis and other reactions can be forced to proceed by use of more drastic reaction conditions. Thus chlorobenzene is hydrolysed by aqueous sodium hydroxide solution at 300° and 150 atmospheres pressure; sodium phenoxide, the sodium salt of phenol (see below) is produced.

$$\underset{}{C_6H_5Cl} + \text{aq. NaOH} \longrightarrow \underset{\text{sodium phenoxide}}{C_6H_5O^{\ominus}\,Na^{\oplus}} \xrightarrow{HCl} \underset{\text{phenol}}{C_6H_5OH} + NaCl$$

The chlorine of chlorobenzene can also be replaced by an amino-group. When the halogen compound is heated with aqueous ammonia at 200° under pressure with a catalytic amount of cuprous oxide, the primary amine, aniline, is formed in good yield.

$$C_6H_5Cl + 2NH_3 \xrightarrow{Cu_2O} \underset{\text{aniline}}{C_6H_5NH_2} + NH_4Cl$$

Electrophilic Substitution. Halogen atoms attached to an aromatic ring show *o*, *p*-directive effects in electrophilic substitution reactions. Chlorination of chlorobenzene in the presence of aluminium chloride affords a mixture of *o*- and *p*-dichlorobenzenes that can be separated by careful fractional distillation (the *p*-isomer is used as a moth repellent):

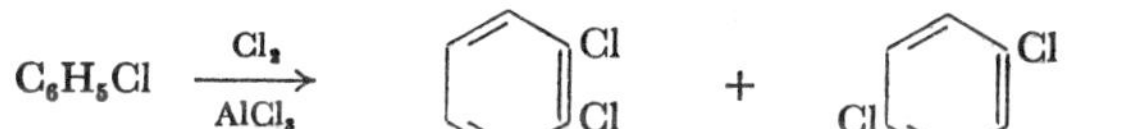

o-dichlorobenzene, b.p. 179°   *p*-dichlorobenzene, b.p. 173° (m.p. 53°)

Nitration of chlorobenzene with nitric and sulphuric acids at 70° proceeds similarly:

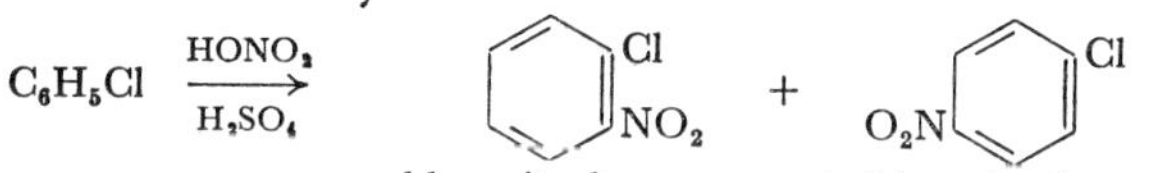

o-chloronitrobenzene, m.p. 32°   *p*-chloronitrobenzene, m.p. 84°

If these reactions are considered in terms of the relative stabilities of the three possible intermediate hybrid cations, the directive effect of the chlorine substituent is seen to be due to its stabilisation of the *o*- and *p*-hybrid cations by electron donation. This effect is most readily seen in a simple chloro-substituted carbonium ion, $>\overset{\oplus}{C}-Cl$. Carbonium ions are normally unstable (difficult to form) because of their electronic configuration (six electrons only in the outer shell, p. 57), but a chlorine substituent stabilises the ion by donation of electrons from one of its unbonded pairs, $>\overset{\oplus}{C}\text{—}\ddot{\underset{\cdot\cdot}{Cl}}:$ This process gives the alternative formula, $>C{=}\overset{\oplus}{\underset{\cdot\cdot}{Cl}}:$ in which both atoms have an octet of electrons. Such a carbonium ion has therefore a hybrid, resonance-stabilised structure (1):

(1)

The three hybrid cations that may be formed in the nitration of chlorobenzene are:

ortho attack

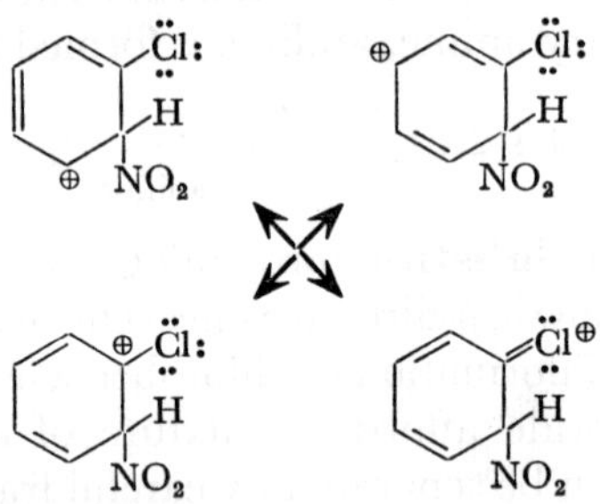

(2)

meta attack

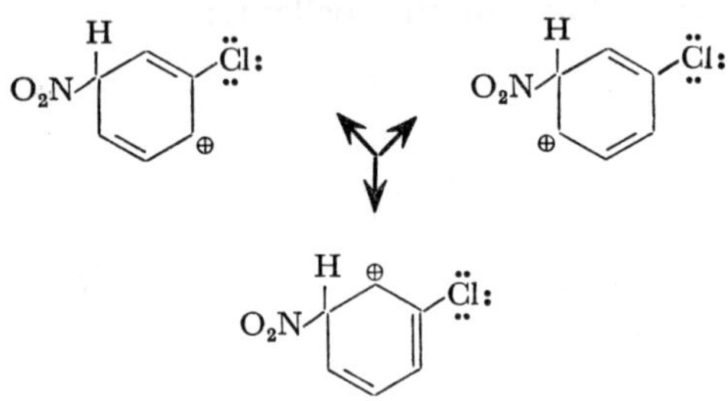

para attack

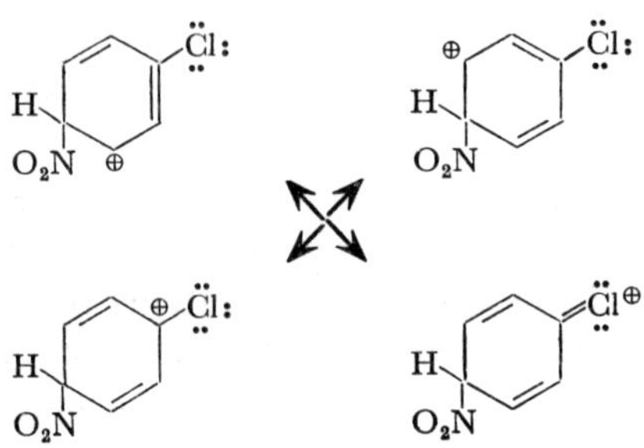

(3)

Four forms contribute to each of the *o*- and *p*-hybrid cations whereas the *m*-intermediate has only three forms. Electron donation from the chlorine substituent is involved in the additional forms (2) and (3), as in the simpler system (1). This additional stabilisation is denied to the *m*-hybrid cation, and therefore *o*-, and *p*-nitration occurs.

As chloro- and nitro-substituents (p. 239) are *o*, *p*- and *m*-directive groups respectively, further nitration of *o*- or *p*-chloronitrobenzene yields 2,4-dinitrochlorobenzene.

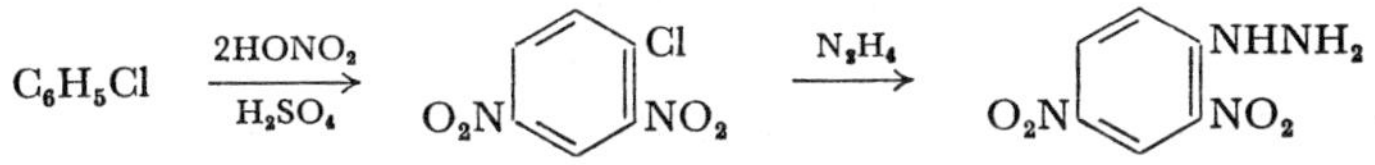

2,4-dinitrochlorobenzene, m.p. 52° 2,4-dinitrophenylhydrazine, m.p. 200°

Reaction of 2,4-dinitrochlorobenzene with hydrazine gives 2,4-dinitrophenylhydrazine, an extremely useful compound for the identification of aldehydes and ketones (p. 77).

DDT, a particularly potent insecticide, is prepared from chlorobenzene and chloral:

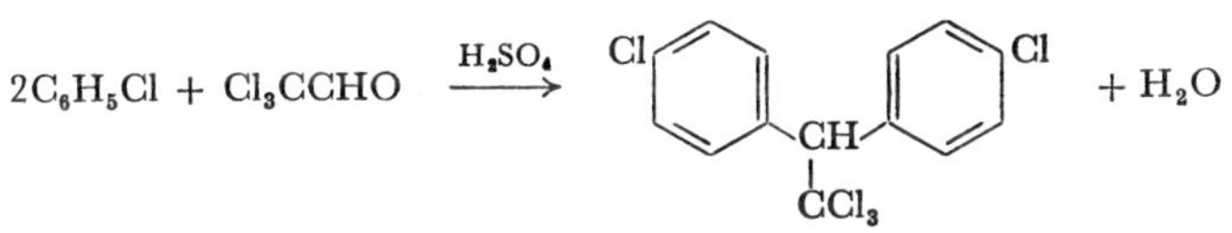

DDT
('dichlorodiphenyl' trichloroethane)

The reaction proceeds by an electrophilic substitution mechanism and some *o*-substituted products are also formed.

BENZYL HALIDES AND RELATED COMPOUNDS

Benzyl halides have the general formula, $C_6H_5CH_2$-halogen; the individual compounds are:

benzyl fluoride, b.p. 140°
benzyl chloride, b.p. 179°
benzyl bromide, b.p. 210°
benzyl iodide, m.p. 24°

Benzyl chloride is made by radical chlorination of toluene (p. 197); when bromine is used benzyl bromide is obtained. The properties of these halides are similar to those of alkyl halides. Nucleophilic substitution reactions occur readily, and benzyl fluoride and iodide are made by the methods used in the aliphatic series (p. 38):

$$C_6H_5CH_2Cl + NaI \xrightarrow[\text{acetone}]{\text{in}} C_6H_5CH_2I + NaCl$$

$$2C_6H_5CH_2Br + HgF_2 \xrightarrow[\text{chloroform}]{\text{in}} 2C_6H_5CH_2F + HgBr_2$$

Benzylidene chloride, $C_6H_5CHCl_2$, is formed when benzyl chloride is chlorinated in the presence of light, and by continuing the process, benzotrichloride, $C_6H_5CCl_3$, is produced.

$$C_6H_5CH_2Cl \xrightarrow[-HCl]{Cl_2} C_6H_5CHCl_2 \xrightarrow[-HCl]{Cl_2} C_6H_5CCl_3$$

Hydrolysis of these readily-available compounds, for example, with calcium hydroxide and water, yields the following oxygenated compounds.

$$C_6H_5CH_2Cl \longrightarrow C_6H_5CH_2OH$$

benzyl alcohol, b.p. 205·5°

$$C_6H_5CHCl_2 \longrightarrow C_6H_5C{\overset{\displaystyle O}{\underset{\displaystyle H}{\lessgtr}}}$$

benzaldehyde, b.p. 179°

$$C_6H_5CCl_3 \longrightarrow C_6H_5C{\overset{\displaystyle O}{\underset{\displaystyle OH}{\lessgtr}}}$$

benzoic acid, m.p. 122°

In most respects benzyl alcohol behaves as a typical primary aliphatic alcohol. Oxidation gives benzaldehyde and then benzoic acid, cf. Chapter 17.

PHENOLS

Aryl alcohols, compounds in which aryl and hydroxyl groups are united, are generally known as *phenols*, from the name of the parent compound. Phenol, C_6H_5OH, first obtained in a pure form from coal tar in 1847, is a crystalline substance, m.p. 41°, with a characteristic smell. Lister (1867) first used a dilute aqueous solution of phenol as an antiseptic for surgical operations, but more active and less toxic compounds are now available.

Phenol is obtained from coal tar, but increasing amounts are being made by synthesis, e.g. by hydrolysis of chlorobenzene (p. 203). Phenol is an important chemical intermediate, e.g. for the preparation of pharmaceutical products, dyestuffs, and plastics. Baekeland submitted the first significant patent on the preparation of a plastic in 1909. Bakelite, used for example as an insulating material in electrical fittings, is made by heating phenol with formalin. Plastic resins are also used for binding plywood.

Other procedures for the preparation of phenols include the fusion of aryl sulphonic salts with alkali (p. 231) and the decomposition of diazonium salts in water (p. 243). These methods are useful for obtaining many substituted phenols.

Reactions at the Hydroxyl Group. Phenol is a weak acid (cf. its early name, carbolic acid) and dissolves in aqueous sodium hydroxide to give the salt, sodium phenoxide, $C_6H_5O^{\ominus}Na^{\oplus}$. The following dissociation constants show that the acid strength of phenol lies between those of typical alcohols and carboxylic acids.

K_a	CH_3CH_2OH	$7{\cdot}3 \times 10^{-20}$
K_a	C_6H_5OH	$1{\cdot}3 \times 10^{-10}$
K_a	CH_3COOH	$1{\cdot}85 \times 10^{-5}$

(For the ionisation of phenol in water at 25°, $\Delta H = 5{\cdot}7$ and $T\Delta S = -8$ kcal./mole.) As a result of such large differences in acidity it is possible to separate a mixture of an alcohol, a phenol and a carboxylic acid by chemical methods. Phenols do not react with sodium bicarbonate, which will therefore extract only the carboxylic acid from a dichloromethane solution of such a three component mixture. The residual mixture of alcohol and phenol can then be separated by extraction of the latter component with aqueous sodium hydroxide. There are physical limitations on the scope of the method; phenol can be separated from benzyl alcohol but not from the water-soluble ethanol.

Phenols resemble alcohols in that ethers, carboxylic esters and sulphonic esters can be prepared. The following reactions are typical.

$$C_6H_5OH \xrightarrow[(CH_3)_2SO_4]{\text{aq. NaOH}} C_6H_5OCH_3$$

anisole, b.p. 154°
(methyl phenyl ether)

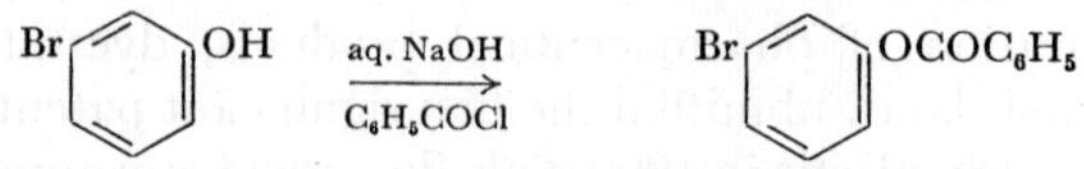

m-bromophenol — *m*-bromophenyl benzoate, m.p. 86°

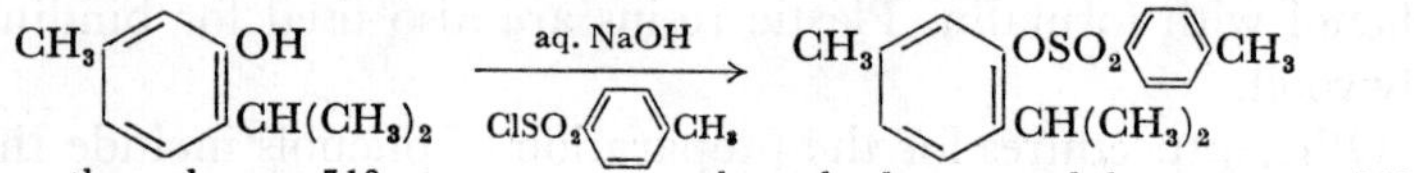

thymol, m.p. 51° (present in oil of thyme) — thymyl toluene-*p*-sulphonate, m.p. 71°

Sodium hydroxide is included in these reaction mixtures in order to convert the phenols into phenoxide ions which are the more reactive nucleophiles (cf. methylation of alkoxides, p. 36). Aryl methyl ethers can also be prepared by reaction of phenols with diazomethane:

$$ArOH + CH_2N_2 \longrightarrow ArOCH_3 + N_2$$

Diazomethane is a yellow poisonous gas, b.p. $-23°$, but can be handled safely as a dilute solution in ether, wherein it reacts rapidly with acidic compounds, for example carboxylic acids, phenols, and enols, and liberates nitrogen.

Aryloxyacetic acids, $ArOCH_2COOH$, are useful compounds for the identification of phenols. These acids, which are generally crystalline and sharp melting, are prepared by heating a mixture of the sodium salts of the phenol and chloroacetic acid in water or aqueous ethanol solution. When reaction is complete, addition of mineral acid liberates the aryloxyacetic acid:

$$ArO^{\ominus}\,Na^{\oplus} + ClCH_2COO^{\ominus}\,Na^{\oplus} \longrightarrow ArOCH_2COO^{\ominus}\,Na^{\oplus} + NaCl$$

$$ArOCH_2COO^{\ominus}\,Na^{\oplus} + HCl \longrightarrow ArOCH_2COOH + NaCl$$

Infrared spectroscopy (hydroxyl absorption band near 3600 cm.$^{-1}$) and colour tests are helpful for the recognition of phenolic groups. Ferric chloride, in aqueous solution gives blue or violet colours with many phenols; catechol (p. 210) is exceptional in giving a green colour.

Reactions in the Nucleus. Phenol is much more reactive than benzene towards electrophilic reagents; halogenation, for example occurs without aid from a catalyst. Reaction between equimolar quantities of phenol and chlorine yields a mixture of *o*- and *p*-chlorophenols, separable by distillation. Further chlorination gives 2,4,6-trichlorophenol, and then pentachlorophenol, a useful wood preservative.

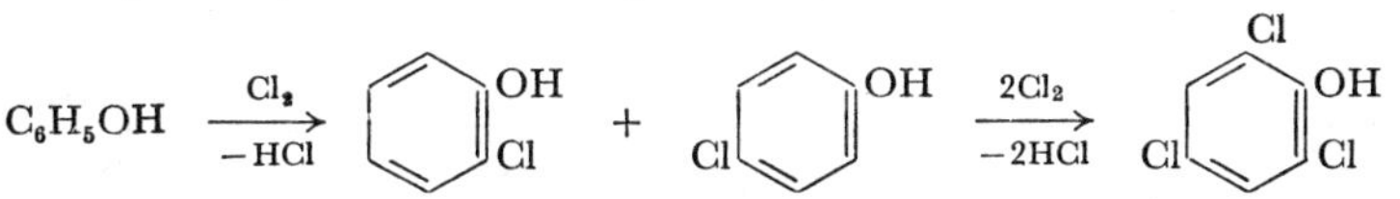

o-chlorophenol, b.p. 175·5° *p*-chlorophenol, b.p. 217° (m.p. 37°) 2,4,6-trichlorophenol, m.p. 68°

Nitration of phenol also proceeds readily. A mixture of *o*- and *p*-nitrophenols is first obtained, from which the exceptionally volatile *o*-isomer can be separated by steam distillation.

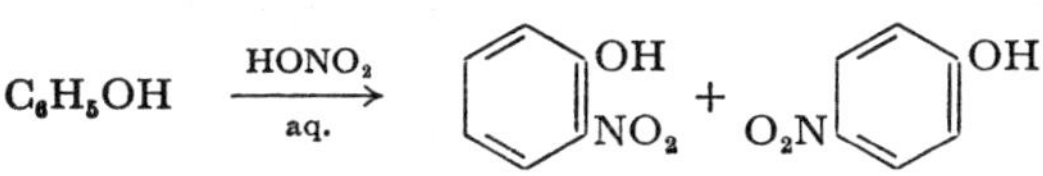

o-nitrophenol, m.p. 45° *p*-nitrophenol, m.p. 114°

An emulsion of phenol (1 mol.) in water at 25° is added gradually to an aqueous solution of nitric acid (1·25 mol). The reaction is exothermic and the temperature is kept below 30° by external cooling. The mixture is then kept for a few hours in order to allow the remainder of the reaction to go to completion. The aqueous layer and the crude product (a dark oil) are separated, and the latter is heated in a current of steam to drive out the volatile *o*-nitrophenol, which collects in the cooled aqueous distillate as a yellow solid (25% yield). *p*-Nitrophenol is extracted from the black residue in the distillation flask by hot water, and is obtained as off-white crystals (15%) on cooling the extract. Colourless material is obtained by sublimation under reduced pressure.

Nitration of either mononitrophenol yields 2,4-dinitrophenol and then 2,4,6-trinitrophenol (picric acid), a useful compound for the identification of amines (p. 116).

The *o*, *p*-directive properties that the hydroxyl group shows in the chlorination and nitration reactions of phenol are similar in origin to those of the chlorine atom in the related reactions of chlorobenzene (p. 203). Only the *o*- and *p*-hybrid cationic intermediates in the phenol reactions are stabilised by electron donation from the oxygen-containing group, $>\overset{\oplus}{C}—\ddot{\underset{..}{O}}—H \leftrightarrow >C=\overset{\oplus}{\underset{..}{O}}—H$.

A carboxyl group can be introduced into phenol by heating sodium phenoxide with carbon dioxide at 120–140° under pressure. Sodium salicylate is formed, from which salicylic acid (*o*-hydroxybenzoic acid) is obtained by acidification:

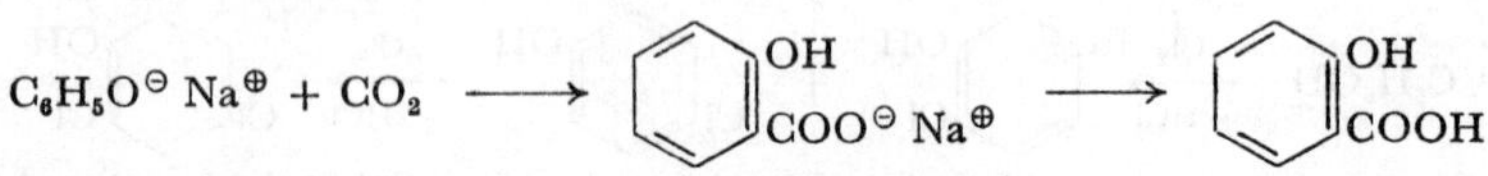

sodium salicylate salicylic acid

This carboxylation reaction is interesting in that the *o*-isomer is obtained. Salicylic acid occurs naturally as the ester, methyl salicylate (oil of wintergreen). Acetylsalicylic acid (aspirin) is the well-known analgesic; about 4000 million tablets of aspirin and related salicylates are consumed annually in the United Kingdom.

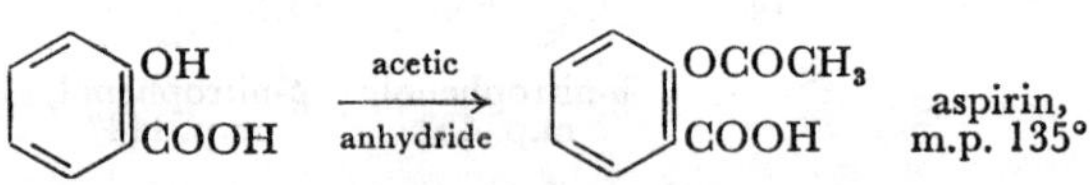

aspirin, m.p. 135°

DIHYDROXYBENZENES AND QUINONES

The three dihydroxybenzenes are colourless crystalline solids:

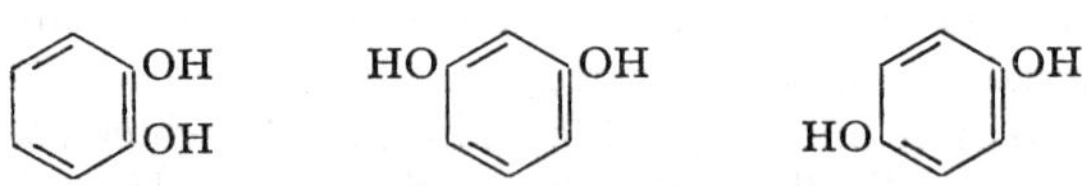

catechol, m.p. 105° resorcinol (p. 232), m.p. 110° hydroquinone (quinol) m.p. 171°

Lignin, a component (20–30%) of wood, has the partial structure (4) in which R represents various complex organic groups. The aqueous part of the distillate from the thermal decomposition of wood contains methanol, formed from the methoxyl (OCH_3) groups of lignin. The oily part of the distillate,

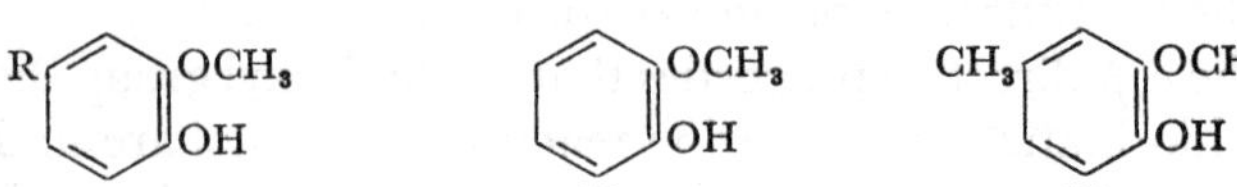

(4) (5) (6)

wood creosote, contains phenols, especially guaiacol (5), the monomethyl ether of catechol, and creosol (6), a methyl-guaiacol. The phenolic constituents of wood, including lignin, may function in part as biological preservatives by inhibiting the attack of microorganisms, cf. the antiseptic properties of simple phenols. Pinosylvin, an antifungal substance, and a styryl derivative of resorcinol, has been isolated from the heartwood of *Pinus sylvestris.*

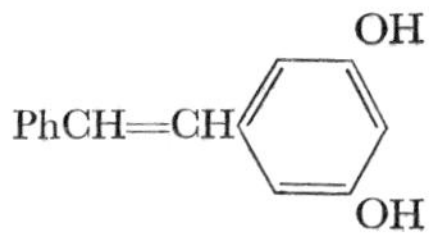

pinosylvin

Quinones are compounds in which two hydrogens of an aromatic nucleus have been replaced by two oxygen atoms. In the benzene series, *o*-benzoquinone and *p*-benzoquinone are known. *o*-Benzoquinone is a rather unstable red compound obtained by dehydrogenation of catechol: a solution of catechol in ether is shaken with lead dioxide (freshly prepared by hydrolysis of lead tetra-acetate) and sodium sulphate which absorbs the water produced in the reaction.

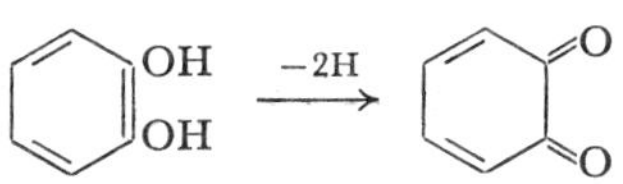

catechol *o*-benzoquinone (decomposes when heated)

p-Benzoquinone was the first quinone to be isolated (1838). Oxidation of the natural compound, quinic acid, gave a yellow, volatile crystalline product that was named 'quinone'. Later this word came to be used in a wider sense (cf. above definition), and the compound from quinic acid was named more specifically (*p*-benzoquinone) as a derivative of benzene.

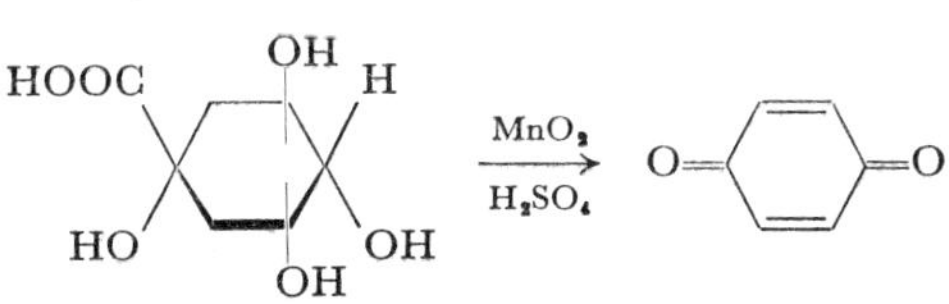

quinic acid *p*-benzoquinone, m.p. 116°

p-Benzoquinone can be produced on a larger scale by oxidation of aniline. Hydroquinone is obtained by reduction of *p*-benzoquinone; conversely, oxidation of hydroquinone yields the quinone. The interconversion of these two compounds by reduction and oxidation reactions proceeds so readily that it can be studied by electrochemical procedures:

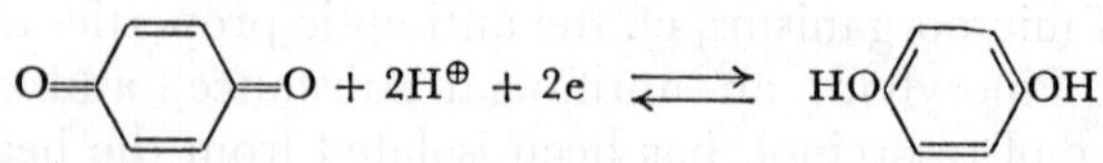

The reducing properties of hydroquinone are employed in photography; as a 'developing' agent it reduces the photo-activated grains of silver halide to silver.

Quinones and compounds with quinone-like (quinonoid) structures are essential components of living cells, where they function as hydrogen acceptors in sequences of oxidation-reductions. Riboflavin (vitamin B_2) has a quinonoid structure; two carbons and one oxygen of the normal quinone ring are replaced by nitrogen. The polyhydroxy side-chain is related in structure to ribose (p. 177):

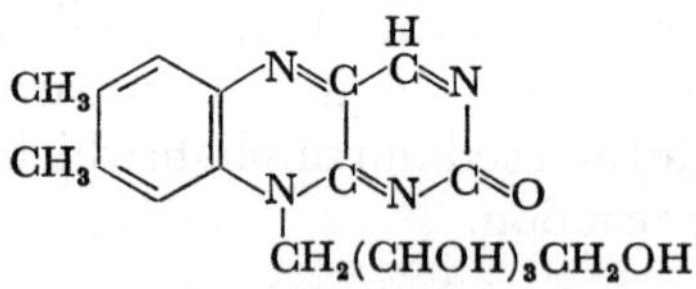

riboflavin

QUESTIONS

1. (*a*) Are there any regular trends in the physical properties of 1-halobutanes (p. 33) and phenyl halides?

(*b*) Arrange the three isomeric dichlorobenzenes in the increasing order of the sizes of their dipole moments.

2. (*a*) How many naphthaquinones, $C_{10}H_6O_2$, can, in theory, exist?

(*b*) Write a balanced equation for the oxidation of quinic acid to benzoquinone (p. 211). The manganese dioxide is converted into manganese (II) sulphate during the reaction.

(*c*) What is the structure of the yellow crystalline compound, $C_6Cl_4O_2$, that separates when an excess of chlorine is passed into a solution of hydroquinone in aqueous hydrochloric acid?

3. Comment on the following observations: (*a*) 1-chlorohexane and benzyl chloride react with warm aqueous alcoholic silver nitrate to give a white precipitate, but chlorobenzene does not; (*b*) treatment of phenol (in water at 20°) with bromine (3 mol.) gives a colourless solid, $C_6H_3Br_3O$, and (*c*) a similar reaction using bromine (4 mol.)

affords a yellow product, $C_6H_2Br_4O$, which shows unsaturated ketone absorption in the infrared.

4. (*a*) Comment on the fact that although phenol has a m.p. near 41°, it forms a separate *liquid* layer when kept with water at 25°.

(*b*) Draw formulae to show how guaiacol could form a hydrogen bond within one molecule, and how it could form a hydrogen bond between two or more molecules.

(*c*) Interpret the following observations. When catechol (1 mol.) and phosgene (1 mol.) are shaken with aqueous sodium hydroxide, a solid, $C_7H_4O_3$, m.p. 118°, is formed in good yield. Under the same conditions, guaiacol gives a compound, $C_{15}H_{14}O_5$, m.p. 85°, in poor yield. Acid hydrolysis of these products gives carbon dioxide and the respective phenol.

5. *o*-Methoxynitrobenzene reacts with acetyl chloride and aluminium chloride to give a product in 84% yield. Predict the structure of the product and suggest a mechanism for the reaction. Draw an energy profile and comment on the role of the aluminium chloride.

6. (*a*) Outline routes from toluene to the following compounds.

(i) $C_6H_5CH_2{}^{82}Br$ (ii) $C_6H_5CH_2SH$, (iii) $C_6H_5CH_2{}^{14}COOH$
(iv) $C_6H_5CH_2{}^2H$ (v) $(C_6H_5CH_2S)_2Hg$

(*b*) Which of the compounds in question (*a*) would be radioactive? By what technique can radioactivity be detected? How could the presence of an unusual isotope in the non-radioactive compounds be detected?

(*c*) Give essential practical details for carrying out any one of the reactions in question (*a*) and for isolating the organic product.

7. (*a*) A selective weedkiller is made by refluxing a solution of sodium 2,4-dichlorophenoxide and sodium chloroacetate in aqueous methanol, followed by acidification. Formulate the reaction and suggest a mechanism for it.

(*b*) The greater acidity of phenols compared to alcohols can be attributed to stabilisation of the phenoxide ion. Write the hybrid structure of the anion in terms of four contributing forms, and draw comparative energy profiles for the ionisation of alcohols and phenols. The pK_a values (where $pK_a = -\log K_a$) for phenol, *p*-nitrophenol, and 2,4-dinitrophenol are 9·9, 7·2, and 4·0. Discuss these values in relation to the hybrid structures of the anions, and the possibilities for solvation (e.g. with water) at various positions in each of the anionic forms.

(*c*) Employing general reactions given in the book, propose a several stage route for the conversion of phenol into hexane-1,6-diol.

The faculty of imagination is the principal source of human improvement.

DUGALD STEWART (1753–1828)

CHAPTER 17

ARYL ALDEHYDES, KETONES, CARBOXYLIC ACIDS AND SULPHONIC ACIDS

ALDEHYDES AND KETONES

The simplest aromatic aldehyde is benzaldehyde, C_6H_5CHO, a liquid with a pleasant almond-like odour. In fact, almond seeds contain a substance, amygdalin, which on hydrolysis with dilute acid yields benzaldehyde, hydrogen cyanide and glucose:

$$C_6H_5CH(CN)OC_{12}H_{21}O_{10} \xrightarrow{2H_2O} C_6H_5CHO + HCN + 2C_6H_{12}O_6$$

'Oil of bitter almonds' provided the first source of benzaldehyde, but the compound is now prepared by hydrolysis of benzylidene chloride or by controlled oxidation of benzyl alcohol.

Aromatic ketones can be divided into aryl alkyl ketones, ArCOR, and diaryl ketones, ArCOAr. A method for the preparation of the parent compound of each series, acetophenone, $C_6H_5COCH_3$, and benzophenone, $C_6H_5COC_6H_5$, has been discussed.

Aromatic aldehydes are named as derivatives of the parent benzaldehyde, although specific names for some compounds are also available:

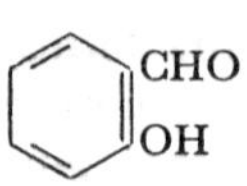

o-hydroxybenzaldehyde
or salicylaldehyde
(present in essential
oils of *Spiraea*)

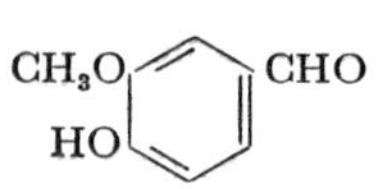

4-hydroxy-3-methoxy-
benzaldehyde
or vanillin
(present in vanilla beans)

Ketones derived from acetophenone or benzophenone are usually named accordingly, but the general ketone system of nomenclature is also used.

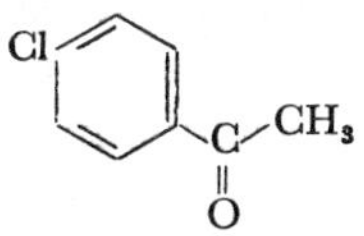

p-chloroacetophenone (*p*-chlorophenyl methyl ketone)

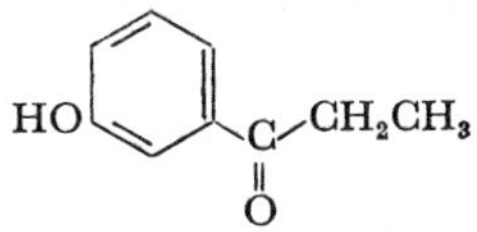

ethyl *m*-hydroxyphenyl ketone

Compounds in which the aryl and carbonyl groups are joined *via* one or more carbon atoms, e.g. phenylacetaldehyde ($C_6H_5CH_2CHO$) and benzyl methyl ketone ($C_6H_5CH_2COCH_3$) exhibit most of the properties of aliphatic carbonyl compounds.

Reactions in the Side-Chain. Aryl aldehydes and ketones behave like their aliphatic analogues towards nucleophilic reagents, addition (or addition-elimination) reactions occurring. For example, Grignard reagents add to aryl carbonyl compounds and, by the use of the appropriate reagents, acetophenone or benzophenone can each be converted into the tertiary alcohol, 1,1-diphenylethanol:

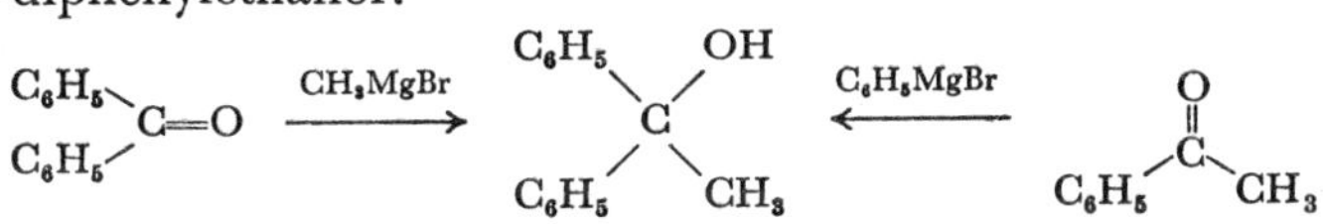

1,1-diphenylethanol, m.p. 81°

As in the aliphatic series, the formation of crystalline oximes, semicarbazones and 2,4-dinitrophenylhydrazones assists in the identification of aryl aldehydes and ketones.

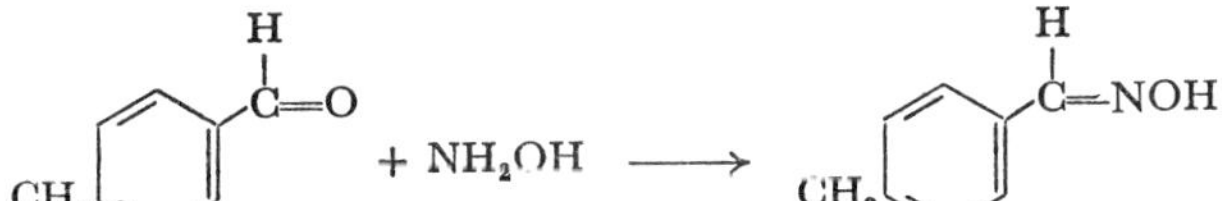

p-tolualdehyde (*p*-methylbenzaldehyde)

p-tolualdehyde oxime, m.p. 80°

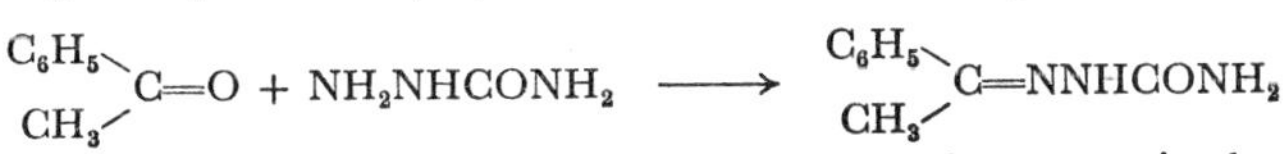

acetophenone semicarbazone, m.p. 198°

Two molecules of benzaldehyde cannot combine by an aldol reaction due to the absence of the group

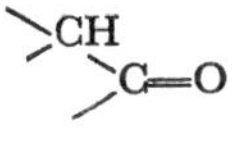

(cf. p. 81). Benzaldehyde does react, however, with acetaldehyde in alkaline solution to give cinnamaldehyde, formed *via* the hydroxy-aldehyde (1). This 'mixed' aldol reaction involves the nucleophilic addition of the anion $^{\ominus}CH_2CHO$ (from acetaldehyde) to benzaldehyde:

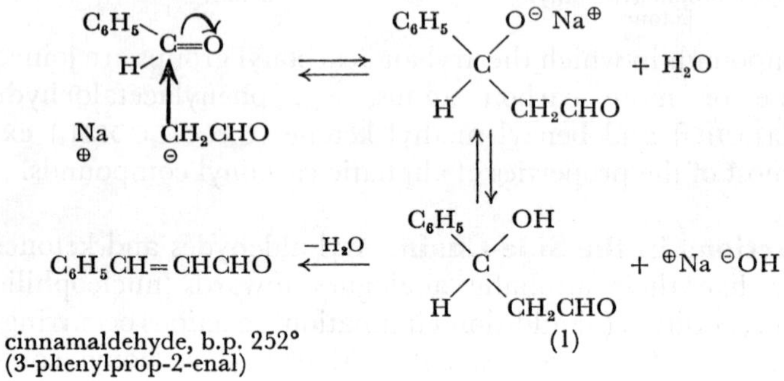

The final stage, elimination of water from intermediate hydroxy-aldehyde (1), is similar to the formation of crotonaldehyde from aldol.

Analogous base-catalysed reactions occur between benzaldehyde (and other aromatic aldehydes) and methyl ketones, including acetophenone. As aryl aldehydes and ketones are not very soluble in water these reactions are often performed in ethanol containing sodium ethoxide. Ethoxide anions behave like hydroxide anions (p. 80), and initiate each reaction by removing a proton from a C—H group next to a carbonyl bond.

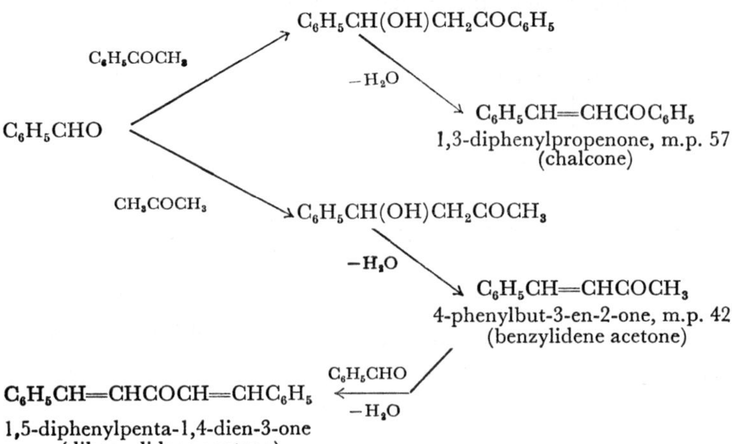

One or two benzylidene groups can be introduced into acetone; the unsaturated dibenzylidene acetone forms pale yellow crystals, m.p. 112°. The methyl groups of acetophenone and acetone show similar behaviour in their reactions with benzaldehyde. A further similarity is that acetophenone, like acetone, undergoes acid-catalysed halogenation at the methyl group:

$$\underset{\substack{\text{chloroacetophenone,}\\ \text{m.p. } 59^\circ\\ \text{(phenacyl chloride)}}}{C_6H_5COCH_2Cl} + HCl \xleftarrow[H^{\oplus}]{Cl_2} C_6H_5COCH_3 \xrightarrow[H^{\oplus}]{Br_2} \underset{\substack{\text{bromoacetophenone,}\\ \text{m.p. } 50^\circ\\ \text{(phenacyl bromide)}}}{C_6H_5COCH_2Br} + HBr$$

Cinnamic acid, $C_6H_5CH{=}CHCOOH$, is produced when a mixture of benzaldehyde, acetic anhydride and potassium acetate is heated under reflux. The function of the acetate anion is to detach a proton from the methyl group of acetic anhydride to give the resonance-stabilised anion (2), which then adds to the carbonyl group of benzaldehyde. These steps are mechanistically analogous to those occurring in the aldol reaction between acetone and benzaldehyde. The final step, conversion of the intermediate (3) into cinnamic acid, will not be considered in detail.

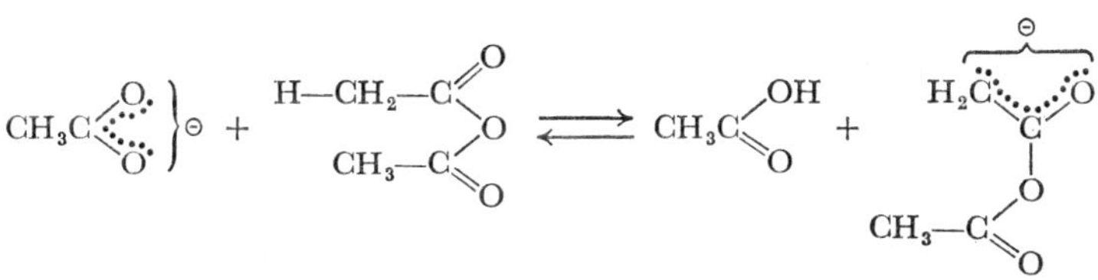
(2)

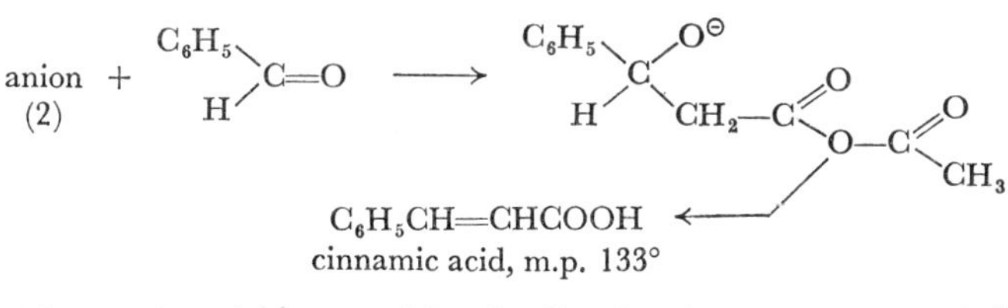
(3)

Cinnamic acid has a wide distribution in nature, especially in the form of esters.

Addition of hydrogen cyanide to benzaldehyde, catalysed by an alkali cyanide, yields the cyanohydrin (mandelonitrile). Hydrolysis of this nitrile with concentrated hydrochloric acid gives racemic mandelic acid, m.p. 118°:

$$C_6H_5CHO + HCN \xrightarrow{^{\ominus}CN} \underset{\text{mandelonitrile}}{C_6H_5CH(OH)CN} \xrightarrow[H^{\oplus}]{H_2O} \underset{\text{mandelic acid}}{C_6H_5CH(OH)COOH}$$

Hydrolysis of the natural glycoside, amygdalin (p. 214) with concentrated hydrochloric acid affords (–)-mandelic acid, m.p. 133°. As with the aliphatic analogue, lactic acid, the melting points of the racemic and optically active forms of mandelic acid are different.

Aryl aldehydes are readily oxidised to acids, and Tollens reagent (p. 82) can be used to differentiate them from the unreactive aryl ketones. Aliphatic and aromatic aldehydes give a band near 2800 cm.$^{-1}$ in the infrared spectrum which is not shown by the ketones of either series (p. 8).

Reactions in the Nucleus. Aldehydic and alkyl ketonic groups both exert *m*-directive effects on electrophilic substitution reactions in an aryl ring. Nitrations of benzaldehyde and acetophenone yield *m*-nitro-compounds as main products:

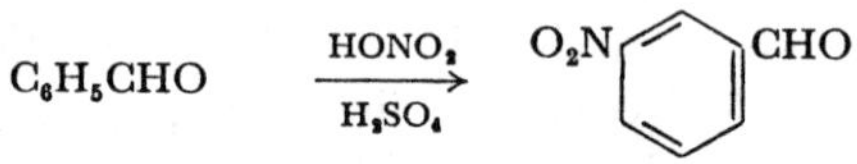

m-nitrobenzaldehyde, m.p. 58°

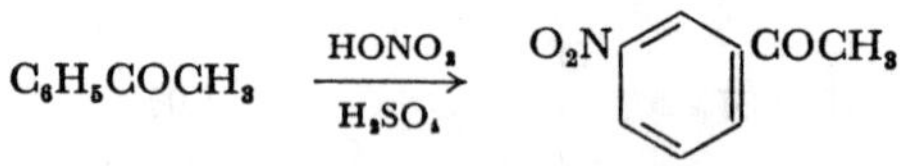

m-nitroacetophenone, m.p. 81°

In order to understand why *m*-nitration predominates in the reactions of these and related carbonyl compounds (p. 223) it is necessary to compare the hybrid cations obtained by addition of nitronium ions at *o*, *m* and *p* positions:

ortho attack

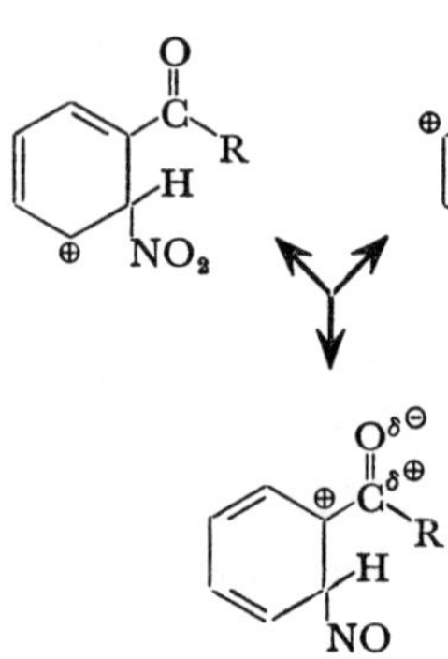

(4)

meta
attack

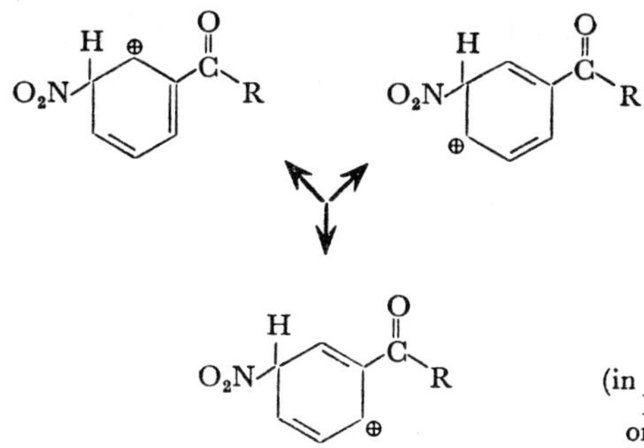

(in these formulae R=H or an organic group)

para
attack

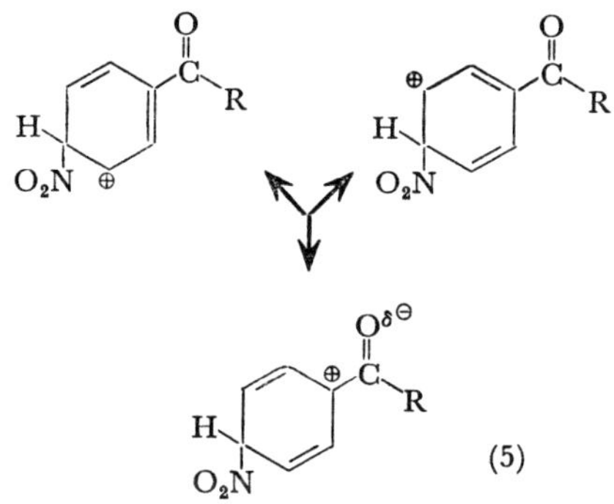

(5)

The *o*-hybrid and the *p*-hybrid each contain a form [(4) and (5) respectively] in which a positive charge is next to the $\delta^{\oplus}$ carbon of the carbonyl group, an electrostatically unfavourable situation because of the juxtaposition of positive charges. This situation does not arise in the hybrid formed by attack of the electrophile at the *m*-position. All three forms of this hybrid will contribute to its stabilisation by resonance. The *m*-hybrid is therefore the easiest to form and loss of a proton from it affords a *m*-nitro-compound.

Kinetic Effects in Electrophilic Aromatic Substitution Reactions. The rates of electrophilic aromatic substitution reactions are conveniently expressed relative to benzene, and the table gives some approximate results for the mono-nitration of benzene derivatives, C_6H_5R. The rate of nitration of phenol is 1000 times greater than that of benzene, whereas nitrobenzene nitrates much less readily than benzene. In terms of the energy profile (p. 190), the value of E_A for the nitration of phenol is lower than that for the corresponding reaction of benzene. The hydroxyl group must therefore stabilise the transition state

R	*Rates relative to benzene* = 1	%*o*-	%*m*-	%*p*-
OH	1000			
CH_3	24	56	4	40
Cl	$3{\cdot}3 \times 10^{-2}$	30	1	69
$COOC_2H_5$	3×10^{-4}	28	69	3
NO_2	*ca.* 10^{-7}	6	93	1

leading to the cationic intermediate of the reaction. The intermediate can be stabilised by resonance involving a lone pair of the hydroxyl group, cf. (*a*), and it appears that the transition state resembles the intermediate in this respect.

$$(a)\quad \gt\overset{\oplus}{C}-\overset{..}{\underset{..}{O}}-H \longleftrightarrow \gt C=\overset{\oplus}{\underset{..}{O}}-H$$

The opposite argument applies to carbonyl and to nitro-groups (p. 218).

This discussion of relative rates takes no account of the distribution of *o*-, *m*-, and *p*-isomers in nitration of monosubstituted benzenes. The percentage yields of products (see table) are clearly very different from the statistical values (40:40:20 for *o*:*m*:*p*). In fact the orientation depends on the relative rates of formation of the three isomers, *o*-, *p*-directive groups, for example halogen (p. 204), stabilising the carbonium ion involved in *o*- and in *p*-substitution (and lowering the energy of transition state A, p. 190) at the expense of that leading to *m*-substitution. *m*-Directive groups, e.g. $COCH_3$ (p. 218) reduce the stability of the intermediates responsible for *o*- and *p*-substitution, resulting in faster reaction at the *m*-position (but still decreasing the reactivity of the ring relative to benzene). The reactions of benzene derivatives, ArR, towards electrophilic reagents follows the order:

$$(R=)\qquad OH > OCH_3 > CH_3 > H > Cl > m\text{-directive groups}$$

The differences in reactivity have profound practical consequences; compare, for example, conditions employed for mono- and poly-nitration of benzene (p. 191, 239), of phenol (p. 209) and of chlorobenzene (pp. 203, 205) and for halogenation of phenol (p. 209), aniline (p. 245), benzene (p. 188), and nitrobenzene (question 5 on p. 248).

The rates of electrophilic reactions are determined by the type of reagent as well as by the nature of the substituent. For example, the rate of bromination of toluene at 25° is 2580 times that of benzene, whereas the figure is 28 for nitration. Activation values for the toluene–bromine reaction are $\Delta H^{\ddagger} = 11{\cdot}3$ and $T\Delta S^{\ddagger} = 9{\cdot}6$ kcal./mole. These results were obtained in trifluoroacetic acid, which provides a suitable medium for the more quantitative study of some electrophilic aromatic substitution reactions; good second-order plots are obtained, whereas the kinetic order is complex in many other solvents, including acetic acid. The $o:m:p$ ratio determined by gas chromatography also depends on the electrophilic reagent used:

$CH_3C_6H_4X$ *isomers from*	*Bromination and*	*Nitration of toluene*
o-	17·6	61·6
m-	<0·1	2·6
p-	82·4	35·8%

Molecular Complexes. When a solution of 1,2,4,5-tetramethylbenzene and 1,3,5-trinitrobenzene in hot acetic acid is cooled, a yellow 1:1 molecular adduct, m.p. 92–98°, is formed. In this type of adduct, often referred to as a π-complex, the

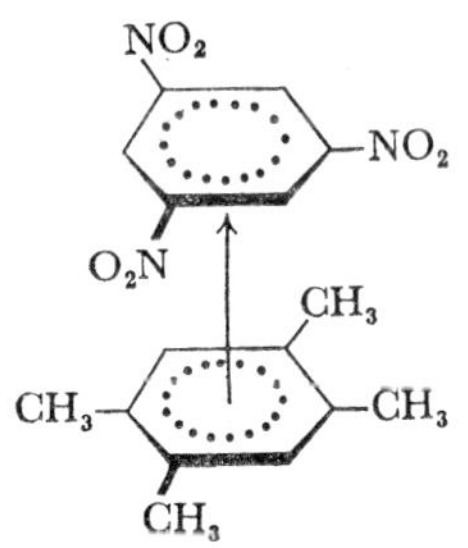

benzene rings lie in parallel planes as shown in the diagram and are held together by a transfer of electron density from one ring to the other. In the example the latter ring is the one that contains electron attracting nitro-groups. Molecular complexes formed from 1,3,5-trinitrobenzene and from picric acid (p. 209), are used to characterise and thence to purify aromatic hydrocarbons.

Related adducts may be formed initially in electrophilic substitution reactions. In nitration this initial step may be the slowest and hence the rate-determining one in the reaction, whereas in bromination the conversion of the π-adduct into the cationic intermediate seems to be the slowest step:

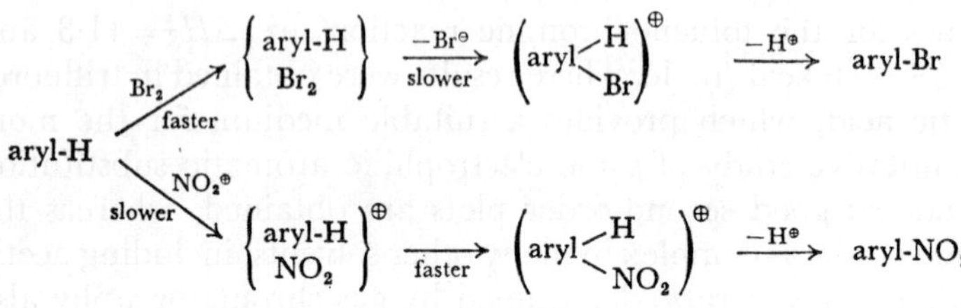

The contrast between the rates of bromination and nitration of toluene (relative to benzene) (p. 221) may be partly due to these factors.

MONOCARBOXYLIC ACIDS

Benzoic acid, C_6H_5COOH, occurs naturally in the form of esters and as the free acid, for example, in gum benzoin from which it was isolated (1608) by sublimation. It is now prepared by oxidation of toluene either directly or *via* chlorination to benzotrichloride followed by hydrolysis. General methods for the introduction of a carboxyl group into an aromatic system include the reactions of aryl Grignard reagents with carbon dioxide (p. 129), and the hydrolysis of aryl nitriles obtained for example, from amines *via* diazonium salts (p. 244). Readily available compounds such as benzoic acid and salicyclic acid (p. 210) can serve as starting materials for the synthesis of related compounds with additional substituents.

Aromatic acids are crystalline solids with melting points mostly in the range 100–250°.

Benzoic Acid and its Reactions. The properties of benzoic acid are typical of those of aromatic acids. Benzoic acid, m.p. 122°, is sparingly soluble in water but its dissociation constant, $K_a = 6{\cdot}6 \times 10^{-5}$, reveals that its acid strength is somewhat greater than that of acetic acid, $K_a = 1{\cdot}85 \times 10^{-5}$.

By the following reactions, benzoic acid can be transformed into esters and other acid derivatives:

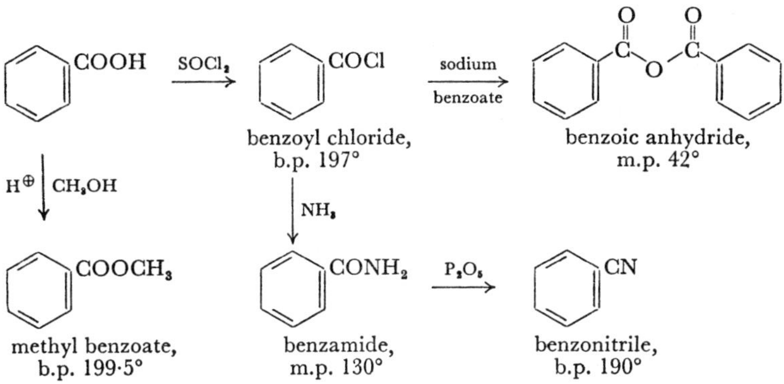

Like the carbonyl group a carboxyl substituent has a *m*-directive effect towards electrophilic reagents. Nitration of benzoic acid yields *m*-nitrobenzoic acid and then, more slowly, 3,5-dinitrobenzoic acid. Esters of the latter acid have reasonably high melting points and are often prepared as derivatives for the characterisation of alcohols, e.g. ethyl 3,5-dinitrobenzoate has m.p. 93°.

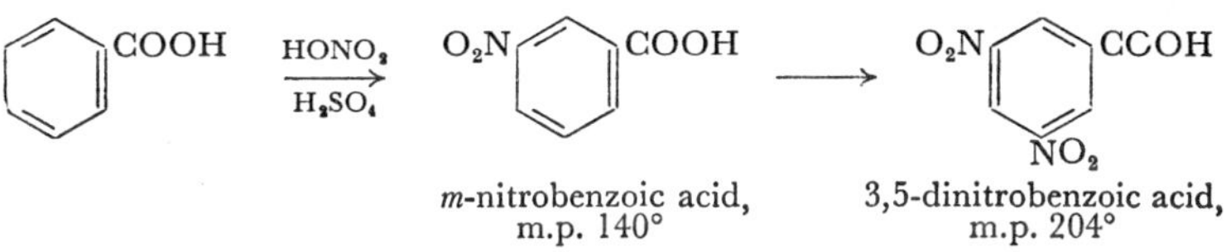

The derivatives of benzoic acid, methyl benzoate and benzonitrile, are also nitrated predominantly at *m*-positions.

Aryl Amino-acids. Two classes may be considered: compounds that contain the NH_2 and COOH groups as nuclear substituents and those that have both groups in a side-chain. Examples of the former type of compound are:

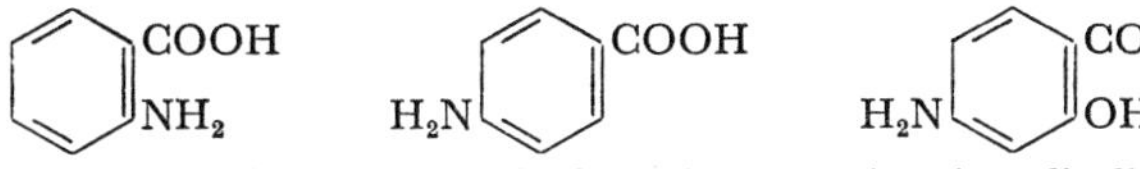

o-aminobenzoic acid (anthranilic acid) — *p*-aminobenzoic acid — 4-aminosalicylic acid (6)

p-Aminobenzoic acid is prepared by reduction of p-nitrobenzoic acid, readily available from the oxidation of p-nitrotoluene (p. 196). The local anaesthetic, procaine, is the β-diethylaminoethyl ester of p-aminobenzoic acid, and is made by the following route:

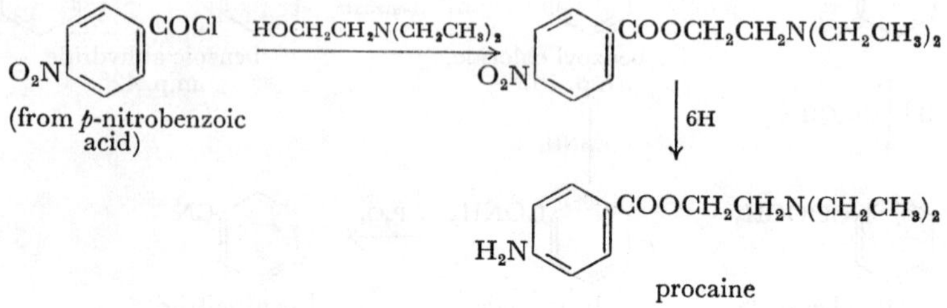

4-Aminosalicyclic acid (6), PAS, has been used since 1948 in the treatment of tuberculosis.

Two of the α-amino-acids that provide the building units of proteins contain benzene rings:

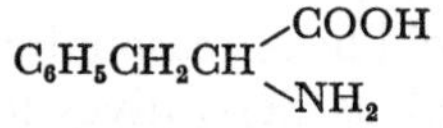

(−)-phenylalanine, $[\alpha]_D -4°$

(−)-tyrosine, $[\alpha]_D -10°$

(−)-Phenylalanine is an essential amino-acid for humans. Tyrosine, first obtained by Liebig in 1834, is a constituent of silk fibroin and of the hormone, oxytocin (p. 161).

C_5 monosaccharides (pentoses) are biological precursors of nucleic acids (p. 177), and in higher plants are also converted into many aromatic compounds, including phenylalanine and tyrosine. A C_3 component (7) (a phosphate ester of pyruvic acid enol, pp. 167–168) and a C_4 fragment (8), both derived from pentose phosphate, combine in a multi-stage process to give shikimic acid (9). Addition of a further molecule of the pyruvate derivative (7) furnishes prephenic acid (10) and thence phenylalanine and tyrosine.

The thyroid gland (situated in the neck) secretes an iodine-containing hormone, thyroxine, which is necessary for the maintenance of good health. Thyroxine was isolated from thyroid gland tissue in 1915, and its structure (11) was confirmed by synthesis in 1927. Thyroxine is synthesised in the body from tyrosine *via* diiodotyrosine. The compound actually responsible for thyroid activity in body cells is triiodothyronine, which is

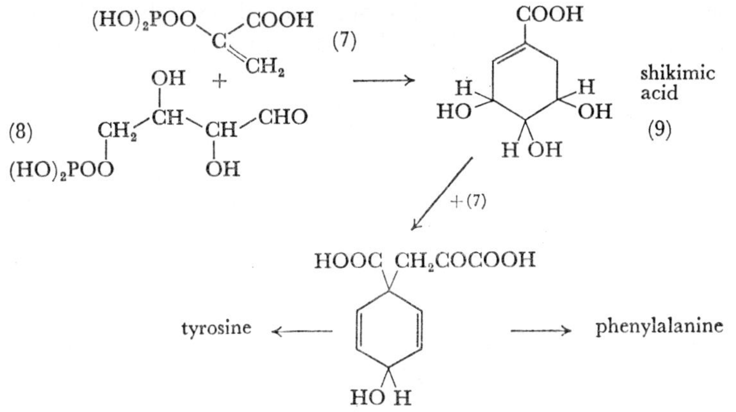

prephenic acid (10)

formed in the cell by deiodination of the thyroxine supplied by the thyroid gland.

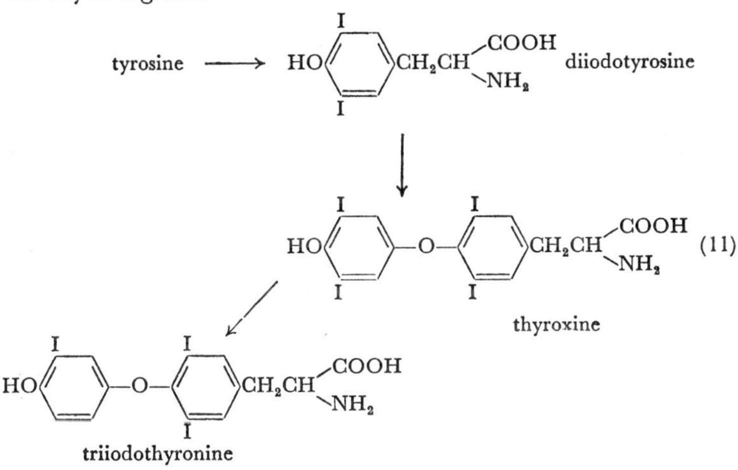

DIBASIC ACIDS

The isomeric benzene dicarboxylic acids are:

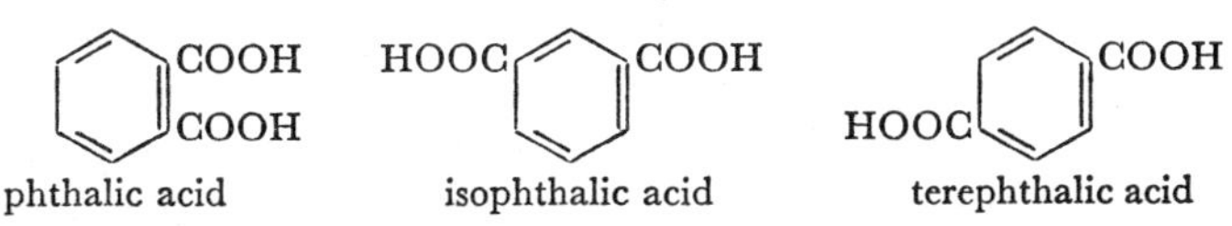

Phthalic Acid must be the *o*-isomer as it is the only one that forms a cyclic anhydride (12). This anhydride is produced on a large scale by aerial oxidation of naphthalene at 400–500° with vanadium pentoxide as catalyst. The amount of naphthalene available from coal tar is limited, and an alternative route to the anhydride involves the similar oxidation of *o*-xylene obtained from petroleum (p. 188). Reaction between the anhydride and water at 100° yields phthalic acid. Conversely, when phthalic acid is heated above its melting point, water is evolved and the anhydride is formed.

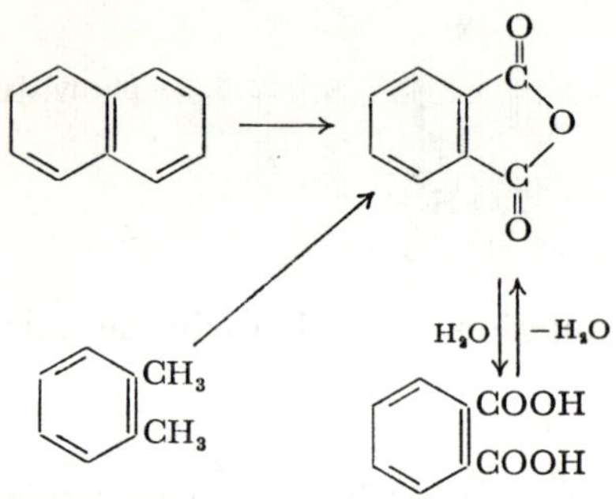

phthalic anhydride, m.p. 131·5° (12)

phthalic acid, m.p. 200–206° (with decomposition)

Reaction between phthalic anhydride and methanol affords the monomethyl ester of phthalic acid (methyl hydrogen phthalate). When sulphuric acid is present, esterification becomes complete with the formation of dimethyl phthalate.

(12) 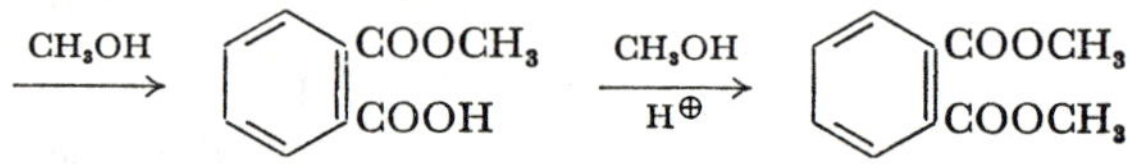

methyl hydrogen phthalate

dimethyl phthalate

Complex polyesters known as glyptal or alkyd resins are produced by reactions between phthalic anhydride and glycerol or other polyhydroxy-compounds. These polyesters are used extensively in the manufacture of paints and surface coatings. The tough, light materials formed by hardening of a mixture of glass-fibre and a glyptal resin by further polymerisation are finding an increasing number of constructional uses. The cyclic imide, phthalimide, is produced when phthalic anhydride is heated with ammonia at 200°. Phthalimide is sparingly soluble in water but dissolves readily in a dilute aqueous solution of potassium hydroxide. The acidic character of phthalimide is due to resonance stabilisation in the anion formed by removal of the proton from the NH group.

(12) 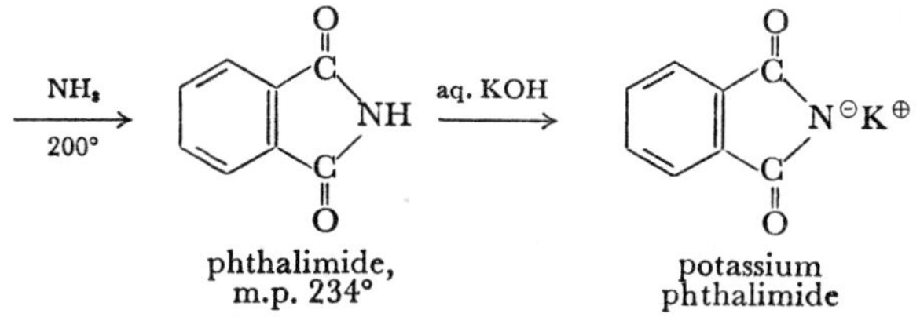

The potassium salt of phthalimide is used in the synthesis of aliphatic primary amines from alkyl halides *via* N-alkylphthalimides (question 8 on p. 249).

Phthalimide dissolves in a cold aqueous solution of ammonia, but when the solution is kept the diamide, phthalamide, separates. Dehydration of the diamide by heating it with acetic anhydride yields the dinitrile, phthalonitrile.

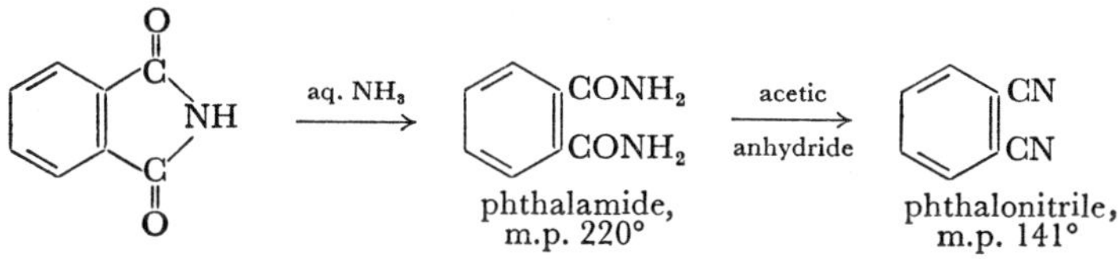

Phthalonitrile can be converted remarkably easily into phthalocyanine (13).

A mixture of the nitrile (1 g.) and glycerol (*ca.* 0·1 g.) is heated at 240–250° for 1 hour. After cooling, the residue is triturated with 10 ml. portions of hot ethanol until the extracts are pale in colour. The remaining crystals are then washed twice with 10 ml. portions of a 1:1 mixture of dichloromethane and ether. Phthalocyanine, permanganate-like in its crystalline appearance, is slightly soluble in some organic solvents to give greenish-blue solutions.

Metal derivatives of phthalocyanine can be obtained in which the two central hydrogens are replaced by the metal atom. The blue pigment, copper phthalocyanine, is prepared from phthalonitrile and copper powder at 200°. Its stability to many chemical reagents and to heat (it can be sublimed at 500° in a vacuum) is noteworthy.

The central ring of the phthalocyanine molecule is composed

of sixteen atoms, eight each of carbon and nitrogen. Sixteen-membered rings, consisting of twelve carbons and four nitrogens, are present in the biological compounds, haem (14) (cf. p. 160) and chlorophyll.

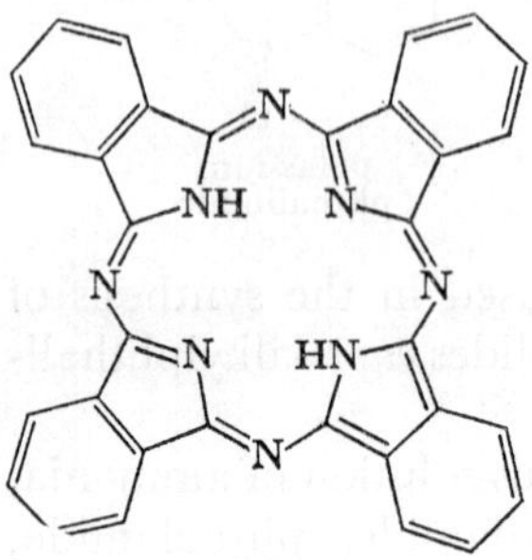

(13)

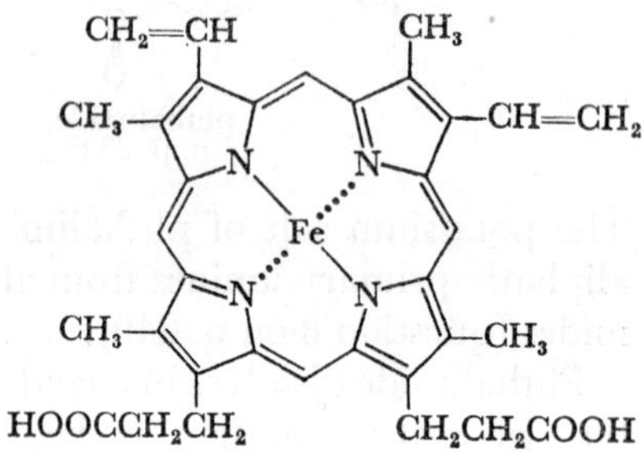

(14)

Fluorescein, a compound that gives an intense green fluorescence in dilute alkaline solution, is prepared by heating a mixture of resorcinol and phthalic anhydride at 200°.

Terephthalic Acid. This acid is required for the production of the synthetic polyester fibre, Terylene (1954). In one process, *p*-xylene (from coal tar or petroleum) is oxidised in acetic acid by air, the reaction being catalysed by metal bromides. In another method, the potassium salt of the readily available phthalic acid is isomerised thermally (300°) to the potassium salt of terephthalic acid, from which the free acid is obtained by acidification.

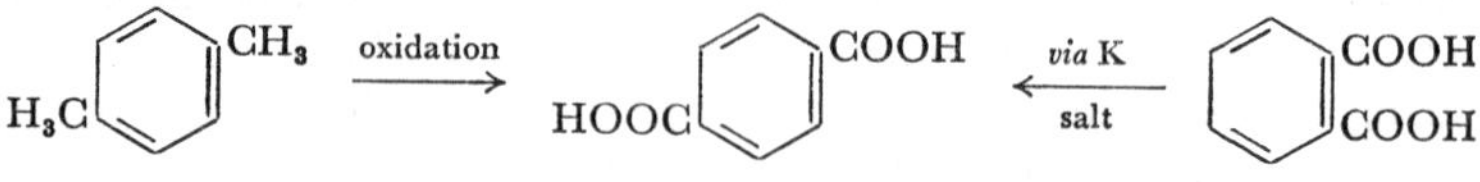

In the production of Terylene, the dimethyl ester of terephthalic acid is heated with ethylene glycol:

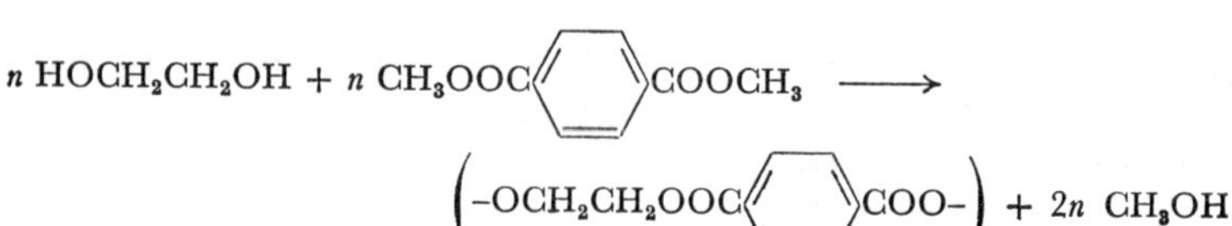

The following polymer, derived from benzene 1,2,4,5-tetracarboxylic acid, is stable for useful periods in air above 300°, whereas most organic polymers cannot be used above 200° in the presence of oxygen.

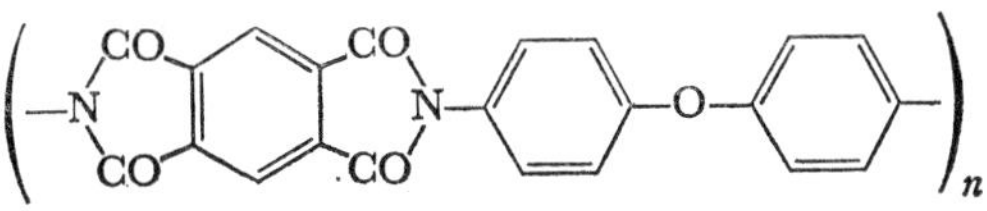

SULPHONIC ACIDS

If the carbonyl group of a carboxylic acid is replaced (in theory) by a sulphonyl group ($-SO_2-$), a sulphonic acid is obtained:

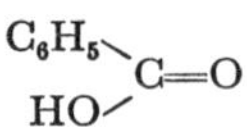

benzoic acid

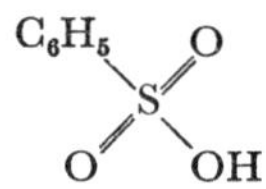

benzenesulphonic acid

In this book the sulphonyl group will be written with two sulphur–oxygen double bonds, but the following hybrid dipolar structure probably expresses the electronic state of the group more accurately.

sulphonyl group

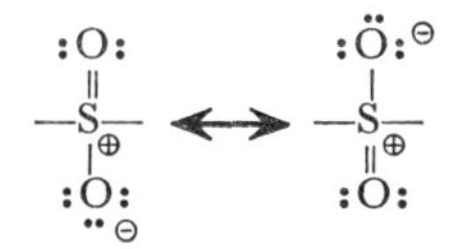

The most important method for the preparation of aromatic sulphonic acids is by sulphonation of the aromatic ring using sulphuric or chlorosulphonic acid (p. 193). Benzenesulphonic acid, $C_6H_5SO_2OH$, is also formed by oxidation of thiophenol, C_6H_5SH (cf. the oxidation of ethanethiol, p. 51).

Sulphonic acids are strong acids and form stable salts:

$$2C_6H_5SO_2OH + K_2CO_3 \longrightarrow 2C_6H_5SO_2O^{\ominus}\,K^{\oplus} + CO_2 + H_2O$$

potassium benzenesulphonate

The benzenesulphonate anion has a hybrid structure to which three equivalent forms contribute:

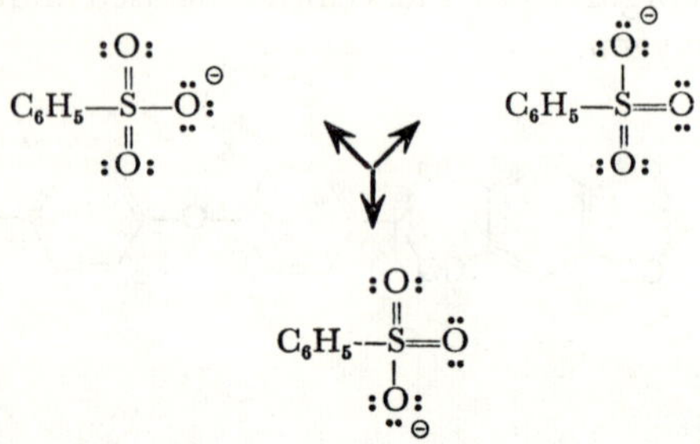

Sulphonic Acid Derivatives. Replacement of the hydroxyl group of a sulphonic acid by other substituents affords a series of compounds that are related structurally and chemically (below) to derivatives of carboxylic acids.

sulphonic ester 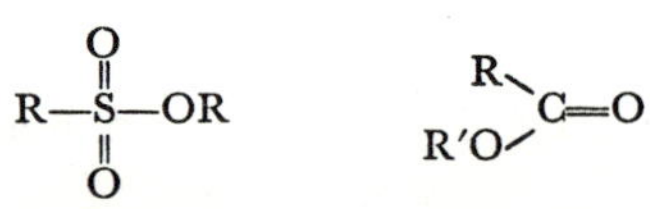 ester

sulphonamide 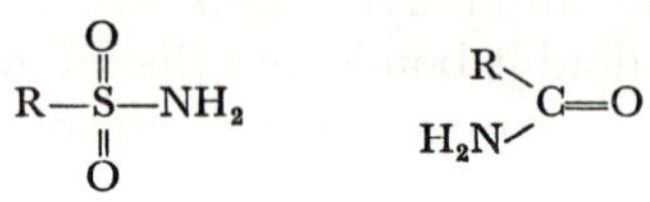 amide

sulphonyl halide  acyl halide

The series can be extended by comparing sulphones with ketones:

sulphone  ketone

(R and R′ in all these formulae represent alkyl or aryl groups.)

Many of the reactions by which the sulphonyl compounds are formed or interconverted show marked similarities to those of the carboxylic series. The formation of benzenesulphonyl chloride from the action of phosphorus pentachloride on the sulphonic acid is analogous to the production of an acyl chloride from a carboxylic acid:

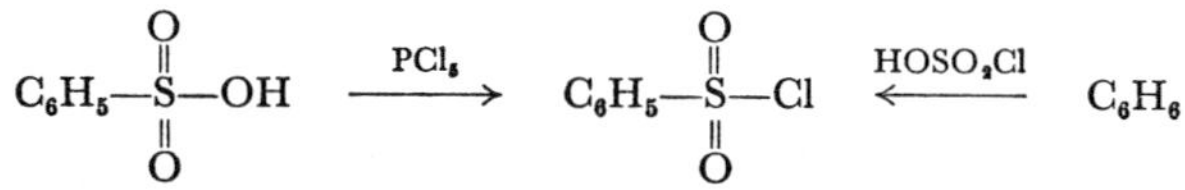

benzenesulphonyl chloride

(The alternative method of making the sulphonyl chloride is also noted.)

Arenesulphonyl chlorides, like acyl chlorides, react with nucleophilic reagents, e.g. alcohols and ammonia:

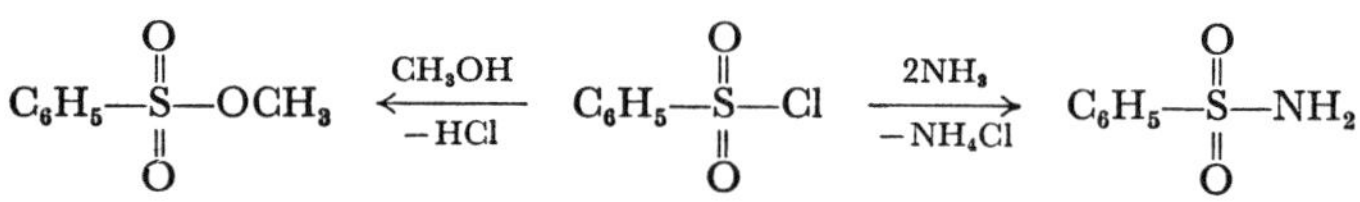

methyl benzenesulphonate, m.p. 28°

benzene sulphonamide, m.p. 153°

Arenesulphonyl chlorides undergo aluminium chloride-promoted reactions with aromatic compounds to give diaryl sulphones, $ArSO_2Ar'$ (cf. the formation of benzophenone from benzene and benzoyl chloride).

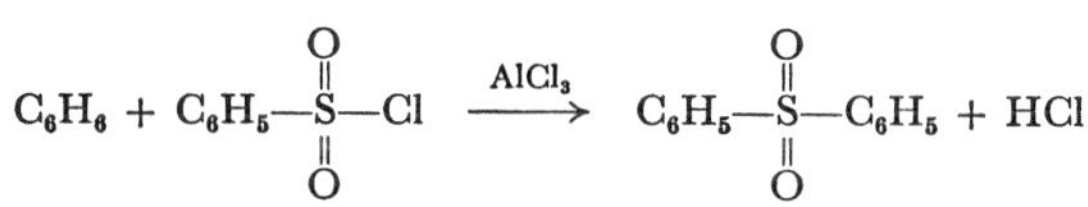

diphenyl sulphone, m.p. 128°

A derivative (4,4′-diaminodiphenyl sulphone or 'DDS') of this last compound was found (1943) to have a curative effect on leprosy; this hitherto intractable disease is being eradicated steadily by the use of DDS and related drugs.

Replacement of the Sulphonic Acid Group by Hydroxyl. The following change occurs when sodium benzenesulphonate is fused with sodium hydroxide at 300°.

$$C_6H_5SO_2O^{\ominus}\,Na^{\oplus} + 2Na^{\oplus}\,OH^{\ominus} \longrightarrow$$

$$Na_2SO_3 + H_2O + C_6H_5O^{\ominus}\,Na^{\oplus} \xrightarrow{H^{\oplus}} C_6H_5OH$$

The initial organic product is sodium phenoxide which, on acidification, yields phenol. In the sequence of reactions, $C_6H_6 \rightarrow C_6H_5SO_2OH \rightarrow C_6H_5OH$, the overall result is that benzene is oxidised to phenol, and sulphuric acid is reduced (to the sulphite level).

Although phenol itself is not made by this method, the reaction is very useful for the preparation of certain other phenols, including resorcinol and the two naphthols. For the preparation of resorcinol, benzene is sulphonated with an excess of fuming sulphuric acid at 250° to give the *m*-disulphonic acid (a sulphonic acid group has a *m*-directive effect like a carboxyl group). The sodium salt of the *m*-diacid is then fused with sodium hydroxide, and resorcinol is obtained by acidification.

$$C_6H_6 \longrightarrow C_6H_5SO_2OH \longrightarrow$$

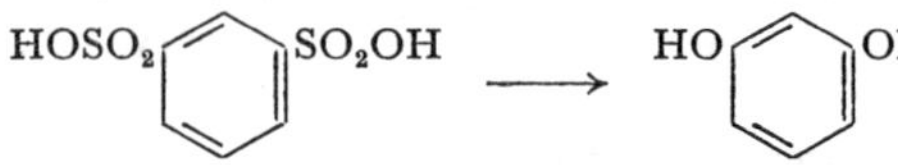

resorcinol

Fusion with alkali is also used to convert the salts of naphthalene 1- and 2-sulphonic acids into 1- and 2-naphthols respectively; the equation for the latter reaction is

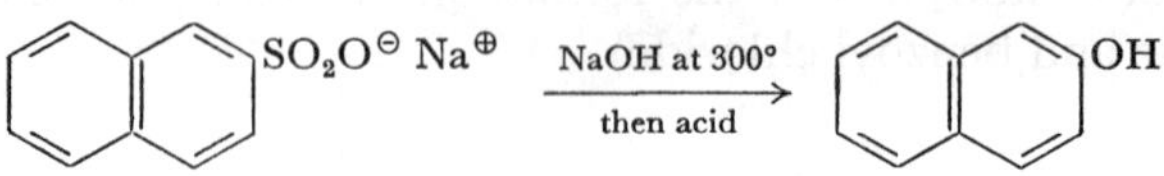

2-naphthol, m.p. 123°
(β-naphthol)

Saccharin. This compound is about 500 times as sweet as sucrose for which it is used as a substitute, for example, by diabetic patients. Synthesis of saccharin begins with toluene, which on treatment with chlorosulphonic acid gives a mixture of the *o*- and *p*-sulphonyl chlorides:

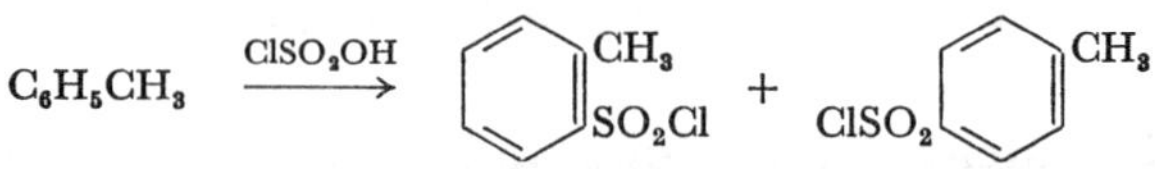

toluene-*o*-sulphonyl chloride, m.p. 10°   toluene-*p*-sulphonyl chloride, m.p. 69°

The liquid and solid isomers are separated, and the former is converted into the sulphonamide (15) by reaction with ammonia. Alkaline permanganate oxidises the methyl group of this sulphonamide, and the resultant sulphonamido-acid loses water to give the cyclic imide, saccharin.

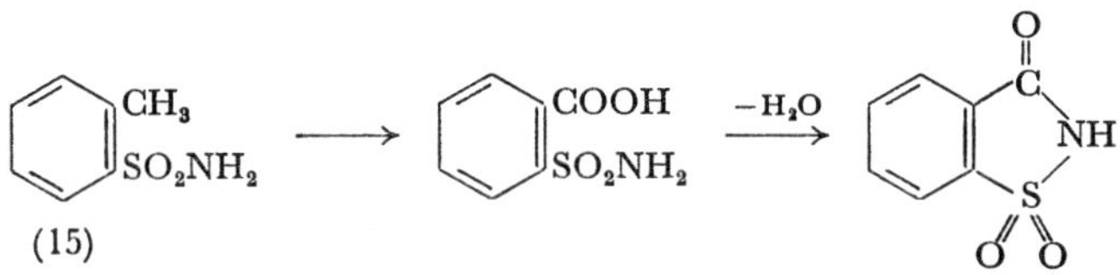

(15)

saccharin

Sulphonamides, Chemotherapy and Antibiotics. A chemotherapeutic agent is a compound that, ideally, will kill or inhibit the growth or reproduction of pathogenic organisms without exerting any harmful effect upon the host. The term *chemotherapy* was coined by Ehrlich, who made one of the earliest discoveries in this field when he demonstrated that injections of an organoarsenic compound 'salvarsan' (16) would cure syphilis.

The search for chemotherapeutically useful compounds accelerated considerably in the 1930's with the discovery that sul-

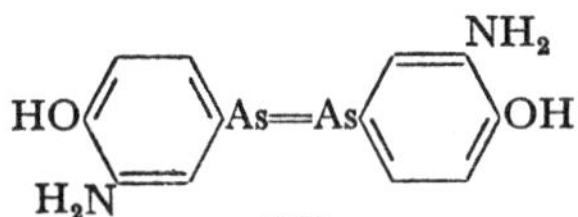

(16)

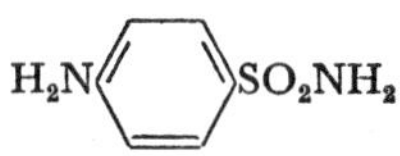

(17)

phanilamide (17) and many of its derivatives were active against streptococci, pneumococci, meningococci and gonococci. Although these developments were overshadowed later by discovery of antibiotics, derivatives of sulphanilamide (generally called 'sulphonamides' or 'sulpha-drugs') still have many uses, for example, for the treatment of bacterial meningitis, and bacillary dysentery.

An example of an unexpected development in the quest for clinically useful compounds was the finding that the reaction of the sulphonamide (18) with formic acid gave a product, chlorothiazide (19) which, although it it did not display antibacterial properties, proved to be a valuable oral diuretic.

(18) (19)

Sulphonamides control disease by preventing the multiplication of the bacteria responsible, but penicillin-G (20) is a more bacteriocidal agent. Recent advances in penicillin chemotherapy were initiated by the discovery in 1959 that useful

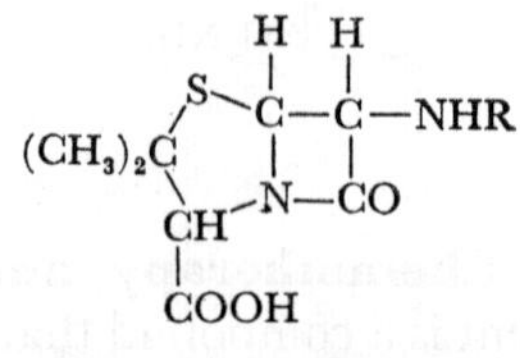

(20) penicillin-G ($R = COCH_2C_6H_5$)
(21) 6-aminopenicillamic acid ($R = H$)
(22) phenethicillin [$R = COCH(C_2H_5)OC_6H_5$]

quantities of 6-aminopenicillamic acid (21) can be isolated by adsorption techniques from the fermentation broth of *Penicillium chrysogenum.* The nucleophilic reactivity of the free amino-group was then employed to prepare penicillin derivatives with novel side-chains, phenethicillin (22) being one example. Penicillin-G is unstable to acid, and is therefore destroyed in the stomach, but phenethicillin is relatively acid-stable, and can be administered orally.

Parallel to these developments, a related antibiotic was detected in cultures of a *Cephalosporium* species of fungus. The structure (23) of the active compound, cephalosporin-C, is very similar to that of the penicillins except for the presence of a six- instead of a five-membered ring.

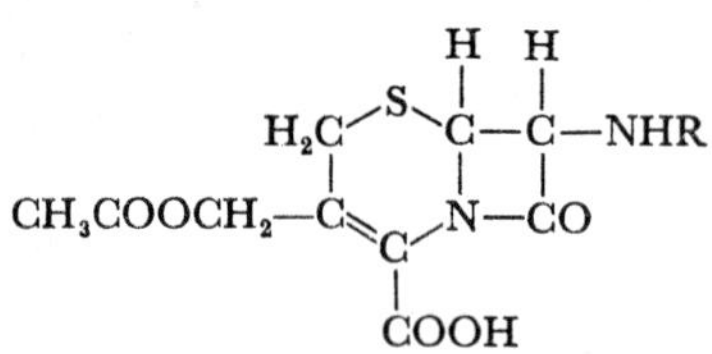

(23) $R = CO(CH_2)_3CH(\overset{\oplus}{N}H_3)COO^{\ominus}$

(24) $R = COCH_2$—

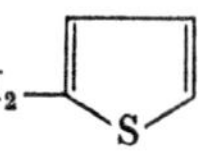

The clinically more effective cephaloridine (24) is made by treating cephalosporin-C with nitrosyl chloride in formic acid to give an amino-compound, which is then acylated with the acid chloride of 2-thiophene acetic acid. Cephaloridine has a wide range of antibacterial activity, and is resistant to the enzyme, penicillinase, which catalyses the decomposition of many of the penicillins.

QUESTIONS

1. Write structural formulae for (i) 1,4-dimethylbenzene, (ii) *o*-hydroxybenzoic acid, and (iii) 2-amino-3-phenylpropanoic acid. What are the more usual names of these compounds?

2. Oxidation of graphite (or charcoal) gives a benzene derivative, $C_{12}H_6O_{12}$, which on heating is converted into a colourless substance, $C_{12}O_9$. Suggest structures for the two compounds.

3. (*a*) The enzyme, tyrosinase, which catalyses the aerial oxidation of tyrosine to melanine, the brown pigment in hair and skin, contains *ca.* 0·2% copper. How could the presence of this element in the enzyme be established?

(*b*) Tracer experiments have shown that penicillin-G is derived biosynthetically from three building blocks: phenylacetic acid, cysteine and valine, $(CH_3)_2CHCH(\overset{\oplus}{N}H_3)CO\overset{\ominus}{O}$. Represent this finding on the formula of penicillin-G by dividing the structure into the three parts that most closely correspond to the building units.

4. (*a*) When *Pseudomonas aeroginosa* is grown in the presence of naphthalene and air, an acidic compound, $C_7H_6O_3$, giving a violet colour with iron (III) chloride, is produced. Suggest possible structures for this product.

(*b*) The predominant substance, A, in an oil obtained from a plant source showed a strong infrared absorption band at 1735 cm.$^{-1}$ When it was heated with aqueous alkali a volatile compound, subsequently identified as *cis*-hex-3-en-1-ol, distilled with the water. Acidification of the alkaline solution gave a compound, B, which was isolated by extraction with dichloromethane. B had the molecular formula, $C_8H_8O_2$, and was converted into benzoic acid on oxidation. What was the original compound, A?

5. Give structural formulae for the major products of the following reactions:

(*a*) 3,5-dimethylbenzenesulphonyl chloride and 1-amino-3-methylbutane;
(*b*) *m*-methylacetophenone and 2,4-dinitrophenylhydrazine;
(*c*) *p*-chlorobenzaldehyde and propionaldehyde in alkaline solution.

Suggest mechanisms for (*b*) and (*c*).

6. (*a*) Express by means of three formulae, the hybrid structure of the resonance-stabilised anion of potassium phthalimide.

(*b*) If an iron vessel is used for the preparation of phthalimide (by passing ammonia into molten phthalic anhydride), traces of a blue insoluble crystalline substance are sometimes obtained. This compound contains C, 62·4%; H, 2·9%; N, 19·2%; Fe, 12·6%; what is it?

7. (*a*) Discuss the following results. Treatment of benzonitrile with nitric acid (at 35° in 72% perchloric acid solution) gave 15·5% *ortho*-, 80% *meta*-, and 4·5% *para*-nitrobenzonitrile, and no dinitrobenzonitrile. By what technique can a mixture of *o*-, *m*-, and *p*-nitrobenzonitriles be analysed to this precision?

(*b*) Comment on the following prices (per 100 g. for each compound).

o-chlorobenzoic acid	30/–
m-chlorobenzoic acid	130/–
p-chlorobenzoic acid	17/–

8. Starting from toluene, suggest a practical method of synthesis of each of the following carboxylic acids: *p*-methylbenzoic acid, phenylacetic acid, and *p*-nitrobenzoic acid. Specify reagents, and indicate briefly the reaction conditions.

9. Suggest reagents (A and B) for, and intermediates (X and Y) in, the synthesis of sulphanilamide from acetanilide

$$\text{acetanilide} \xrightarrow{2ClSO_3H} X \xrightarrow{A} Y \xrightarrow{B} \text{sulphanilamide}$$

Of divers voices is sweet music made.

DANTE ALIGHIERI (1265–1321)

CHAPTER 18

ARYL NITRO-COMPOUNDS, AMINES AND RELATED COMPOUNDS

NITRO-COMPOUNDS

Organic nitro-compounds contain a nitro-group (NO_2), linked *via* the nitrogen to a carbon atom. The high dipole moments (3 to 4 D) of these compounds show that the nitro-group has a polar structure; it is now represented as a resonance hybrid in which a negative charge is shared equally between the two oxygen atoms while the nitrogen carries the positive charge:

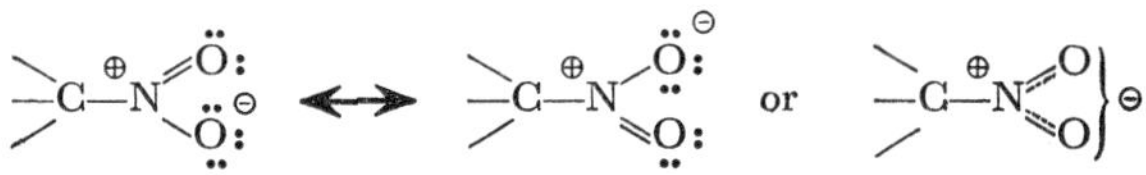

Nitration with nitric acid (or a mixture of nitric and sulphuric acids) is the most general method for the introduction of a nitro-group into an aromatic compound:

$$ArH + HONO_2 \longrightarrow ArNO_2 + H_2O$$

Nitrations of nitrobenzene and acetanilide are discussed below, and the nitrations of benzene, toluene, chlorobenzene, phenol, benzaldehyde, acetophenone and benzoic acid have been described in previous chapters. Some of the possible isomeric nitro-compounds are produced only in low yields from nitration reactions because of the directive effects of one or more nuclear substituents, and less direct methods of synthesis are then required (cf. p. 246).

Reduction of Nitro-compounds. The outstanding chemical property of the nitro-group is that it is readily reduced. The following account will be confined to the reduction of nitrobenzene, $C_6H_5NO_2$, but analogous products are obtained from many other aromatic nitro-compounds.

Powerful reducing agents convert nitrobenzene into the primary aromatic amine, *aniline*, $C_6H_5NH_2$.

$$C_6H_5NO_2 + 6H \longrightarrow C_6H_5NH_2 + 2H_2O$$

The many methods for effecting this reaction include hydrogenation over a catalyst, reduction by metal/acid systems, and reduction in acid solution by salts in which the cations can move to a higher valence state. Stannous chloride and titanous chloride are examples of such salts, and the latter is used as a reagent for the quantitative estimation of nitro-groups:

$$ArNO_2 + 6TiCl_3 + 6HCl \longrightarrow ArNH_2 + 6TiCl_4 + 2H_2O$$

Other reducing agents may be used in special cases; for example, sodium disulphide reduces only one of the nitro-groups of *m*-dinitrobenzene:

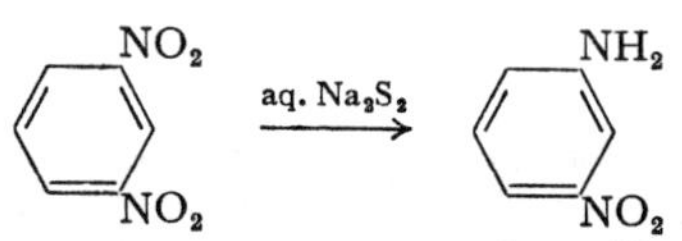

m-nitroaniline, m.p. 114°

Electrochemical studies of the reduction of nitrobenzene to aniline have shown that three steps are involved:

$$C_6H_5NO_2 \xrightarrow[-H_2O]{2H} \underset{\text{nitrosobenzene}}{C_6H_5N{=}O} \xrightarrow{2H} \underset{\text{phenylhydroxylamine}}{C_6H_5N\begin{smallmatrix}H\\OH\end{smallmatrix}} \xrightarrow[-H_2O]{2H} C_6H_5NH_2$$

Nitrosobenzene is not easy to isolate because it is more readily reduced than nitrobenzene. Phenylhydroxylamine, m.p. 81°, is obtained by stirring nitrobenzene with zinc dust and an aqueous solution of ammonium chloride:

$$C_6H_5NO_2 + 4H \longrightarrow C_6H_5N\begin{smallmatrix}H\\OH\end{smallmatrix}$$

Oxidation of phenylhydroxylamine with the correct amount of chromium trioxide gives nitrosobenzene:

$$C_6H_5N\begin{smallmatrix}H\\OH\end{smallmatrix} \xrightarrow{\text{aq. }CrO_3} C_6H_5N{=}O$$

Nitrosobenzene so obtained is a colourless solid that has the striking property of forming a green liquid above its melting point (68°), or a green vapour when distilled with steam. The colourless form is a dimer, $(C_6H_5NO)_2$, and the green monomeric compound, C_6H_5NO, is formed from it by a readily reversible dissociation process.

The formation of the nitrogen–oxygen double bond of nitrosobenzene by the removal of two hydrogens from phenylhydroxylamine is reminiscent of the production of the carbonyl bond by dehydrogenation of an alcohol; chromium trioxide is an excellent oxidising agent for either reaction. A further analogy is that nitrosobenzene* undergoes carbonyl-like reactions with nucleophilic reagents, for example with aniline:

$$C_6H_5N{=}O + H_2NC_6H_5 \longrightarrow \underset{\text{azobenzene}}{C_6H_5N{=}NC_6H_5} + H_2O$$

cf.

$$C_6H_5CH{=}O + H_2NC_6H_5 \longrightarrow \underset{\text{benzylidene aniline, m.p. 52°}}{C_6H_5CH{=}NC_6H_5} + H_2O$$

Azobenzene produced in the former reaction is the *trans*-isomer, m.p. 68°, but the *cis*-isomer, m.p. 71·5°, is also known:

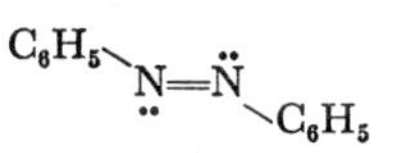

trans-azobenzene

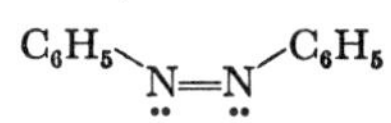

cis-azobenzene

Both geometrical isomers are orange-red. Azo-dyestuffs are derivatives of *trans*-azobenzene, the additional substituents, often complex in structure, modifying and altering the colour of the parent compound. Another method of making compounds that contain an *azo-bond*, –N=N–, is given on p. 243.

Reactions in the Nucleus. A high yield of nitrobenzene is obtained when benzene is nitrated at room temperature, because the product (nitrobenzene) is less reactive than benzene towards electrophilic reagents. A second nitro-group is, however, introduced when a mixture of nitrobenzene, nitric acid and sulphuric acid is heated at 90°. The nitro-group has a *m*-directive effect (like acyl, COOH, and CN) and, since the introduction of a third nitro-group is slow, *m*-dinitrobenzene can be obtained in good

* Dipole moments confirm that the polarities of the carbonyl and nitroso-bonds ($\overset{\delta\oplus}{N}{=}\overset{\delta\ominus}{O}$) are similar. The moments of benzaldehyde and nitrosobenzene are 2·90 and 3·15 D respectively.

yield. Nitration under more vigorous conditions gives 1,3,5-trinitrobenzene:

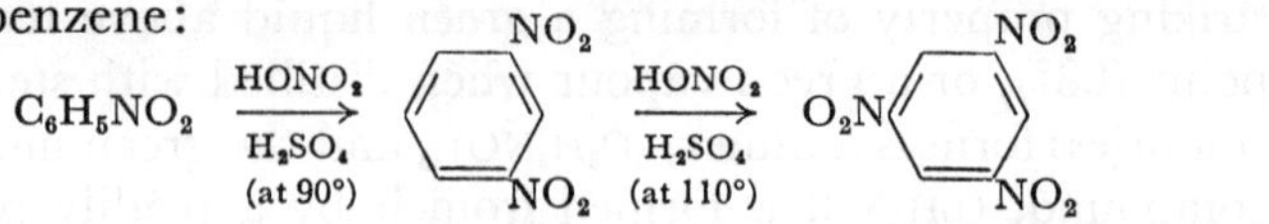

m-dinitrobenzene, m.p. 90° — 1,3,5-trinitrobenzene, m.p. 122°

ARYLAMINES

Aromatic primary amines contain an amino-group attached directly to the ring. Some of the longer known compounds have specific names, e.g. aniline and toluidine. Aniline is used as a stem name for many substituted compounds, e.g. *m*-fluoroaniline. Systematic names employing amino- as a prefix are rarely used but are given below in parentheses.

$C_6H_5NH_2$ aniline (aminobenzene)

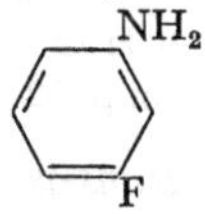

m-fluoroaniline (1-amino-3-fluorobenzene)

p-toluidine (*p*-aminotoluene)

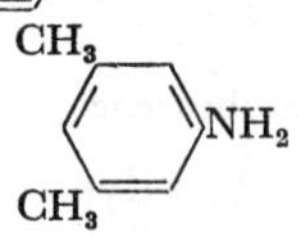

3,5-dimethylaniline (1-amino-3,5-dimethylbenzene)

Compounds in which the amino-group is separated from the aromatic nucleus by an aliphatic group, e.g. benzylamine, $C_6H_5CH_2NH_2$, behave in many respects as alkylamines and will not be discussed further.

The two most important general methods for the preparation of primary arylamines, the reduction of nitro-compounds and the reactions of aryl halides with ammonia, were considered earlier (pp. 203, 238).

Secondary and tertiary amines are obtained when one or two hydrogens in the primary arylamines are replaced (see below) by other alkyl groups:

$C_6H_5NHCH_3$	(*N*-)methylaniline*	b.p. 196°
$C_6H_5N(CH_3)_2$	(*NN*-)dimethylaniline*	b.p. 193°
$C_6H_5N(CH_2CH_3)_2$	(*NN*-)diethylaniline*	b.p. 218°

* The *N*- or *NN*- in front of these names signifies that the alkyl group or groups are linked to the nitrogen of aniline (not to a nuclear position). These prefixes can be omitted on most occasions.

Aromatic primary, secondary or tertiary amines are sparingly soluble in water but dissolve in dilute acids to form salts. Extraction with acid is a very effective method for separating amines from non-basic components of a mixture.

Acylation and Alkylation Reactions. Arylamines have nucleophilic properties and, in the cases of primary and secondary amines, react with acyl halides and acid anhydrides to form amides. These derivatives are crystalline and useful in identification work. Acetylation to give N-substituted acetamides (N-acetyl derivatives of the amines) can be performed by adding acetic anhydride to a solution of the primary or secondary amine in aqueous acetic acid:

$$C_6H_5NH_2 \xrightarrow[\text{aq.}CH_3COOH]{(CH_3CO)_2O} C_6H_5NHCOCH_3$$

acetanilide, m.p. 114°
(*N*-acetylaniline)

$$C_6H_5NHCH_3 \xrightarrow[\text{aq. }CH_3COOH]{(CH_3CO)_2O} C_6H_5N(CH_3)COCH_3$$

N-methylacetanilide, m.p. 102°
(*N*-acetyl-*N*-methylaniline)

Benzoylation to give N-substituted benzamides can be carried out by shaking a mixture of the amine, benzoyl chloride and aqueous alkali:

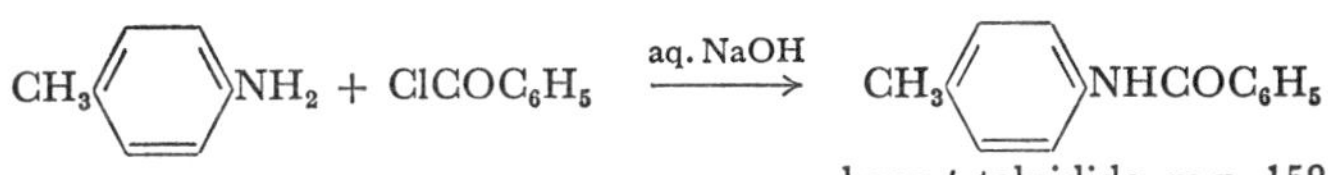

benz-*p*-toluidide, m.p. 158°
(*N*-benzoyl-*p*-toluidine)

N-Acetyl- and N-benzoyl-derivatives are hydrolysed back to the original amines when heated with aqueous acid; the preparation, crystallisation and hydrolysis of these N-acyl compounds is a general method for the purification of amines:

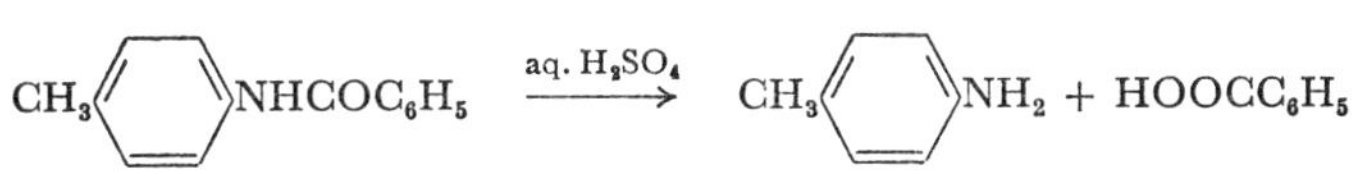

An N-alkyl bond is formed when a primary arylamine reacts with an alkyl halide:

$$ArNH_2 + RX \longrightarrow Ar\overset{\oplus}{N}H_2R\ X^{\ominus}$$

salt of secondary amine

As in the aliphatic series (p. 116), a mixture of the salts of secondary and tertiary amines (and sometimes quaternary salts) may be obtained from further transformations:

$$Ar\overset{\oplus}{N}H_2R\ X^{\ominus} + ArNH_2 \rightleftharpoons ArNHR + Ar\overset{\oplus}{N}H_3\ X^{\ominus}$$

$$ArNHR + RX \longrightarrow \underset{\text{salt of tertiary amine}}{Ar\overset{\oplus}{N}HR_2\ X^{\ominus}}$$

Nevertheless, the ratio of amine: alkyl halide used initially gives considerable control over the ratio of secondary: tertiary amine obtained.

To take a specific example, aniline can be methylated with methyl iodide at 20° or with methanol at 200° in the presence of acid. The ratio of methylaniline to dimethylaniline produced depends on the reaction conditions including the ratio of reactants. The boiling points of these two amine products are very close but the mixture can be separated by acetylation followed by distillation. Dimethylaniline distils out and the residual *N*-methylacetanilide gives the pure secondary amine on acid hydrolysis:

$$\underset{\text{(in mixture)}}{C_6H_5NHCH_3} \xrightarrow[\text{anhydride}]{\text{acetic}} C_6H_5N(CH_3)COCH_3 \xrightarrow{\text{aq. acid}} \underset{\text{(as a salt)}}{C_6H_5NHCH_3} + CH_3COOH$$

Dialkylanilines react with methyl iodide to give quaternary ammonium salts; e.g. phenyltrimethylammonium iodide, $C_6H_5\overset{\oplus}{N}(CH_3)_3\ I^{\ominus}$, starting from dimethylaniline.

Formation and Reactions of Diazonium Salts. Salts of the unusual structure, $Ar\overset{\oplus}{N}{\equiv}N\ X^{\ominus}$, are formed in solution when sodium nitrite is added to an acidic aqueous solution of a primary amine. Nitrous acid is generated from the alkali nitrite and the following reaction then occurs:

$$ArNH_2 + HX + HNO_2 \longrightarrow Ar\overset{\oplus}{N}{\equiv}N\ X^{\ominus} + 2H_2O$$

These salts are unstable and difficult to isolate. Instability is, however, equivalent to considerable reactivity, and these *diazonium salts* undergo a tremendous number of useful reactions. Most of these transformations can be performed using the aqueous solution obtained by 'diazotisation' of the primary amine. In the first two of the following reactions, the nitrogens are retained in the molecule; in the remainder, elemental nitrogen is ejected.

(a) *Formation of arylhydrazines.* Diazonium salts are reduced to arylhydrazines, $ArNHNH_2$, on treatment with sodium sulphite:

$$C_6H_5NH_2 \xrightarrow[HCl]{HNO_2} C_6H_5\overset{\oplus}{N}{\equiv}N\ Cl^{\ominus} \xrightarrow{Na_2SO_3} C_6H_5NHNH_2$$

benzenediazonium chloride — phenylhydrazine, m.p. 19°

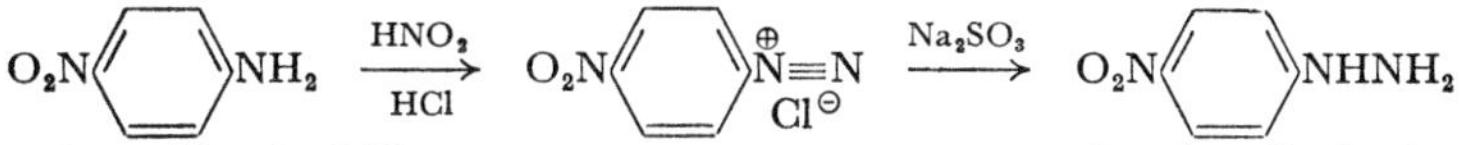

p-nitroaniline (p. 245) — *p*-nitrophenylhydrazine, m.p. 157°

Aldehydes and ketones yield coloured crystalline *p*-nitrophenylhydrazones on reaction with the latter arylhydrazine (cf. 2,4-dinitrophenylhydrazine, p. 77).

(b) *Formation of azo-compounds.* Diazonium cations behave as electrophilic reagents towards phenols in alkaline solution and towards certain amines. In each of these reactions a new bond is

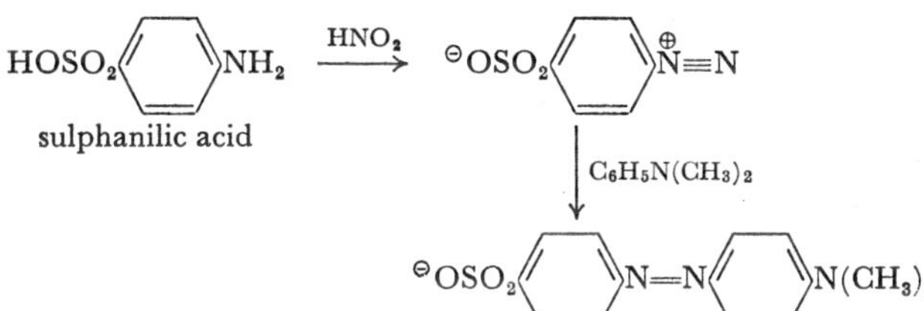

sulphanilic acid

methyl orange (anion)

formed to the terminal nitrogen of the diazo-group, and compounds containing an azo (–N=N–) link are produced. The reaction of 'diazotised' sulphanilic acid with dimethylaniline, yields the indicator, methyl orange, as shown above.

(c) *Formation of phenols.* When an aqueous solution of a diazonium salt is heated, nitrogen is eliminated and a phenol is produced:

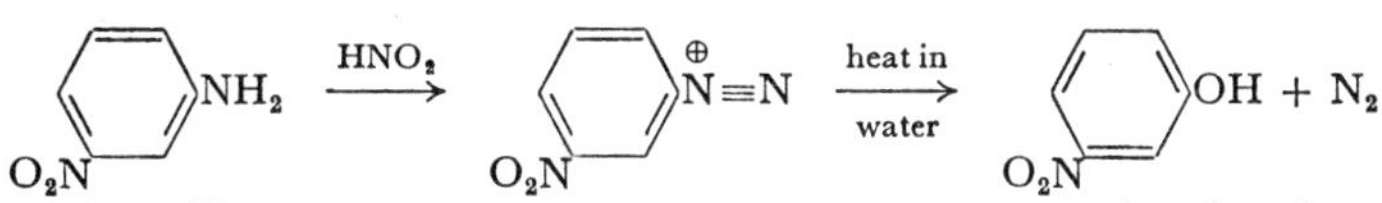

m-nitroaniline — *m*-nitrophenol, m.p. 97°

It will be recalled that only *o*- and *p*-nitrophenols are produced in the mono-nitration of phenol (p. 209).

(d) *Replacement of diazonium group by hydrogen.* Treatment of a diazonium salt with hypophosphorous acid (H_3PO_2) leads to replacement of the diazo-group by hydrogen. A good example of the utility of this reaction is the removal of the amino-group from 2,4,6-tribromoaniline (p. 245) to give 1,3,5-tribromobenzene, m.p. 120°, a compound not formed by bromination of benzene:

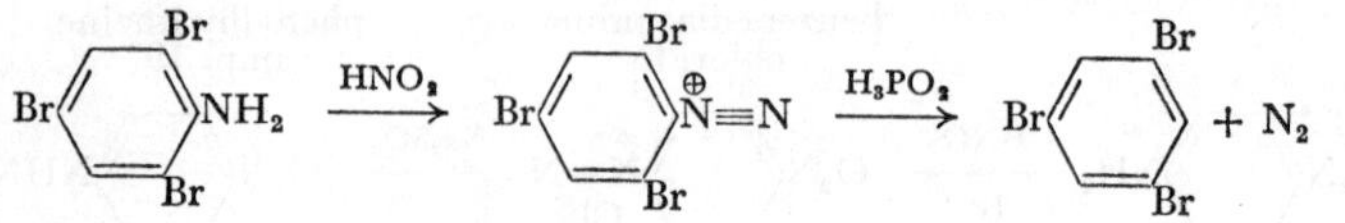

(e) *Formation of aryl halides.* The diazonium group can be replaced by iodine merely by the addition of a solution of sodium iodide:

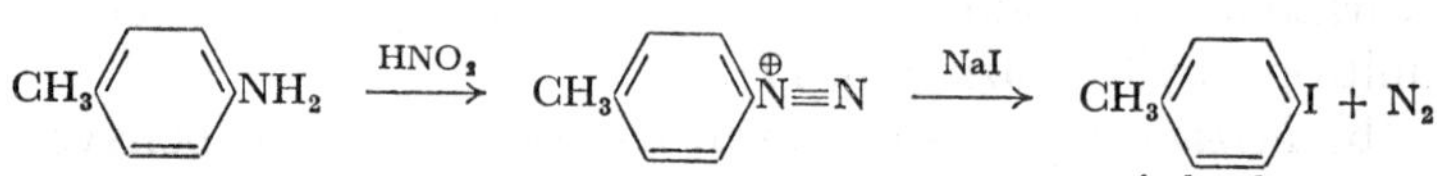

p-iodotoluene, m.p. 35°

When a solution of cuprous chloride in hydrochloric acid is added to the diazonium salt solution and the mixture is heated to 60°, the diazonium cation is replaced by chlorine (Sandmeyer reaction). Bromine is introduced similarly when cuprous bromide in hydrobromic acid is used:

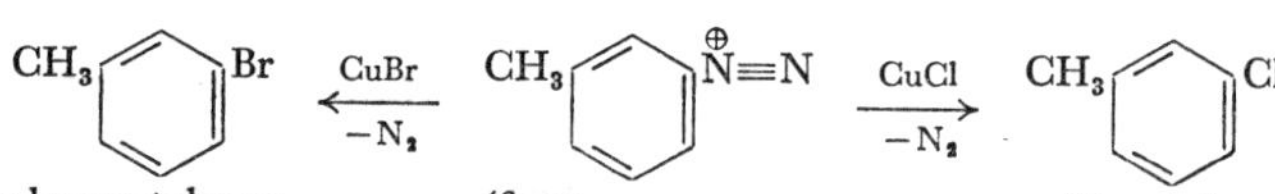

m-bromotoluene, b.p. 184° (from *m*-toluidine) *m*-chlorotoluene, b.p. 162°

Substitution of the diazonium group by fluorine is carried out by dry distillation of the diazonium fluoroborate. These salts are stable at room temperature and often separate from solution when fluoroboric acid (HBF_4) is used in the initial diazotisation reaction.

$$C_6H_5NH_2 \xrightarrow[HBF_4]{HNO_2} C_6H_5\overset{\oplus}{N}{\equiv}N \; {}^{\ominus}BF_4 \xrightarrow{heat} C_6H_5F + N_2 + BF_3$$

fluorobenzene, b.p. 85·5°

(f) *Formation of aryl nitriles.* A nitrile group can be introduced by adding the diazonium salt to a warm solution of potassium cuprocyanide (prepared from cuprous and potassium cyanides):

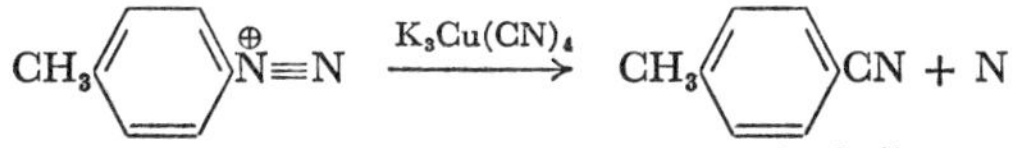

p-tolunitrile,
m.p. 29°

Nuclear Substitution. Aromatic amines react extremely readily with electrophilic reagents. Thus, 2,4,6-tribromoaniline rapidly precipitates when bromine is added to a cold solution of aniline in dilute aqueous acid.

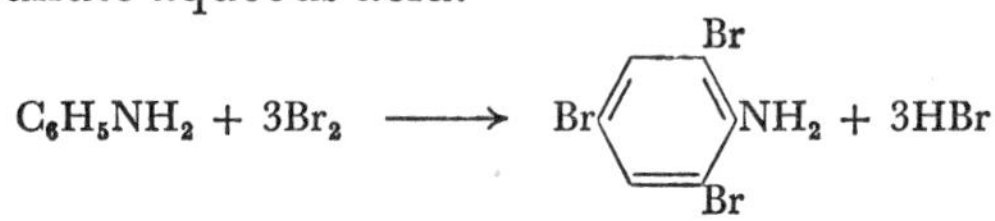

2,4,6-tribromoaniline,
m.p. 122°

*Mono*substitution of aniline by electrophilic reagents is often best achieved by protecting (diminishing the reactivity of) the amino-group by acylation. In order to prepare *p*-bromoaniline, aniline is converted into acetanilide and this, on treatment with bromine in acetic acid, gives *p*-bromoacetanilide with only a small amount of its *o*-isomer. Purification of the *p*-compound by crystallisation, followed by hydrolysis, yields *p*-bromoaniline:

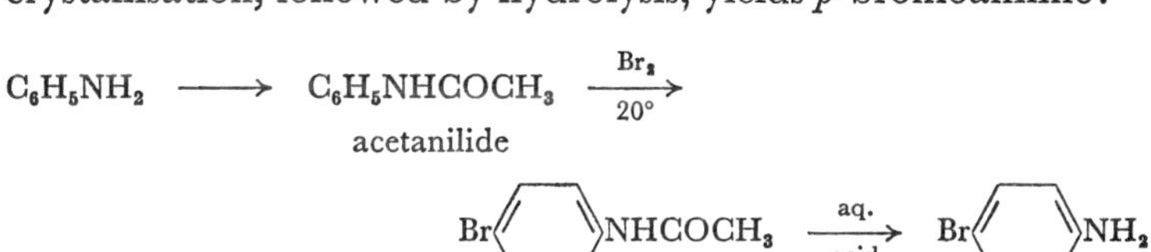

p-bromoacetanilide, m.p. 168° *p*-bromoaniline, m.p. 66°

Nitration of acetanilide proceeds similarly to give *p*-nitro-acetanilide together with a small proportion of the more soluble *o*-nitroacetanilide. *p*-Nitroaniline is obtained on hydrolysis of the *p*-isomer:

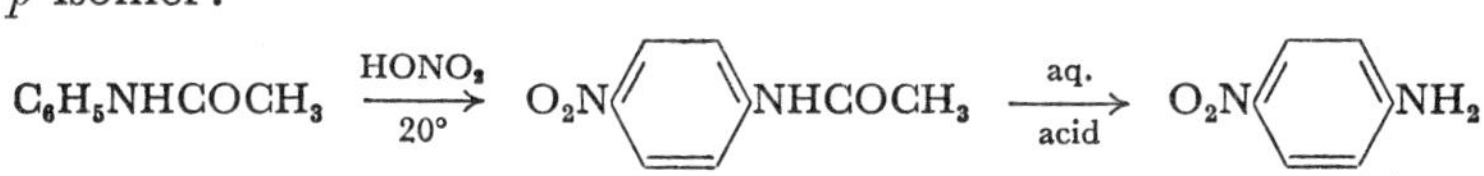

p-nitroacetanilide, m.p. 215° *p*-nitroaniline, m.p. 147°

Whereas the reaction between aniline and nitrous acid yields diazonium salts, nitrous acid reacts with dimethylaniline at the *p*-position to give *p*-nitrosodimethylaniline, a green solid.

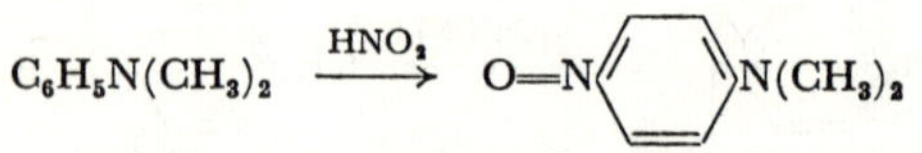

***p*-nitrosodimethylaniline, m.p. 87°**

Oxidation. Aromatic amines undergo a great variety of oxidation reactions and only a small selection can be considered here.

Oxidation of primary amines with an excess of peroxytrifluoroacetic acid yields nitro-compounds, a useful reaction when the amine is more readily available than the nitro-compound:

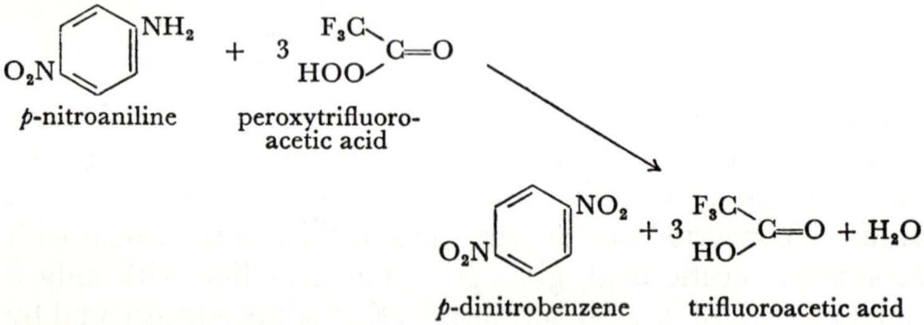

In this case, *p*-nitroaniline is available from aniline whereas *p*-dinitrobenzene is only a minor by-product in the nitration of nitrobenzene.

The oxidation of *p*-dialkylaminoanilines is an important stage in colour photography. The following equation summarises the chemistry involved in the formation of a blue dye from reactions involving *p*-diethylaminoaniline, photoactivated silver bromide and 1-naphthol.

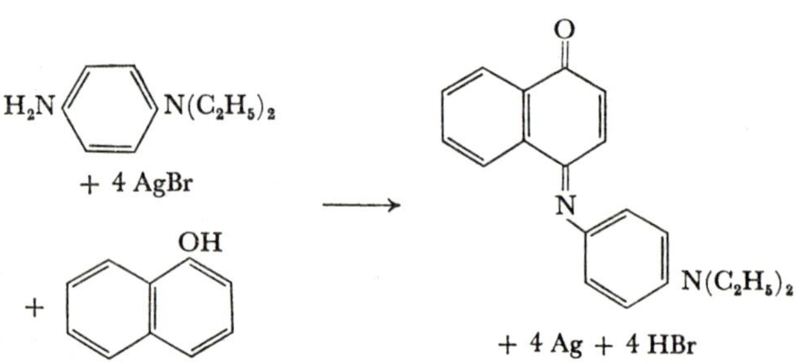

The dependence of a reaction path on the reagent used is illustrated by the oxidation of dimethylaniline:

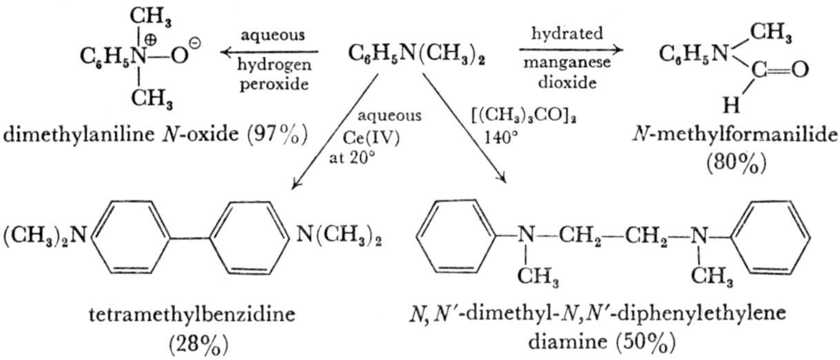

Further oxidation of tetramethylbenzidine in acid solution gives a blue salt, shown by its e.s.r. spectrum (p. 198) to contain a radical cation:

$$(CH_3)_2\overset{\bullet\oplus}{N}C_6H_4{-}C_6H_4\ddot{N}(CH_3)_2 \longleftrightarrow (CH_3)_2\ddot{N}C_6H_4{-}C_6H_4\dot{N}^{\oplus}(CH_3)_2$$

Radical cations are also believed to be involved in the mechanism of action of the herbicides, diquat and paraquat (1962), which kill plants (preferably weeds) in the presence of light. Seeds can be safely planted in ground that has been cleared in this way.

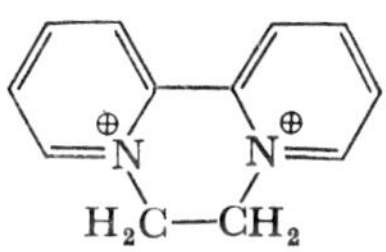

diquat cation

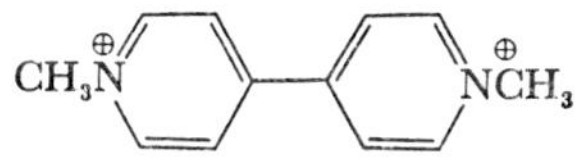

paraquat cation

Each of these dications is easily reduced (addition of one electron), and this reaction competes with the reductive stage of photosynthesis which is required for the survival of the plant.

QUESTIONS

1. Suggest methods for carrying out the following several-stage preparations:

m-nitrophenol from nitrobenzene;
m-nitrobromobenzene from benzene;
m-cyanoacetophenone from acetophenone.

2. Nitration of carboxylic acid (A), $C_8H_6O_4$, gives only one compound (B), $C_8H_5NO_6$. (B) is converted by tin and hydrochloric acid into compound (C), $C_8H_7NO_4$. A solution of compound (C) in hydrochloric acid, when treated first with sodium nitrite and then with sodium iodide, gives compound (D), $C_8H_5IO_4$. Suggest structures for compounds (A)–(D).

3. The dyestuff, mauveine, was discovered by Perkin in 1856 when he treated a sample of the impure aniline then available with aqueous

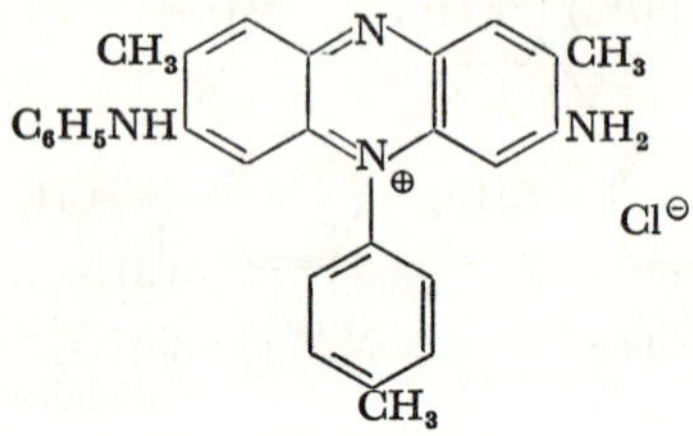

mauveine

dichromate solution. Mauveine is only formed when *o*- and *p*-toluidine are present in the aniline. How could a pure sample of aniline (for a control experiment) be obtained from impure aniline? Suggest a reason for the presence of *o*- and *p*-toluidines, but not the *m*-isomer, in the material that Perkin used.

4. (*a*) As far as possible, discuss the following physical properties of the nitroanilines.

	m.p.	*Solubility (g. per 100 ml. at 25°)* *in water*	*in ethanol*
o-	72°	0·12	27·9
m-	114°	0·09	7·9
p-	148°	0·06	6·05

(*b*) Comment on the fact that the dipole moment of *p*-nitro-*N*,*N*-dimethylaniline is higher (6·87 D) than the sum of the moments of nitrobenzene (4·01 D) and *N*,*N*-dimethylaniline (1·58 D).

5. Chlorine (1 mol.) is passed into nitrobenzene (26·5 g.) containing iron (III) chloride (1·3 g.) at 75–85°. Fractional distillation gives a compound (18 g.), m.p. 43–44°, of composition, $C_6H_4ClNO_2$. What is this compound? Write an equation, mechanism and energy profile for its formation, and explain the function of the iron (III) chloride. Calculate the yield of the main product, and suggest reasons for the yield being less than 100%.

How can two other compounds, also of formula $C_6H_4ClNO_2$, be prepared starting from benzene?

6. Adrenalin (epinephrine), produced from tyrosine (p. 224) in the adrenal gland, increases blood pressure, and was the first crystalline hormone to be isolated (1897). Adrenalin has the structure, 3,4-$(HO)_2C_6H_3CH(OH)CH_2NHCH_3$. In how many isolable stereoisomeric forms can a molecule of this structure exist? Would you expect the biological product to be optically active or not?

7. Briefly suggest what kind of work may be described in the patents whose titles are: (i) ethylene recovery systems; (ii) process for the preparation of 1,3,5-trivinylbenzene; (iii) water-soluble organic dyestuffs; (iv) triarylacrylic acids; (v) method of preparing tetra-fluoroethylene yarn.

8. Identify each of the compounds (A to E) involved in the following synthesis of methionine, write a balanced equation for each reaction and, as far as possible, discuss the chemistry of each stage.

$$CH_2(COOC_2H_5)_2 + A(1{\cdot}03\text{ mol.}) \xrightarrow[\text{light}]{\text{in boiling }CCl_4} BrCH(COOC_2H_5)_2$$

b.p. 134–136°/33 mm. (73–75%)

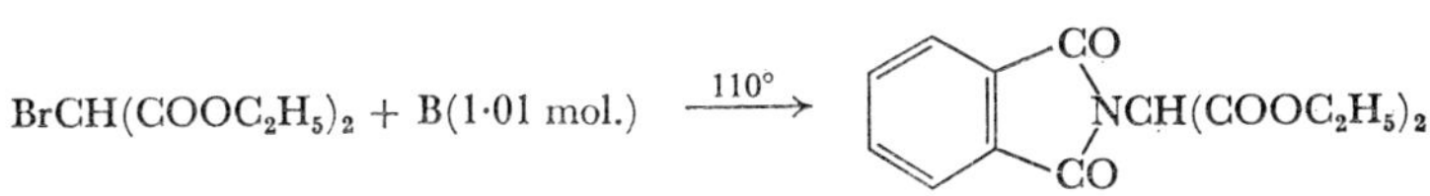

m.p. 73–74° (67–71%)

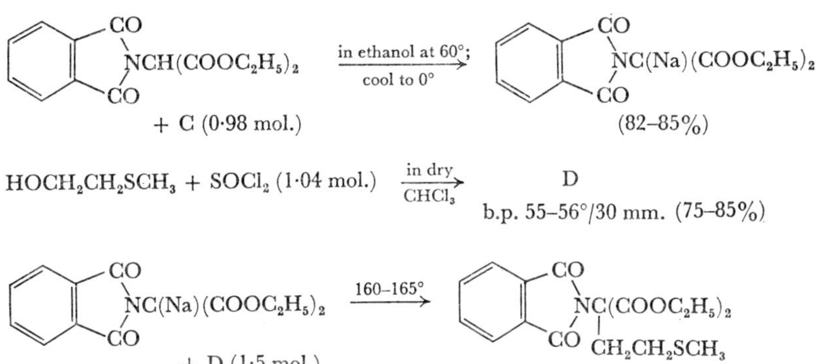

(82–85%)

$$HOCH_2CH_2SCH_3 + SOCl_2\ (1{\cdot}04\text{ mol.}) \xrightarrow[CHCl_3]{\text{in dry}} D$$

b.p. 55–56°/30 mm. (75–85%)

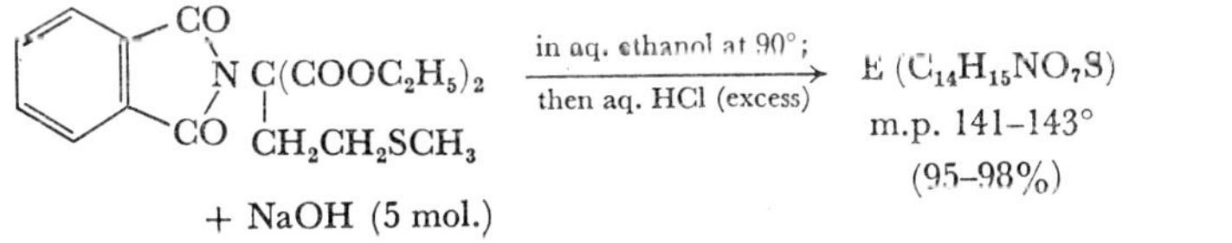

m.p. 66–67° (76–80%)

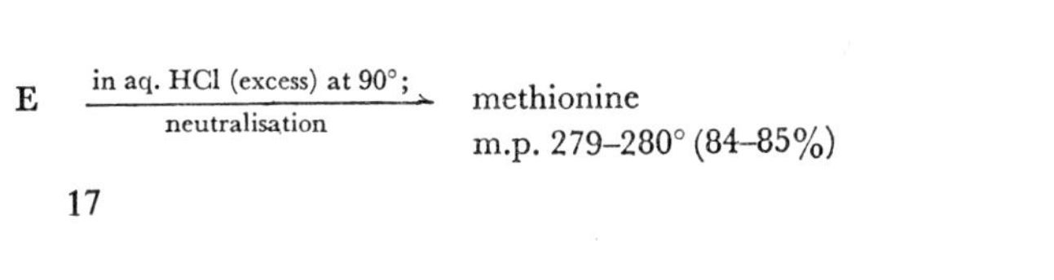

m.p. 141–143° (95–98%)

$$E \xrightarrow[\text{neutralisation}]{\text{in aq. HCl (excess) at }90^\circ;} \text{methionine}$$

m.p. 279–280° (84–85%)

9. Comment on the fact that alkanes are unchanged when treated with a mixture of nitric and sulphuric acids under the conditions in which benzene is converted into nitrobenzene.

However when a mixture of propane and nitric acid (vapour) is passed through a zone at 400°, a condensate is obtained, which contains two compounds: A of b.p. 120°, and B of b.p. 132°. A and B each have the molecular formula $C_3H_7O_2N$, and both give basic compounds on complete hydrogenation ($3H_2$ absorbed) over a metal catalyst. What are A and B?

ANSWERS TO QUESTIONS

Chapter 1 **1(a)** diamond; **2(a)** $C_{16}H_{10}N_2O_2$; **2(c)** $C_9H_9ClO_3$; **4** $4C_3H_5N_3O_9 \longrightarrow 6N_2 + 10H_2O + 12CO_2 + O_2$, 469 ml.; **5(a)** by extraction with light petroleum.

Chapter 2 **1** hexane, 2-methylpentane, 3-methylpentane, 2,2-dimethylbutane, 2,3-dimethylbutane; **2(a)** condensed formulae are

$$ClCH_2CH_2CH_2CH(CH_3)CH_2CH_2CH_3,$$
$$CH_3CHClCH_2CH(CH_3)CH_2CH_2CH_3,$$
$$CH_3CH_2CHClCH(CH_3)CH_2CH_2CH_3,$$
$$CH_3CH_2CH_2CCl(CH_3)CH_2CH_2CH_3,$$
$$CH_3CH_2CH_2CH(CH_2Cl)CH_2CH_2CH_3;$$

(3a) $CH_3CH_3 + HNO_3 \longrightarrow CH_3CH_2NO_2$ (nitroethane, actually) $+ H_2O$; **4(a)** the equation is not balanced, and other products, e.g. CH_3Br, are also formed; **4(b)** *combustion* analysis only gives the empirical formula, and $C_5H_{10}Cl$ is an impossible formula; **7** ΔH in kcal./mole for the three successive stages of fluorination are +36 (endothermic), $101 - 134 = -33$ (exothermic), $36 - 107 = -71$ (exothermic) and for bromination +46 (endothermic), $101 - 87 = +14$ (endothermic), $46 - 67 = -21$ (exothermic); **8(a)** isobutane has a moment; **8(b)** for example, cool the mixture strongly or use a molecular sieve; **8(c)** 11,700 ml.

Chapter 3 **2** 1,1-dichloro-2-methylbutane, 1,2-dichloro-2-methylbutane, 1-chloro-2-(chloromethyl)butane, 1,3-dichloro-2-methylbutane, 1,4-dichloro-2-methylbutane; none has a zero dipole moment; **3(a)** $O^{2\ominus}$, $CH_3O^{\ominus}$, $H_2\overset{\oplus}{O}CH(CH_3)_2$, ${}^{\ominus}CH_2CH_3$, ICl, $\overset{\oplus}{N}H_4$, $\overset{\ominus}{N}HCH_2CH_3$, $CH_3SCH_2CH_3$; **3(b)** $CH_3CH_2\cdot$, $CH_3CH_2^{\oplus}$; **4(d)** 63%; **4(e)** 1·1:7·8:1, excess of butanol acts as a solvent; **5(b)** evaporate acetone, add ether and water, separate, evaporate water to give LiBr + Li^{82}Br, evaporate ether and distil to give $CH_3(CH_2)_3{}^{82}Br + CH_3(CH_2)_3Br$.

Chapter 4 **1(c)** covalent; **1(d)** water gives $Al(OH)_3$ which being insoluble coats the surface of the metal;

3($CH_3CH_2OH \xrightarrow{HBr,\ H_2SO_4} CH_3CH_2Br \xrightarrow{NaCN} CH_3CH_2CN$, $CH_3CH(OH)CH_3$
$\xrightarrow{Na} CH_3CH(O^{\ominus})CH_3 \xrightarrow{CH_3I} CH_3CH(OCH_3)CH_3$;

... p. 33; **4(b)** ethanol is more associated by hydrogen bonding than ...thiol and therefore has a higher b.p.; in the absence of hydrogen bonding the ...s a higher b.p. than the ether because of its higher molecular weight; ...efractive index increases as the molecular weight of the substance in... ...re many exceptions to this rule, e.g. bromobenzene has a higher ... 1-bromohexane); **5(b)** $CH_3(CH_2)_4OH + HBr + H_2SO_4 \longrightarrow$ ... $+ HSO_4^{\ominus}$; **5(d)** cf. p. 45; **6** ethanol $\longrightarrow$ ethyl cyanide.

... *cis*- and *trans*-pent-2-ene, 2-methylbut-1-ene, 2-...ut-1-ene; **3(a)** A is $CH_3(CH_2)_{30}CH{=}CH_2$ and B is ...H_3; **4(a)** since $-2{\cdot}303 \log K = \Delta G/RT$, $\Delta G =$ ... 0·15 kcal.; **4(b)** 2-methylpropene; **6** dichloro-...ene is planar with all bond angles approximately ... C≡C—H is linear and there is a tetrahedral arrangement of substituents of the methyl carbon atom; in dimethyl ether there is a tetrahedral arrangement (of substituents or unbonded pairs) around the carbon and oxygen atoms.

Chapter 6 **1(a)** condensed formulae are

$$CH_3CH_2CH_2COCH_3,$$
$$CH_3(CH_2)_6CHO,$$
$$CH_3CH_2C(CH_3)_2COCH_2CH_3,$$
$$ClCH_2CHO;$$

1(b) pentanal, 4-bromobutan-2-one, 2,2-dimethylhexan-3-one; **3(a)** labelling occurs by reversible hydration (p. 76), separate by distillation; **3(c)** loss of an electron from propanal gives the molecular ion $CH_3CH_2CHO^{\oplus}$ (M/e 58) which decomposes into $CH_3CH_2\cdot$ and $H—C{\equiv}O^{\oplus}$ (M/e 29); propanal partly labelled with ^{18}O would show additional peaks at 60 and 31; **4** note that hydration of acetone is reversible with a small equilibrium constant (p. 76), whereas hydration of isobutene is a two-step reaction which in practice can go to completion (p. 63);

5(a) (i) $(CH_3)_2CHCH_2Br \xrightarrow{NaOH} (CH_3)_2CHCH_2OH \xrightarrow{CrO_3} (CH_3)_2CHCHO$
(avoid an excess of oxidising agent)

(ii) $CH_3CH_2CH{=}CH_2 \xrightarrow{aq.\ H_2SO_4} CH_3CH_2CH(OH)CH_3 \xrightarrow{CrO_3} CH_3CH_2COCH_3$

(iii) use NaCN with less than 1 mol. of aq. H_2SO_4; **6(b)** 5·4%.

Chapter 7 **2(b)** approx. 0·4%; **3(b)** in methyl formate the C=O bond is shorter than C—O as expected; in each molecule the single bond to the sp^2 carbon is shorter than that to the methyl group; **3(d)** $CO_3^{2\ominus}$, which contains equivalent carbon–oxygen bonds of length intermediate between single and double bonds (cf. $RCOO^{\ominus}$, p. 90); **5** organic products are

$$(CH_3)_2CHCOCHBr(CH_3)_2 + (CH_3)_2CBrCOCBr(CH_3)_2,$$
$$CH_3CH(CH_3)CH_2CH_2CH_2OH + HOCH(CH_3)_2,$$
$$CH_3CH_2CH(CH_2CH_3)CH_2CH(CH_3)CH_2CH_2NH_2,$$

and names of starting compounds are 2,4-dimethylpentan-3-one, isopropyl 4-methylpentanoate and 5-ethyl-3-methylheptanamide; infrared indicates the type of carbonyl group present (p. 100) and only the amide absorbs at 3600–3100 cm.$^{-1}$ (NH); **7(i)** an excess of NaOH would hydrolyse the ester; **7(ii)** octadecane is derived from $CH_3(CH_2)_8COO^{\ominus}$ alone; **8(a)** potassium ethyl sulphate is not hydrolysed rapidly at 20° but on heating $SO_4^{2\ominus}$ is formed and this gives $BaSO_4$; **10** consider the effect of pressure on K (cf. Le Chatelier's principle).

Chapter 8 **1(c)** $CH_3(CH_2)_{16}COOH$ (stearic acid), $CH_3(CH_2)^{14}COOH$ (palmitic acid), $HOCH_2CH(OH)CH_2OH$ (glycerol), H_3PO_4 (phosphoric acid), $HOCH_2CH_2\overset{\oplus}{N}(CH_3)_3\ \overset{\ominus}{O}H$ (choline); the acids are obtained initially as sal[illegible] **2(i)** 192; **(ii)** 227; **4(a)** e.g. remove non-polar compounds with light petro[illegible] test residue for Mg, suspend in water, acidify and extract the stearic acid w[illegible] petroleum. Compare it (m.p. and infrared) with an authentic sample;

5(a) $CH_3I + Na^{14}CN \longrightarrow CH_3{}^{14}CN \xrightarrow[aq.\ H_2SO_4]{reflux} CH_3$[illegible]

take precautions for handling radioactive materials; **5(b)** [illegible] COOH).

Chapter 9 **1(a)** $CH_3(CH_2)_3NH_2$, $(CH_3)_2CHC$[illegible] $CH_3CH_2CH_2NHCH_3$, $(CH_3)_2CHNHCH_3$, $(CH_3CH_2$[illegible]

3(a) triethylamine since only one reaction occurs (p. 118); follow the reaction, e.g. by withdrawing samples and estimating $I^{\ominus}$ with $AgNO_3$; **4** main reaction:

$$CH_3NHCH_3\ (B) + C_6H_5COCl\ (\text{benzoyl chloride, A}) + NaOH \longrightarrow C_6H_5CON(CH_3)_2\ (C) + NaCl + H_2O;$$

additional reaction: $C_6H_5COCl + 2NaOH \longrightarrow PhCOO\ ^{\ominus}Na^{\oplus} + NaCl + H_2O$;

acidification: $PhCOO^{\ominus} + H_3O^{\oplus} \longrightarrow PhCOOH(D) + H_2O$: without NaOH some $(CH_3)_2NH_2{}^{\oplus}Cl^{\ominus}$ would be formed and benzoylation would be incomplete; an excess of A is employed because its hydrolysis to sodium benzoate is a competing reaction; **5(a)** acetylene.

Chapter 10 **1(a)** CH_2O; **1(b)** $XMg(CH_2)_3CH_3$; **1(d)** $(CH_3)_3CCH_2MgX$ (where X=halogen); **2(b)** 1-chlorobutane, its b.p. (78°) is 43° higher than that of ether and 48° lower than that of n-octane (cf. pp. 13–33); **3(a)** $C_{10}H_{10}Fe$; **3(b)** mass spectrometry; depression of f.p. or rise in b.p. of a solvent (cryoscopic or ebullioscopic method); isothermal distillation method (vapour pressure osmometer). The method can depend on the properties (e.g. solubility, volatility and stability) of the compound; **5** $C_7H_{16}MgO$, $CH_3(CH_2)_3Mg{-}O{-}CH(CH_3)_2$.

Chapter 11 **1(a)** CHI_3 (iodoform); $CH_3COCH_3 + 3NaI + 3NaOCl \longrightarrow CHI_3 + CH_3COONa + 3NaCl + 2NaOH$; **1(b)** 1,1-, 1,2-, 1,3-, 1,4-, 2,2- and 2,3-dibromobutanes, 1,1-, 1,2- and 1,3-dibromo-2-methylpropanes; **3(a)** glycerol; **4** two; **5(a)** thick-walled sealed glass tube; gas chromatography; **5(b)**

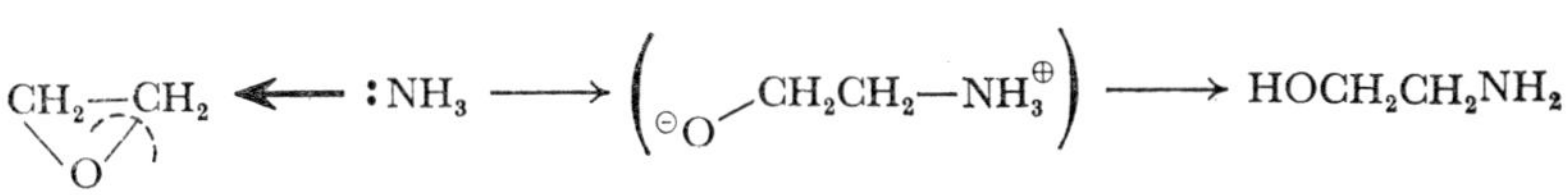

Chapter 12 **1(a)** (i) 5-bromohex-2-ene; (ii) 4,4-dimethylpent-2-en-1-ol; (iii) pent-2-enal; (iv) 2-chloropenta-1,3-diene; (v) 5-methylhex-4-enoic acid; (vi) 1-iodohepta-2,5-diene. Conjugated bonds are present in (iii) and (iv); **2(a)** a bond from Cl to *sp*^{2} carbon is shorter than one to *sp*^{3} carbon, cf. Chap. 7, question 3(b); **4(a)** the amide gives a (carbonyl) band near 1650 cm.$^{-1}$ whereas the cyanohydrin gives a (nitrile) band near 2300 cm.$^{-1}$ Most amides are solid (p. 94) whereas simple alcohols and nitriles are liquid; **4(b)** pass a mixture of isobutene, ammonia and air at *ca.* 400° over a catalyst containing molybdenum oxide (cf. p. 144); **5(b)** poly(methyl methacrylate); **5(c)** acid-catalysed hydrolysis of the ester groups in the polymer gives the water-soluble poly(vinyl alcohol), $({-}CH_2{-}CH(OH){-})_n$, and acetic acid. The latter is removed during steam distillation.

Chapter 13 **1** (b), (c), (e); **2** leucine, glycine and lysine, the first and the third compounds can be optically active; **4(a)** four (*cis*- and *trans*-isomers each of which exists in enantiomeric forms); **5(a)** negative, the flexibility of the molecule being reduced during cyclisation; **5(b)** $({-}O(CH_2)_4CO{-})_n$; **7** crystalline solid is urea nitrate ($CH_5N_3O_4$); **8(a)** nylon 6.10, because this contains a higher proportion of hydrocarbon groups (e.g. densities$^{20}_{4}$ of methyl octanoate and methyl decanoate are 0·8775 and 0·8733); **9** NaCN (in aq. ethanol), H_2/metal catalyst, hot aq. acid; **11** electron diffraction; **12(a)** $C_4H_{10}ClNO_2$ (glycine ethyl ester hydrochloride, $C_2H_5OCOCH_2NH_3{}^{\oplus}Cl^{\ominus}$); **12(b)** thiourea.

Chapter 14 **1(a)** sucrose in water, and its octa-acetate in dichloromethane; **2** 16%, add aq. sodium acetate (enough to neutralise the remaining nitric acid) and then aq. calcium chloride, insoluble calcium oxalate precipitates and can be weighed; **3(a)** *cis* and *trans*-1,4-dimethoxycyclohexane (cf. p. 174 for the related diols); one from each for the other two compounds; **5(a)** cellulose (e.g. *via* ester links); **5(c)** the name, though convenient, is inaccurate because each molecule contains between 6000 and 10,000 acetate groups.

Chapter 15 **1** 1,2,4-trichlorobenzene, *m*-diethylbenzene, *m*-nitrobenzoic acid, *p*-fluorotoluene; **3(a)** nitrobenzene is readily soluble; **3(b)** the percentage yield should be given as 85% because the masses from which it is derived each contain only two significant figures; **4(a)** no; **4(b)** (i) methylcyclohexane, (ii) CHO.CHO, $CH_3CO.CHO$ and $(CH_3CO)_2$ (or reduced or oxidised products depending on the procedure used, cf. p. 65); **6(a)** isopropylbenzene; **7(b)** the tetrahydronaphthalene ('tetralin') in which the benzenoid structure of one ring is present; **7(c)** HBr is missing from right-hand side of equation.

Chapter 16 **2(a)** six; **2(b)** $C_7H_{12}O_6+2MnO_2+2H_2SO_4 \longrightarrow C_6H_4O_2+CO_2+2MnSO_4+6H_2O$; **2(c)** tetrachloro-*p*-benzoquinone ('chloranil'); **4(a)** water dissolves in the phenol and lowers its m.p.; **4(c)** the two *o*-hydroxy-groups in catechol react with one molecule of phosgene to give a cyclic carbonate, $C_7H_4O_3$; a similar reaction cannot occur for guaiacol, two molecules of which react to give *o*-$CH_3OC_6H_4OCOOC_6H_4OCH_3$-*o*;

6(a) (i) $C_6H_5CH_3+Br_2 \xrightarrow{\text{light}} C_6H_5CH_2Br+HBr$

$C_6H_5CH_2Br+Li^{82}Br \longrightarrow C_6H_5CH_2{}^{82}Br+LiBr$ (p. 39)

(ii) $C_6H_5CH_2Br+NaSH \longrightarrow C_6H_5CH_2SH+NaBr$ (p. 36)

(iii) $C_6H_5CH_2Br+Mg$ then $^{14}CO_2$ then acid $\longrightarrow C_6H_5CH_2{}^{14}COOH$ (p. 129)

(iv) $C_6H_5CH_2MgBr+{}^2H_2O \longrightarrow C_6H_5CH_2{}^2H+BrMgO^2H$ (p. 127)

(v) $2C_6H_5CH_2SH+\text{aq. }Hg(OCOCH_3)_2 \longrightarrow (C_6H_5CH_2S)_2Hg+2HOOCCH_3$ (p. 49)

6(b) $C_6H_5CH_2{}^{82}Br$ and $C_6H_5CH_2{}^{14}COOH$, geiger counter, mass spectrometry; **7(b)** one form with negative charge on oxygen and three with the charge on the two *o*- and one *p*-positions respectively; **7(c)**

phenol $\xrightarrow{\text{(p. 195)}}$ cyclohexanol $\xrightarrow{\text{(p. 47)}}$ cyclohexanone $\xrightarrow{\text{(p. 83)}}$ adipic acid

hexane-1,6-diol $\xrightarrow{\text{(p. 96)}}$ adipic acid

Chapter 17 **1** *p*-xylene, salicylic acid, phenylalanine; **2** benzene hexacarboxylic acid and its trianhydride; **4(a)** *o*-, *m*-, or *p*-hydroxybenzoic acid (*o*-isomer in fact); **4(b)** $C_6H_5CH_2COOCH_2CH_2CH\overset{\text{cis}}{=}CHCH_2CH_3$; **6(a)** one form with the negative charge on nitrogen and two with the charge on oxygens; **6(b)** iron phthalocyanine;

9 X=*p*-$CH_3CONHC_6H_4SO_2Cl$, A=NH_3, Y=$CH_3CONHC_6H_4SO_2NH_2$, B=aq. HCl.

Chapter 18 **2** A is terephthalic acid; **3** prepare acetanilide, purify it by crystallisation, and then hydrolyse it back to aniline. The nitrobenzene used for preparing aniline contained *o*- and *p*-nitrotoluenes derived from toluene in the not very well purified benzene then available; **4(b)** electron donation (resonance interaction) from $N(CH_3)_2$ to the NO_2 group in the nitro-amine; **6** (+) and (−)-enantiomers; yes ($\alpha_D -53°$ in aq. HCl). **9** A and B are isomeric nitropropanes.

INDEX

(Page numbers printed in **bold type** indicate principal references)

Absolute configuration, 158, 159
Acetaldehyde, 69, **83**, 216
Acetals, 173
Acetanilide, **241**, 245
Acetic acid, **101**, 110, 167
Acetone, 80, **84**, 216
Acetophenone, 193, 216, 218
Acetylacetone, 167, **169**
Acetylcholine, 135
Acetylene, 70
Acetylenes, *see* Alkynes
Acid anhydrides, 86, **92**, 95–97, 101, 119, 151, 153, 217, 223, 226
Acidity of
  alcohols, 45, 90
  aldehydes, 80, 91
  alk-l-ynes, 67
  amino-acids, 158
  carboxylic acids, **89–91**, 222
  phenols, 207, 213
  thiols, 50
Acrolein, 145
Acrylic acid, 139
Acrylonitrile, 139, 144, 165
Acylation, 96, 193, 241
Acyl group, 86
Acyl halides, 86, **92**, **95–96**, 118, 150, 193, 223
Additions to
  alkenes, 59–66
  alkynes, 68–70
  carbonyl groups, **75–79**, 97
  epoxides, 134
Adipic acid and adiponitrile, 149, 164
ADP, 103, 180
Adrenalin and adrenal gland, 248
Aerosol spray, 38
Alanine, 158
Alcohols, 33, **43–49**, 79, 96, 128, 206
Aldehydes, 47, **73–83**, 128, 173, **214–219**
Aldohexoses, 176
Aldol reaction, **79**, 216
Alkaloids, 122–123
Alkanes, **12–29**, 99, 127
Alkanoic acids, 87
Alkenes, 40, 46, **55–56**, 129, 138–147
Alkoxides, 45
Alkylammonium salts, 37, **115**, 241
Alkylation, 116–118, 191, 241
Alkyl chloroformates, 165
Alkyl group, 15
Alkyl halides, **31–41**, 116, 157
Alkynes, **66–71**, 127
Allenes, 108, **139**
Allyl alcohol and chloride, 146
Allylic compounds, 146
Amides, 86, **94**, **96**, 119, 223, 241
Amines, 97, 102, **114–124**, 156, 227, **240–247**
Amine salts, 35, **115**, 241
Amino-acids, **158**, 223
2-Aminoethanol, 110, **135**
Ammonia, 36, 94, 117, 135, 166, 203
Ammonium carbamate, 166
Amygdalin, 176, **214**, 218
Amylopectin and amylose, 179
Anaemia, 6
Anaesthetics, 51, 55, 123, 132, 224
Analysis, 3–5, 143
Aniline, 186, 203, 211, 238, **240**
Anisole, 208
Antibiotics, 2, 108, **234**
Antiseptics, 207
Aromaticity, 191, 195, 199
Arrhenius equation, 25
Arylamines, 240–247
Aryl group, 184
Aryl halides, 188, **202–205**, 244
Arylhydrazines, 243
Aryloxyacetic acids, 208
Asphalt, 12, 27
Aspirin, 210
Asymmetry, 108, **155–158**, 159, 175
Atomic charge and structure, 39
Atomic number, 9
ATP, 103, 180
Automation, 4, 162
Autoxidation, 109
Azo-compounds, 239, 243

Bakelite, 83, 207
BAL, 136
Basicity of
  amines, **115**, 241
  amino-acids, 159
Benzaldehyde, 206, **214–218**
Benzene, 182–195
Benzene hexachloride, 195

Benzenesulphonic acid, 193, **229**
Benzenesulphonyl chloride, 194, **231**
Benzoic acid, 202, 206, **222–223**
Benzonitrile, 223, 236
Benzophenone, 193
o-Benzoquinone, 211
p-Benzoquinone, 211
Benzotrichloride, 206
Benzoyl chloride, 119, **223**, 241
Benzyl compounds, 197, **205**
Benzylidene acetone, 216
Benzylidene chloride, 206
Biosynthesis of
  aromatic amino-acids, 224
  fatty acids, 110
  hydrocarbons, 26
  penicillin, 235
Bitumen, 12
Bombykol, 147
Bond angles, 7, 15, 58, 75, 134, 183
Bond dissociation energies, **23**, 27, 82, 88, 197
Bond lengths, 7, 58, 66, 75, 90, 100, 104, 182, 199
Bond order, **66**, 182, 199
Bond repulsion, 18–19, 43
Bromination of
  acetone, 81–82
  alkanes, 22, 30
  benzene, 188, 221
  ethyl acetoacetate, 167
  naphthalene, 199
  toluene, 196, 221, 222
Bromobenzene, 188, **202**
Bromonium ions, 61, 81
Buta-1,3-diene, 138, 140, 145
Butanes, 13, 19
Butan-1-ol, 45, 48, 146
But-1-ene and but-2-ene, 56, 71
t-Butyl alcohol, 48, 63, 64, 122

Carbanions, **57**, 126
Carbohydrates, 111, 134, **172–180**
Carbon dioxide, 48, 129, 152, 165
Carbon-metal bonds, 126, 144
Carbonic acid, 165
Carbonium ions, **57**, 62–64, 189, 191, 193, 203, 219
Carbon tetrachloride, 21
Carbonyl bond, 73, **75**, 91
Carboxylation, 129, 210
Carboxylic acids, 82, **86–102**, 129, **222–226**
  aryl, 222
  halogenated, 91, 150, 152, 208
  olefinic, 139
β-Carotene, 141
Cartilage, 103
Catechol, 208, 210, 213
Cellobiose, 177
Cellulose, 179
Cell wall, 110, 179
α-Cephalin, 110
Cephalosporidine, 234
Cephalosporin, 234
Cerebrosides, 106, 110
Chain reactions, 20, 197
Chalcone, 216
Chemical language, 147
Chemotherapy, 231, **233–234**
Chloral, **76**, 205
Chlorination of
  alkanes, 19
  benzene, 188
  chlorobenzene, 203
  phenol, 209
  toluene, 197
Chlorobenzene, 190, **202**
2-Chloroethanol, 135
Chloroform, 21, 132
Chloromethane, 20, 191
Chloronitrobenzenes, 203
Chlorophyll, 228
Chloroprene, 71, 145
Chlorotoluenes, 244
Choline, 110, **135**
Chromatography
  column, 3
  gas, 3, 93, 221
  ion-exchange, 161
  thin-layer, 3
Cinnamaldehyde, 216
Cinnamic acid, 217
Cis-isomers, **58–59**, 69, 107, 147, 152, 174, 239
Coal, 12, 187, 198
Cocaine, 123
Collagen, 103
Colour, 141, 211, 227, 239
Colour photography, 246
Computers, 6, 137
Conformations, **16–17**, 162, 176
Conjugation, 139–141, 190
Contact lens, 148
Co-ordination catalysts, 143
Copolymer, 191
Cosmetic powder, 113
Cost, 6, 29, 201, 236
Covalent bond, 9, 17–18
Cracking, 27, 145
Creosote, 211
Cresol, 185
Crotonaldehyde, 146
Crotonic acid, 58
Crystallisation, 3
Cumene, 184
Cyanides, *see* Nitriles
Cyanohydrins, **77**, 150, 172, 217
Cyclic anhydrides, **151**, 153, 225
Cyclic imides, **151**, 227, 249

Cyclohexane, 26, 195
Cyclohexane-diols, 174
Cyclohexanol, 46, 47
Cyclohexanone, 47, 84
Cyclohexene, 46, 60
Cysteine and cystine, 50, 136, **163**

DDS, 231
DDT, 205
Decarboxylation, 152
Dehydration, **46**, 122, 146
Dehydrogenation, 60, 191
Densities, 33, 202
Depot fats, 106
Desiccants, 137
Design of experiments, 41, 52, 125, 131, 148, 180
Desoxyribonucleic acid, 177
Detergents, 109
Deuterium labelling, 83, 113, 194
Diacetone alcohol, 80
Dialkyl carbonates, 166
Diamines, 136
Diastereomers, **58**, 152, 174
Diazomethane, 208
Diazonium salts, 242–245
Dicarboxylic acids
  aromatic, 225–227
  olefinic, 152
  saturated, 149–152
Dichlorobenzenes, 203
Dichlorodifluoromethane, 38
Dichloromethane, 21, 132
Dienes, 138–141, 145
Diethyl ether, 51, 127
1,2-Dihalides, 133
Diketones, 166–169
Dimethylaniline, 240, 242, 243, 246, 247
Dimethylformamide, 71, 94, 119
Dimethyl sulphate, 102
Dinitrobenzenes, 238, 240, 246
3,5-Dinitrobenzoic acid, 223
2,4-Dinitrophenylhydrazine, 78, 205
2,4-Dinitrotoluene, 196
Diols, 133
Dioxan, 135
Dipole moments, 6, **32**, 59, 75, 102, 158, 237, 239, 248
Diquat, 247
Directive effects, 196, 203, 209, 218–221, 223, 231, 239
Disaccharides, 177
Dissociation constants, **89–91**, 207, 222
Disulphides, 50
Diuretics, 130, 233
Drying oil, 109
Dyestuffs, 29, 207, 239, 246, 248
Dysentry, 233

EDTA, 159
Electrolyses, 99
Electronegativity, 32
Electronic configurations, 9, 18, 56, 183
Electron spin resonance spectroscopy, **198**, 247
Electrophiles and electrophilic reagents, **34**, 188, 192, 193, 199, 244
Electrophilic substitution, **188–196**, 199 209, **218–221**, 222, 223, 239, 244
Elimination reactions, 40, **120–122**
Empirical formulae, 5
Enantiomers, 108, **155**, 174
Energy of activation, **25**, 38, 46, 189, 219
Energy profiles, **17**, 34, 98, 190
Enols, **81**, 146, 167
Entropy, **37–38**, 89, 169, 207, 221
Enzymes, 111, 234, 235
Epoxides, 134, 146
Equilibrium constants, **22–25**, 93, 198
Essential amino-acids, 159, 224
Esters, **93**, **95–99**, 128, 223, 226
  keto-esters, 167–169
  phosphate esters, **102**, 109, 177, 225
  sulphate esters, 102
  thiolesters, 96
Ethanol, 43, **48**, 63, 73
Ethers, **51–52**, 103, 208
2-Ethoxyethanol, 135
Ethyl acetoacetate, 167
Ethyl alcohol, *see* Ethanol
Ethylbenzene, 191
Ethyl chloride, 32
Ethylene, 55, 63, 83, 129, 142
Ethylene glycol, 133–135, 228
Ethylene oxide, 134
Ethyl hydrogen sulphate, 62, 102
Excited state, 10, 139
Explosives, 136, 196
Extraction, 2, 51, 71, 132, 207

Fats, *see* Lipids
Fatty acids, 106
Fehling's solution, **82**, 173, 176
Fermentation, 29, 48, 153
Fibre glass material, 226
Flammability, 51, 133
Fluorescein, 228
Fluorination of alkanes, 22, 30
Fluorobenzene, **202**, 244
Formaldehyde, 76, 81, **83**, 120
Formalin, 83
Formic acid, **100**, 120
Free energy, **22**, 72, 91, 180
Fructose, 176
Fumaric acid, 152
Fungus, 71
Fusion test, 3

Galactose, 175
Gallic acid, 185
Gasoline, 27
Gelatin, 170
Geometrical isomerism, *see* Isomerism
Geraniol, 147
Glucose, 103, **172**–**176**
Glucose 6-phosphate, 111, **180**
Glutamic acid, 158, 161
Glyceraldehyde, 145
Glycerides, 106
Glycerol, 106, **136**
Glycine, 158, 160
Glycogen, 179
Glycols, *see* Diols
Gycosides, 173
Graphite, 200, 235
Grignard reagents, **127**–**129**, 202, 215
Ground state, 10, 34
Guaiacol, 211

Haem and haemoglobin, 163, 228
Hair, 163
Halogenation of
  aldehydes and ketones, 81, 217
  alkanes, 19–26
  benzene, 188, 221
Halogens, addition reactions, 60, 195
Halothane, 132
Hardening of oils, 108
Heat of reaction, **23**, 30, 60, 82, 133, 142, 195, 207
Helix, 162
Hemiacetals, 173
Herbicides, 247
Heterocyclic amines, 114
Heterogeneous catalysis, 60
Heterolytic reactions, 34
Hexamethylene diamine, 136, 164–165
Homogeneous catalysis, 130
Homolytic reactions, 19–25, **34**, 197
Hormones, 160, 161, 224, 248
Hydration of
  alkenes, 62–64
  alkynes, 69
  carbonyl compounds, 76
Hydrazones, **77**, 243
Hydrogenation, 59, 68, 108, 195, 238
Hydrogen bonds, 44, 76, **88**, 163
Hydrogen exchange, 194
Hydrogen halides, addition reactions, 61, 69, 122
Hydrolysis of
  alkyl halides, 33
  allyl halides, 146
  aryl halides, 203
  benzyl halides, 206
  carboxylic acid derivatives, 95, 97–99
  di- and poly-saccharides, 177–179
  nitriles, 102
Hydrolysis of—*cont.*
  proteins, 161, 162
  triglycerides, 106, 109
Hydronium ion, 46
Hydroperoxides, 109
Hydroquinone, 211–212
Hydroxy-acids, 149–150
Hydroxylation, 145

Imides, 151, 227, 249
Inclusion compounds, 166
Industrial production, 29
Infrared spectroscopy, 8, 45, 50, 78, 100, 119, 208, 218
Insecticides, 195, 205
Insulin, 160
Inversion, 157
Invert sugar, 179
Iodine value, 107, 112
Isobutene, 56, 63, 122
Isoelectric points, 159, 160
Isoelectronic, 40
Isolation, 2, 147, 201, 224, 248
Isoleucine, 161, 162
Isomerism,
  geometrical, 58, 147, 174
  optical, 49, 108, **153**–**157**, 174, 218
  structural, 13, 44
Isoprene, 145
Isotopes, 39, 42, 83, 95, 111, 194

Keratin, 160, 163
Kerosene, 27
Keto-acids and esters, 167–170
Ketohexose, 176
Ketones, 47, 69, **73**–**84**, 128, 214–219,
Kinetics, 38, 81, 118, 121, 221

Labelling, *see* Isotopes
Lactic acid, 150, 153–156
Lactones, 151
Lactose, 178
Lattice energy, 9
Lead tetra-acetate, 134
Leaf alcohol, 147, 235
Leprosy, 231
Leucine, 158
Ligands, 126
Light petroleum, 27
Lignin, 210
Ligroin, 27
Linoleic acid, 107, 109
Linolenic acid, 107, 109
Lipids, 106–113
Lithium tetrahydroaluminate, 79, 96 97, 102
Local anaesthetics, 123–124, 224
Lubricating oil, 27
Lysine, 29, 158, 159

Maleic acid and anhydride, 152–153
Malonic acid, 148, 152
Maltose, 177, 179
Mandelic acid and nitrile, 217
Manganese dioxide, 146, 247
Mannose, 175
Markovnikov rule, 62
Mass spectrometry, 6, 7, 30, 71, 84, 95
Mauveine, 248
Melting, 8, 37, 132, 213, 248
Meningitis, 233
Mercaptans and mercaptides, 49
Mesomerism, 90
Metabolism, 111
Meteorite, 11
Methane, 12, 15, 26, 127
Methanol, 43, 48, 85
Methionine, 1, 5–7, 164, 249
Methyl alcohol, *see* Methanol
Methylaniline, 240, 242
Methyl orange, 243
Methylation, 36, 120, 191, 208, 242
Methylene group, 15
Microbiological chemistry, *see* Fermentation
Molecular complexes, 221
Molecular formula, 5, 11, 131
Molecular models, 17
Molecular sieves, 19
Monosaccharides, 172
Morphine, 123
Mucins, 180
Mutarotation, 176
Mycomycin, 108
Myoglobin, 163

Naphtha, 27, 70
Naphthalene, 198
2-Naphthol, 232
Natural gas, 12, 26
Neoprene, 145
Neuraminic acid, 180
Nitration of
  acetanilide, 245
  aryl carbonyl compounds, 210
  benzene, 191
  chlorobenzene, 203
  naphthalene, 199
  nitrobenzene, 239
  phenol, 209
  toluene, 196
Nitriles, 36, **101**, 223, 245
Nitroanilines, 238, 245
Nitrobenzene, 191, 237
Nitroglycerine, 136
Nitronium ion, 192
Nitrophenols, 209, 243
p-Nitrophenylhydrazine, 243
Nitrosobenzene, 239
p-Nitrosodimethylaniline, 246
Nitrotoluenes, 196
Nuclear magnetic resonance, 6, 186, 201
Nucleic acids, 177
Nucleophiles, 34
  carbon, 128
  halogen, 38–39
  nitrogen, 36, 116
  oxygen, 33–36
  sulphur, 36
Nucleophilic substitution, **33–37**, 116 134, 157
Nylon, 164–165, 170

Octane number, 27
Octet rule, 10
Odour, 49, 93, 147, 184
Olefinic alcohols, 146–147
Olefins, *see* Alkenes
Oleic acid, 107
Optical inversion, 157
Optical isomerism, *see* Isomerism
Orbitals, 18, 56–57, 66, 88, 183
Organometallic compounds, 126–131
Origin of life, 164
Oxalic acid, 149
Oxidation of
  aldehydes and ketones, 82, 218
  alkanes, 101
  alkenes, 64
  amines, 246–247
  diols, 134
  formic acid, 101
  hydroxy-compounds, 47, 73, 83, 133 146, 206
  thioethers, 52
  thiols, 50, 163, 229
Oxidative cleavage, 64, 134, 180
N-Oxide, 247
Oximes, 77, 215
Oxonium cation, 46, 52
Oxo-reaction, 48
Oxytocin, 161
Ozonolysis and ozonides, 65–66, 195

Paint, 226
Palmitic acid, 107
Paraffin, 27
Paraformaldehyde, 83
Partition, 3, 125
PAS, 224
Patent, 249
Penicillin, 3, 234, 235
Pentanes, 14
Pentanols, 48
Pentoses, 177
Pepsin, 160
Peptides, 161
Peroxy-acids, 146, 246

Petroleum, 12, 26, 187
Pharmacology of amines, 122–124
Phenacyl halides, 217
Phenethicillin, 234
Phenols, 203, **206–209**, 232, 243
Phenylalanine, 243
Phenylhydrazine, 243
Phenylhydroxylamine, 238
Pheromones, 147
Phosgene, 165
Phosphates, *see* Esters
Phospholipids, 109
Photography, 212, 246
Photosynthesis, 111
Phthalic acid and anhydride, 225–226
Phthalimide and potassium salt, 227, 249
Phthalocyanine, 228
Picrates and picric acid, 116, 209
Pinosylvin, 211
pK_a values, 89, 91, 213
Plasticisers, 93, 144
Polarimeter, 154
Polyalcohols, 133
Polyester resins, 226
Polyethylene, 142
Polyhalides, 132
Polymers and polymerisation, 142, 170, 228–229
Polypropylene, 143
Polysaccharides, 179
Polystyrene, 191
Potassium permanganate, 83
Prephenic acid, 225
Pressure, 63, 105, 143
Prices, 201, 236
Procaine, 123–124
Propylene, 29, 56, 63, 84
Proteins, 29, 160–163
Purity, 3, 143
Putrescine, 136
PVC, 144
Pyridine, 114
Pyrogallol, 185
Pyroligneous acid, 43
Pyruvic acid, 167

Quaternary ammonium salts, 118, **120**, 124
Quinones, 211–212

Racemate, 156
Radical cations, 247
Radicals, **19**, 197, 198, 247
Reducing sugars, 176, 178
Reduction of
  acids and acid derivatives, 96–97
  aldehydes and ketones, 79
  alkenes, 59
  alkyl halides, 127
Reduction of—*cont.*
  alkynes, 68
  diazonium salts, 243, 244
  disulphides, 50, 163
  nitriles, 102
  nitro-compounds, 237-238
  ozonides, 65
Refractive indices, 33, 53, 202
Refrigeration, 38
Reservoir, 148
Resolution, 156
Resonance, **90–91**, 183, 189, 218
Resorcinol, 210, 228
Rhodium catalysts, 130, 195
Riboflavin, 212
Ribonucleic acid, 177
Ribose, 177, 212
Ricinoleic acid, 107, 113
Rope trick, 165
Rubber, 72, 145

Saccharin, 232
Salicylaldehyde, 214
Salicylic acid, 210
Salvarsan, 233
Saponification value, 106, 112
Sawflies, 181
Sedative, 76, 85
Semicarbazones, 77, 167, 215
Shark liver oil, 145
Shikimic acid, 225
Skin, 103
Silicon tetrachloride, 131
Silk, 224
S_N1 reactions, 64
S_N2 reactions, **34**, 121, 158
Soap, 108-109
Sodium tetrahydroborate, 65, 79
Solubility, 8, 85, 105, 180, 200, 248
Solvation, 89
Specific rotation, 154
Spectroscopy, *see* Infrared, etc.
Spermaceti, 43
Spirits of wine, 51
Squalene, 145
Starch, 179
Stearic acid, 100, 107
Stereo chemistry, 15–17, 58–59, 153-158, 159, 162, 174
Stereoregular reaction, 143
Stoichiometry, 77, 79, 131
Structure determination, 6
Styrene, 191
Succinic acid and anhydride, 149, 151
Succinimide, 151
Sucrose, 178
Sulphanilamide and sulpha-drugs, 233, 236
Sulphates, *see* Esters
Sulphonamides, 119, 231, 233

Sulphonation of
  benzene, 193
  naphthalene, 199
Sulphones, 52, 231
Sulphonic acids, 51, 193, 229
Surfactants, 109, 113
Sweeteners, 180, 232
Synthetic fibres, 144, 164, 179, 228
Synthetic rubber, 142, 145, 170
Syphilis, 233

Tartaric acid, 156–157
Teflon, 144
Terephthalic acid, 225, 228
Terylene, 228
Tetraethyl-lead, 126, 130
Tetrafluoroethylene, 133
Tetrahedral carbon, **15**, 153
Tetrahedral nitrogen, 114, 156
Textiles, 144, 179, 181, 228
Thioethers, 51, 164
Thiolesters, *see* Esters
Thiols, 49, 229
Thionyl chloride, 46, 92
Thymol, 208
Thyroxine and thyroid gland, 224–225
Tollens reagent, 82, 101, 218
Toluene, 187, **196–197**
Toluene-*p*-sulphonyl chloride, 119, 232
Tracer experiments, 95, 111
Trans-isomers, **58**, 152, 174
Transition states, 11, **25**, 34, 98, 190
Trifluoroacetic acid, 221, 246
Triglycerides, 106
Trinitrobenzene, 240
Trinitrotoluene, 196
Triphenylmethyl, 198
Tuberculosis, 108, 224
Tuberculostearic acid, 108
Tyrosine, 224

Ultraviolet spectroscopy, 139
Umbrella inversion, 156
αβ-Unsaturated carbonyl compounds 139, 216
Urea, 2, 166
Urine, 170

Vaccenic acid, 112
Vanillin, 214
Vinegar, 101
Vinylacetylene, 71
Vinyl chloride, 144
Vinylic compounds, 141, 144
Viscosity, 105, 136
Visible spectroscopy, 139
Vitamins, 6, 141, 212

Waxes, 71, 106
Whiskey, 43
Wood, 43, 210–211

Xylenes, 185, 188, 226
X-ray diffraction, 6–7, 163, 199

Yields, viii, 29, 30, 35, 40, 66, 135, 220–221

Zinc dialkyls, 126

THE FOLLOWING PAGES MAY BE USED FOR NOTES